Materials in
BIOLOGY AND
MEDICINE

GREEN CHEMISTRY AND CHEMICAL ENGINEERING

Series Editor: Sunggyu Lee
Ohio University, Athens, Ohio, USA

Materials in
BIOLOGY AND
MEDICINE

Edited by
Sunggyu Lee
David Henthorn

CRC Press
Taylor & Francis Group
Boca Raton London New York

CRC Press is an imprint of the
Taylor & Francis Group, an **informa** business

CRC Press
Taylor & Francis Group
6000 Broken Sound Parkway NW, Suite 300
Boca Raton, FL 33487-2742

First issued in paperback 2017

© 2012 by Taylor & Francis Group, LLC
CRC Press is an imprint of Taylor & Francis Group, an Informa business

No claim to original U.S. Government works

ISBN-13: 978-1-4398-8169-9 (hbk)
ISBN-13: 978-1-138-07215-2 (pbk)

Visit the Taylor & Francis Web site at
http://www.taylorandfrancis.com

and the CRC Press Web site at
http://www.crcpress.com

Contents

Green Chemistry and Chemical Engineering

Series Statement

The subjects and disciplines of chemistry and chemical engineering have encountered a new landmark in the way of thinking about, developing, and designing chemical products and processes. This revolutionary philosophy, termed "green chemistry and chemical engineering," focuses on the designs of products and processes that are conducive to reducing or eliminating the use and generation of hazardous substances. In dealing with hazardous or potentially hazardous substances, there may be some overlaps and interrelationships between environmental chemistry and green chemistry. While environmental chemistry is the chemistry of the natural environment and the pollutant chemicals in nature, green chemistry proactively aims to reduce and prevent pollution at its very source. In essence, the philosophies of green chemistry and chemical engineering tend to focus more on industrial application and practice rather than academic principles and phenomenological science. However, as both chemistry and chemical engineering philosophy, green chemistry and chemical engineering derive from and build upon organic chemistry, inorganic chemistry, polymer chemistry, fuel chemistry, biochemistry, analytical chemistry, physical chemistry, environmental chemistry, thermodynamics, chemical reaction engineering, transport phenomena, chemical process design, separation technology, automatic process control, and more. In short, green chemistry and chemical engineering are the rigorous use of chemistry and chemical engineering for pollution prevention and environmental protection.

The Pollution Prevention Act of 1990 in the United States established a national policy to prevent or reduce pollution at its source whenever feasible. And adhering to the spirit of this policy, the Environmental Protection Agency (EPA) launched its Green Chemistry Program to promote innovative chemical technologies that reduce or eliminate the use or generation of hazardous substances in the design, manufacture, and use of chemical products. The global efforts in green chemistry and chemical engineering have recently gained a substantial amount of support from the international community of science, engineering, academia, industry, and governments in all phases and aspects. Some of the successful examples and key technological developments include the use of supercritical carbon dioxide as green solvent in separation technologies, application of supercritical water oxidation for destruction of harmful substances, process integration with carbon dioxide sequestration steps, solvent-free synthesis of chemicals and polymeric materials, exploitation of biologically degradable materials, use of aqueous hydrogen peroxide for efficient oxidation, development of hydrogen proton exchange membrane (PEM) fuel cells for a variety of power generation needs, advanced biofuel productions, devulcanization of spent tire rubber, avoidance of the use of chemicals and processes causing generation of volatile organic compounds (VOCs), replacement of traditional petrochemical processes by microorganism-based bioengineering processes, replacement of chlorofluorocarbons (CFCs) with nonhazardous alternatives, advances in design of energy efficient processes, use of clean alternative and renewable energy sources in manufacturing, and much more. This list, even though it is only a partial compilation, is undoubtedly growing exponentially.

This book series on Green Chemistry and Chemical Engineering by CRC Press/Taylor & Francis is designed to meet the new challenges of the twenty-first century in the chemistry and chemical engineering disciplines by publishing books and monographs based on cutting-edge research and development to effect reducing adverse impacts on the environment by chemical enterprise. To achieve this, the series will detail the development of alternative sustainable technologies that will minimize the hazard and maximize the efficiency of any chemical choice. The series aims at delivering readers in academia and industry with an authoritative information source in the field of green chemistry and chemical engineering. The publisher and its series editor are fully aware of the rapidly evolving nature of the subject and its long-lasting impact on the quality of human life in both the present and future. As such, the team is committed to making this series the most comprehensive and accurate literary source in the field of green chemistry and chemical engineering.

Sunggyu Lee

Preface

The history of human civilization is best understood by tracking the materials and remains thereof that humans have found, created, modified, processed, and used according to their specific life needs at different time periods and geographical regions. Therefore, the prevailing materials of choice of a given era often serve as its defining point, with the Stone Age, Bronze Age, and Steel Age being good examples. Historically, the process of transitioning from one type of principal structural material to another could take up to thousands of years; however, modern development of new materials is much faster paced and more multidirectional in its intended end uses. Some may refer to the current era as the Silicon Age, owing to the profound impact that silicon has had on the way we live and work; however, modern materials are far more diverse in their compositional and structural forms than a single substance and are greatly versatile in their desired applications and deliverable functionality. Although the list of modern materials is truly rapidly increasing, most materials may still be fitted into broader categories of metals, ceramics, semiconductors, polymers, biomaterials, composites, magnetic materials, and exotic materials. Remarkable advances in materials include biomaterials for artificial human body parts and organs, ceramics for bone and tooth replacement, biodegradable and biocompatible polymers, semiconductor materials, solar grade silicon, highly capable energy storage materials, reinforced structural materials, highly functional composites, high-performance alloys, ultra-high temperature-resistant materials, liquid crystals, aerogels, buckyballs and nanotubes, oil-producing microalgae, and much more.

The discipline of materials science and engineering is relatively new and may be defined as an applied and interdisciplinary field concerned with the relationship between the structure and properties of materials. Materials science and engineering is the field where scientists, engineers, and physicians of diverse backgrounds can work together for common targets and goals, thereby developing new materials, enhancing material properties, expanding the boundaries of applications, and devising new technologies of processing or manufacturing. This book is intended to provide the readers with a solid background, including recent successful examples, in the subfield of materials in biology and medicine. Principal foci of the book are placed on biomaterials and bioinspired materials, functional and responsive materials, controlling biology with materials, and development of devices and enabling technologies. All chapters were written by subject experts in a consistent and readable fashion by fully describing the relevant scientific background as well as thoroughly discussing the logical sequences of new development and applications. Although this book is intended for readers who have a background in college-level chemistry, biology, and physics, the value of this book may be more appreciated by graduate students working on diverse scientific and engineering problems involving materials in biology and medicine, researchers and inventors in related fields, and practicing engineers, scientists, and physicians in their chosen areas. It is the editors' wish that this book contribute to scientists, engineers, medical researchers, and industrialists in their technological thinking, tackling challenges in novel materials, carrying out new product and process development projects, and devising and inventing new methodologies in applying novel materials for life quality enhancement.

This book is published as a spin-off volume of the *Encyclopedia of Chemical Processing* and is based on encyclopedia entries recently published and soon to be published in the current theme of materials in biology and medicine. This book is also published as a book in the Green Chemistry and Chemical Engineering book series.

Sunggyu Lee
Athens, Ohio

David B. Henthorn
St. Louis, Missouri

About the Editors

Dr. Sunggyu Lee earned both his Bachelor's and Master's degrees in chemical engineering from Seoul National University, Seoul, Korea, in 1974 and in 1976, respectively. He received his Ph.D. in chemical engineering from Case Western Reserve University, Cleveland, Ohio in 1980. He started his professorial career with The University of Akron in 1980 and was promoted to full professor in 1988. From 1988–1997, he served as Robert Iredell Professor and Chairman of chemical engineering as well as the Founding Director of the Process Research Center. From 1997–2005, he held positions of chairman and C. W. LaPierre Professor of chemical engineering at the University of Missouri. From 2006–10, he was with Missouri University of Science and Technology, where he established the Laboratory for Transportation Fuels and Polymer Processing. Since 2010, he has held positions as the Russ Ohio Research Scholar and professor of chemical and biomolecular engineering, Ohio University, Athens, Ohio. He has established the Sustainable Energy and Advanced Materials (SEAM) Laboratory as a stand-alone off-campus laboratory of excellence.

Dr. Lee has authored 7 books, 9 monographs, 9 book chapters, 144 refereed journal articles, and over 360 proceedings and conference papers. Titles of his published books include *Methanol Synthesis Technology, Oil Shale Technology, Alternative Fuels, Methane and Its Derivatives, Handbook of Environmental Technology*, and *Handbook of Alternative Fuel Technologies*. He is the editor of the *Encyclopedia of Chemical Processing* (5 volumes with 350 chapters), published in 2006. He has received 30 U.S. patents based on his inventions, mainly in the areas of clean alternative energy, functional polymers, and supercritical fluid technology. Most of his inventions are being commercially utilized. He has received and directed over 100 research grants/contracts from both industrial and governmental agencies totaling over $19M as principal investigator and co-principal investigator. His specialties are in the areas of alternative fuels, supercritical fluid technology, chemical process engineering and reactor design, polymer synthesis and processing. He has guided over 90 advanced degree students and 24 postdoctoral fellows as major advisor.

Dr. David B. Henthorn received his Bachelor's (1999) and Doctoral (2004) degrees in chemical engineering from Purdue University in West Lafayette, Indiana. For his doctoral work, Dr. Henthorn studied the formation of densely crosslinked hydrogel networks and their applicability as biomimetic materials under the direction of Professors Nicholas A. Peppas and Kinam Park. In 2004, he joined the Department of Chemical and Biological Engineering at Missouri University of Science and Technology (then the University of Missouri-Rolla) as an assistant professor. In 2010, he was awarded tenure and promoted to associate professor of chemical and biological engineering. In the fall of 2010, Dr. Henthorn moved to the Biomedical Engineering Department at Saint Louis University, where he is currently an associate professor.

Dr. Henthorn's research focuses on polymeric materials in biomaterials, biosensors, bioMEMS devices, and drug delivery devices. A strong proponent of involving students in research, Dr. Henthorn has worked with 21 undergraduate students in the last 7 years. Over that same period of time, he has supervised 8 graduate students.

Contributors

Shubhayu Basu
Department of Chemical and Biomolecular Engineering, The Ohio State University Columbus, Ohio, U.S.A.

Sujata K. Bhatia
Dupont Central Research and Development, Wilmington, Delaware, U.S.A.

Surita R. Bhatia
Department of Chemical Engineering, University of Massachusetts–Amherst, Amherst, Massachusetts, U.S.A.

James Blanchette
University of South Carolina, Columbia, South Carolina, U.S.A.

Frank Davis
Cranfield University, Silsoe, U.K.

Amy S. Determan
Department of Chemical and Biological Engineering, Iowa State University, Ames, Iowa, U.S.A.

Hugo S. Garcia
UNIDA, Instituto Tecnologico de Veracruz, Veracruz, Mexico

David B. Henthorn
Department of Biomedical Engineering, Saint Louis University, St. Louis, Missouri, U.S.A.

Séamus P. J. Higson
Cranfield University, Silsoe, U.K.

Charles G. Hill, Jr.
Department of Chemical and Biological Engineering University of Wisconsin–Madison, Madison, Wisconsin, U.S.A.

Kang Moo Huh
Department of Polymer Science and Engineering, Chungnam National University, Daejeon, South Korea

L. James Lee
Department of Chemical Engineering, The Ohio State University, Columbus, Ohio, U.S.A.

Sunggyu Lee
Department of Chemical and Biomolecular Engineering, Ohio University, Athens, Ohio, U.S.A.

Surya K. Mallapragada
Department of Chemical and Biological Engineering, Iowa State University, Ames, Iowa, U.S.A.

Balaji Narasimhan
Department of Chemical and Biological Engineering, Iowa State University, Ames, Iowa, U.S.A.

Cristina Otero
Departamento de Biocatálisis, Instituto de Catálisis y Petroleoquimica, CSIC, Campus Universidad Autonoma, Cantoblanco, Madrid, Spain

Jae Hyung Park
Department of Advanced Polymer and Fiber Materials, Kyung Hee University, Gyeonggi-do, South Korea

Kinam Park
Departments of Pharmaceutics and Biomedical Engineering, Purdue University, West Lafayette, Indiana, U.S.A.

Clayt Robinson
Department of Chemical and Biomolecular
  Engineering, The Ohio State University,
  Columbus, Ohio, U.S.A.

Gregory T. Rushton
Department of Chemistry and
  Biochemistry, Kennesaw State
  University, Kennesaw, Georgia, U.S.A.

Ken D. Shimizu
Department of Chemistry and
  Biochemistry, University of South
  Carolina, Columbia, South Carolina,
  U.S.A.

Maria P. Torres
Department of Chemical and Biological
  Engineering, Iowa State University,
  Ames, Iowa, U.S.A.

Chun Wang
Department of Chemical Engineering,
  Massachusetts Institute of Technology,
  Cambridge, Massachusetts, U.S.A.

Shang-Tian Yang
Department of Chemical and Biomolecular
  Engineering, The Ohio State University,
  Columbus, Ohio, U.S.A.

Hirotsugu Yasuda
Center for Surface Science and Plasma
  Technology, University of Missouri,
  Columbia, Missouri, U.S.A.

Mingli Ye
Departments of Pharmaceutics and
  Biomedical Engineering, Purdue
  University, West Lafayette, Indiana,
  U.S.A.

1

Introduction: Materials in Biology and Medicine

David B. Henthorn and Sunggyu Lee

CONTENTS

Discoveries in medicine, biotechnology, biology, and biochemistry are occurring at a rate previously unthinkable. The fields of genomics, proteomics, genetic engineering, computational chemistry, etc., have revolutionized our approach to unlocking the mysteries of living organisms and their responses to various stimuli. Development of new materials to complement these advance—whether to serve as replacement limbs, as structures to encourage tissue growth, as sensing elements to determine blood glucose levels, or as micr-scale vessels for determination of serum antibody level—has been equally rapid and diverse.

Materials

The study of biomaterials, materials used in biomedical applications with the designed intent to interact with the surrounding tissue, is one of the most diverse multidisciplinary endeavors, requiring the talents of scientists (biology, chemistry, physics, materials science), engineers (biomedical, chemical, mechanical, electrical, materials), and physicians. In Chapter 2 of this book, Bhatia and Bhatia describe the basic aspects of biomaterials, highlighting the essential property of biocompatibility. Myriad materials have been utilized over the years in medicine and biology, ranging from ceramics, metals, polymers, glasses, and composites. Later chapters highlight aspects of biomaterials research, moving from bulk materials to applications with cells and tissues, and ending with the fabrication of complex and integrated biodevices.

Polymer—long-chain molecules based on repeating monomeric unit—are some of the most versatile biomaterials in use—since the chemistry of the repeating units is so easily varied—leading to diverse physical, chemical, thermal, optical, and mechanical properties. Hydrophilic (water loving) materials are some of the most widely used in medicine and biology, owing to the fact that the physiological environment is largely based in water. Examples of hydrophilic polymer materials include poly(ethylene glycol) (PEG), poly(methacrylic acid), poly(acrylic acid), poly(vinyl alcohol), and poly(2-hydroxyethylmethacrylate). Their uses range from soft contact lenses, burn and wound dressing, membranes for immobilization of proteins and other active biomolecules, drug delivery matrices, and absorbents. The solvation of these chains by water or physiological fluid is an area of intense study, and has led to unique biomedical applications for many materials.(PE) is one of the most widely applied material to increase biocompatibility, for instance. The ability of PEG to be solvated by water has made it an immensely useful coating for various hydrophobic surfaces. Protein deposition, platelet adhesion, and other undesirable processes are slowed through the masking of the surface with hydrophilic polymers, such as PEG.

Cross-linking of these hydrophilic polymers, whether through chemical or physical means, leads to the creation of water-swellable three-dimensional networks known as hydrogels. Long used as a biomaterial due to their high water content, hydrogels are found in applications such as soft contact lenses, hemostatics and absorbents, drug delivery devices, immobilization matrices for enzymes, etc. The degree to which the network swells is dependent upon a number of factors, including the interaction strength between the chains and water, extent of cross-linking, presence of solvent during the cross-linking procedure, temperature, pressure, and concentration of ions. Superabsorbent hydrogels may consist of more than 99% water by weight. Their use as drug delivery matrices has been well-studied, with release rates of drugs showing both traditional diffusional behavior (Fickian) along with more complex (non-Fickian) behavior altered by solvation, chain relaxation, and crystal melting.

Fickian diffusion from these drug delivery matrices, with drug release rates evolving over time with the inverse square root of time ($t^{-1/2}$), has been a hurdle to those interested in designing systems that deliver in constant, pulsatile, or delayed rates. Attempts to control the release of active therapeutics from materials have taken a number of different approaches. A radically different approach to this problem employs materials where the rate of degradation is instead controlled. In this case, the matrix material is sufficiently hydrophobic to halt influx of physiological fluid, preventing solvation and drug from the bulk. Fabrication of such a device from hydrolyzable materials ensures drug release from the surface, allowing the designer to tailor the device's geometry to obtain the desired controlled release rate. Polyanhydrides are often formed from the polymerization of diacid monomers, with copolymerization commonly used to achieve the desired material properties and ultimate degradation rates. By carefully choosing the repeating units of the polymer, hydrolysis of these materials *in vivo* results in the controlled release of the active agent along with monomeric species that are easily metabolized by the body.

Functional and Responsive Materials

For a long time, the search to improve biocompatibility centered on the physical and chemical tailoring of materials and their surfaces to lower protein adsorption, to boost hydrophilicity and wettability, and in general less likely to trigger a foreign-body response. This approach has trade-offs, as exemplified in the evolution of soft contact lenses. While traditional soft contact lenses, developed by Otto Wichterle in the 1960s, were comprised of a hydrogel based around the neutral monomer 2-hydroxyethylmethacrylate and were reasonably hydrophilic and water-swollen, incorporation of monomers that boosted swelling became advantageous. The resulting lenses, with their higher water content, not only were more comfortable to wear but also had higher oxygen permeabilities and therefore lower propensity to trigger hypoxia. The monomers incorporated in this new generation of lenses—the same that are used in disposable diapers for their ability to absorb vast quantities of water—are, however, ionized in aqueous solution and more likely to trigger protein adsorption. The solution, therefore, was to suggest shorter lifetimes for the lenses and rely on improvements in the manufacturing process to maintain price parity. In the late 1980s, it was not uncommon for a single pair of soft contact lenses to be retained by a patient for a full year, with the patient employing various cleaning techniques, including weekly treatment with concentrated protease solutions, to maintain comfort. The introduction of contact lenses with higher water content increased comfort, but the charged materials required more frequent replacement, with monthly, weekly, and even daily replacement possible.

In the design of implantable devices and other applications with long-term use, scientists and engineers are not likely to have options such as these. Other functionalization techniques or utilization strategies must therefore be employed. In a series of seminal articles, leaders in the biomaterials field argued that the passive approach to biocompatibility—building in stealth-like properties that allow a material to hide from the foreign-body response—was not enough. Biomaterials, they argued, need to take an active role, participating in healing. A material, for instance, that resists adsorption of general serum proteins while recruiting specific, beneficial molecules and interacting with the local tissue has a much greater chance of successful integration. Knowledge, built over years of work on surface passivation, is used to create a foundation where no-specific interactions are minimized. Control of specific interactions is then added through the introduction of surface charges, peptidic oligomers, cell adhesion molecules/fragments, carbohydrates, etc.

An example of this approach is used in the technique of molecular imprinting. A hydrogel material, long used as a biomaterial because of its high water content, ability to mimic tissue properties, and passive nature, is tailored to add affinity for a specific target molecule or family of molecules. This affinity is typically imparted during material formation. A template molecule, either the interaction target or a close analog, is added in material formation, driving self-assembly of the monomers species. This assembly is then locked into the material during polymerization and cross-linking. Removal of the template molecule, typically through dialysis, yields analyte-specific binding sites inside the bulk of the normally passive hydrogel structure. In recent time, researchers have expanded this technique to focus on the creation of these binding sites on a material's surface, allowing for interaction with macromolecules and other species that are either too large or immobile to interact with the bulk of the hydrogel.

Molecularly imprinted polymers and other biomimetic materials illustrate the interest in creating synthetic materials for use in medicine, biology, and biotechnology. Natural materials, however, are still the gold standard in terms of signaling, molecular recognition, transport, and catalysis. Enzymes, for instance, have some of the greatest catalytic ability of any known molecular structures. While the specificity and turnover rate of an enzyme may be tailored through genetic engineering, biological organisms are still required for their production. As such, yields are low and expensive; time-consuming separation steps must be employed to capture the proteins in active form. Solid support materials, therefore, are frequently used when enzymatic action is utilized in a chemical or biological process. These support material—microparticles, gels, columns, etc—allow the enzymatic materials to be easily recaptured once the process is finished. The coupling of biologically active molecules, such as enzymes, with nanotechnology and nanomaterials is an exciting field that combines the unique properties at the nanoscale (high surface-to-volume ratio, quantum effects, etc.) with the recognition and catalytic abilities of biomolecules.

Nanomaterials such as carbon nanotubes, normally too hydrophobic to disperse well in the aqueous environment, have many desirable physical, mechanical, and optical properties. Researchers have therefore been interested in the modification of these nanomaterials for use in biomedical and biotechnological applications. First, modification must be done to improve the aqueous dispersibility of the nanotubes, allowing them to interact individually with cells, proteins, etc. This modification, however, is typically done in a manner that allows for simultaneous surface passivation and incorporation of elements to control specific interaction. For instance, hydrophilic polymer chains may be attached to or grown from the nanotube surface, aiding in dispersibility. Functional groups on these chains, e.g., carboxylate or epoxide groups, may serve as sites where adhesion molecules, antigens, antibodies, enzymes, DNA/RNA, etc., may be added. Surface functionalization techniques have been used to fabricate dispersible carbon nanotubes that circulate freely in aqueous media, interact with specific cells, and triggering internalization by targeted ones. These internalized, targeted nanomaterials may act as a vector to identify diseased cells and even to deliver therapeutic agents.

Controlling Biology with Materials

The growth of cells and tissues is inextricably linked with the support and substrate materials provided to them. In cell culture, surfaces covered with peptide oliogmers, cell adhesion molecules, fragments of extracellular matrices, etc., are used to promote growth. This idea of using engineered materials to grow cells in three dimensions has led to the field of tissue engineering. Scaffolds are used to promote and regulate cellular growth in geometric structures, with an ultimate goal of replacing diseased or damaged tissue. Tissue engineering scaffolds have been constructed from a variety of materials, including from both synthetic and naturally derived sources. Bioresorbable materials, such as the copolymer poly(lactic-co-glycolic acid), have found use as scaffolds for the repair of various tissues, with the rate of degradation carefully tuned to allow the tissue to be established before diminishment of mechanical and biochemical support. Other materials, such as bioactive glasses that resorb into hydroxyapatite (the main mineral component of bone), are utilized in the fabrication of scaffolds for hard tissue restoration. The efficient transport of nutrients to the tissue, and waste products away, remains one of the greatest problems facing researchers in the area of tissue engineering.

To protect cells from immune response, allowing for the transplantation of allogenic or xenogenic tissues, it is possible to encapsulate a group of cells in a matrix material that allows for biochemical transport. Transplantation of islet cells into type 1 diabetic patients, for the restoration of insulin production, is one of the most commonly researched applications of this technology. The encapsulating material must allow for transport of nutrients to the transplanted tissue, removal of waste products, and the relatively unhindered release of the therapeutic agent (e.g., insulin in islet cell encapsulation). The material, however, must work to hinder the immune response while retaining the integrity of the cellular aggregates. For example, the molecular size disparity between insulin and immunoglobulins allows researchers to tailor materials with molecular weight cutoffs that allow for diffusion of the peptidic hormone out of the matrix while preventing inflow of antibodies, etc.

Devices

The idea that materials can be made to incorporate, utilize, or release bioactive agents is driving a revolution in implantable device design. Immobilized enzymes, antibodies, or nucleic acids could provide molecular recognition for the device, once implanted. Release of growth factors, cytokines, and other signaling molecules could help recruit desirable tissue growth in the local area of the device, instead of formation of a fibrous capsule and scar tissue. Surface-grafted cell adhesion molecules and biocompatible surfaces would help mask the presence of a device, the operation of which is becoming more complex over time due to improvements in miniaturization and fabrication techniques.

By adopting the fabrication techniques developed in the manufacture of integrated electrical circuits, researchers have been able to create therapeutic and diagnostic devices that operate on the micro- and even nanoscale. Early work in the area of microelectromechanical systems (MEMS) devices led to the development of now-ubiquitous commercial devices such as the accelerometer used in automobiles to trigger activation of airbags. Translation of this work into biomedicine and biotechnology (BioMEMS) has been done to provide for devices that separate, analyze, sense, and respond to various chemical and physiological signals. In addition to research in device creation, interest in BioMEMS has led to advances in materials. While silicon is the most traditional substrate material for integrated circuit manufacture, researchers in the BioMEMS field have pioneered the fabrication of devices using glasses, epoxies, and silicones. Rigid silicon devices with long fabrication times, for instance, are replaced by flexible, rapid-formingpoly(dimethyl siloxane) silicone rubber, allowing for rapid prototyping. In recent years, there has been a substantial push to lower material and fabrication costs, with new disposable devices being contemplated from paper, tape, etc.

2

Biomaterials

Sujata K. Bhatia and Surita R. Bhatia

CONTENTS

Introduction

Biomaterials science is a multidisciplinary endeavor incorporating chemical engineering, medicine, biology, chemistry, materials science, bioengineering, and biomechanics. The past few years have witnessed an explosion in the field of biomaterials, with an expansion of both the compositions and the applications of medical implant materials. As the prevalence of chronic diseases such as diabetes, cardiovascular disease, and neurodegenerative disease increases, there will be an even greater need for innovative biomaterials. This chapter reviews the current status of the field of biomaterials, and highlights new developments in biomaterials. The chapter will provide an overview of medical applications of biomaterials, and will describe current classes of biomaterials, including metals, ceramics and glasses, and polymeric materials. The chapter will then discuss the next generation of biomaterials, including surface-modified biomaterials, smart biomaterials, bioactive materials, biomimetic materials, patterned biomaterials, and tissue engineering and regenerative medicine.

Definitions

A commonly used definition of a biomaterial, endorsed by a consensus of biomaterials experts, is "a nonviable material used in a medical device, intended to interact with biological systems."[1] An essential characteristic of biomaterials is biocompatibility, defined as "the ability of a material to perform with an appropriate host response in a specific application."[1] The goal of biomaterials science is to create medical implant materials with optimal mechanical performance and stability, as well as optimal biocompatibility.

Overview of Biomaterials Applications

Biomaterials are used in diverse clinical applications. Table 2.1 lists several examples of applications of biomaterials in medicine.[2] Note that metals, ceramics, polymers, glasses, carbons, and composite materials are listed.

TABLE 2.1

Examples of Clinical Applications of Biomaterials

Application	Types of Materials
Orthopedic	
Joint replacements (hip, knee)	Titanium, Ti–Al–V alloy, stainless steel, polyethylene
Bone plate for fracture fixation	Stainless steel, cobalt–chromium alloy
Bone cement	Poly(methyl methacrylate)
Bony defect repair	Hydroxyapatite
Artificial tendon and ligament	Teflon™, Dacron™
Cardiovascular	
Blood vessel prosthesis	Dacron™, Teflon™, polyurethane
Heart valve	Reprocessed tissue, stainless steel, carbon
Catheter	Silicone rubber, Teflon, polyurethane
Pacemaker	Polyurethane, silicone rubber, platinum electrodes
Ophthalmologic	
Intraocular lens	Poly(methyl methacrylate)
Contact lens	Silicone-acrylate, hydrogel
Corneal bandage	Collagen, hydrogel
Dental	
Dental implant for tooth fixation	Titanium, alumina, calcium phosphate
Neurologic	
Cochlear implant	Platinum electrode
General surgery	
Skin repair template	Silicone–collagen composite
Sutures	Silk, nylon, poly(glycolide-*co*-lactide)
Adhesives and sealants	Cyanoacrylate, fibrin
Organ replacement	
Heart–lung machine	Silicone rubber
Artificial kidney (hemodialyzer)	Cellulose, polyacrylonitrile
Artificial heart	Polyurethane

(Adapted from Ratner, Hoffman, Schoen, and Lemons.[2])

Types of Biomaterials

Metals

Metals and alloys have long been used in surgical and dental applications where materials with high strength are required. Metals are excellent for providing specific mechanical properties, including strength and ductility; however, corrosion of metallic implants in biological environments remains a concern. Corrosion not only limits device lifetime but also causes release of toxic metal ions that are often carcinogenic or mutagenic.[3,4] Thus, much of the current research focuses on minimizing and reducing corrosion of metallic biomaterials. Blackwood has recently reviewed common types of corrosion encountered in metal implants in vivo, as well as physiological parameters relevant to corrosion.[3] For surgical implants, relevant parameters are chloride content, pH, and dissolved oxygen levels in blood.[3] In vitro tests of corrosion resistance are typically performed in aqueous solutions containing 0.9% NaCl, with a pitting resistance number greater than 26 desirable for implanted materials.[3] Differences between in vitro and in vivo response are often attributed to dissolved oxygen content, sulfur-containing amino acids present in blood, and pathological changes associated with implantation such as generation of hydrogen peroxide and lowered pH (as low as 4) at the implant site.[3] Corrosion is an even greater concern in dental applications, because of the high acidity and chloride ion levels in many foods. Additionally, the corrosiveness of saliva is highly dependent on oral hygiene.[3] Thus, while reasonable in vitro models for saliva are available, it is difficult to predict in vivo corrosive resistance of dental materials.

Most metallic biomaterials fall into one of four categories: stainless steels, titanium and titanium-based alloys, cobalt–chromium alloys, and amalgams.[3] Additionally, research is under way on a number of next-generation metallic biomaterials, including rare earth materials and shape-memory alloys. Of the stainless steels, type 304 had been used previously in medical applications, but problems with localized corrosion and tumor formation were sometimes reported.[3] Type 316L SS is currently the most widely used in biomedical applications.[5] This is an iron–chromium–nickel alloy with a low carbon content, where chromium provides corrosion resistance. The resistance to pitting corrosion can be improved if nitrogen additions are made.[3] More recently, additional corrosion-resistant stainless steels have been developed, including 316LVM grade with a typical composition of 18Cr14Ni3Mo. For dental materials, which must be extremely corrosion resistant, ultraclean high nitrogen austenitic stainless steels are recommended, such as 21Cr10Ni3Mo0.3Nb0.4N.[3] Finally, mixing different grades of stainless steels is not recommended, as this can lead to galvanic corrosion and failure. In general, stainless steels display better mechanical and formability properties but worse corrosion resistance than titanium-based alloys. Release of chromium presents a concern, although the levels of chromium are lower in stainless steels than in cobalt–chromium alloys.[3]

Titanium displays excellent biocompatibility and corrosion resistance; however, it does not have the high strength necessary for several biomedical applications.[3,5] The most popular material for load-bearing orthopedic applications is Ti6Al4V, a dual-phase alloy comprising an Al-stabilized η-phase and a V-stabilized β-phase.[5] Other alloys in use for medical applications include Ti2.5Al2.5Fe, Ti6Al7Nb, and Ti50Ta.[3,5] While all these alloys exhibit a higher tensile strength than titanium, their corrosion resistance is not as high.[3] They also display poor shear strength and thus should not be used for applications such as screws.[3]

Common cobalt–chromium alloys include CoCrMo, used in dentistry and artificial joint applications, and CoCrNiMo, used as a part of replacements for heavily loaded joints because of its very high tensile strength.[3] Other cobalt–chrome biomaterials include CoCrWNi, MP35N, and ASTM F1058 (40Co12Cr15Ni7Mo).[3,5] In all these materials, chromium is present at fairly high levels. Thus, the release of chromium upon corrosion is a concern, as chromium is a carcinogen.[3]

Amalgams are typically used in dental applications and are multiphase alloys. Silver–tin amalgam carries a risk of mercury release through corrosion of the Sn_7Hg $\gamma 2$ phase.[3] Newer high copper amalgams reduce the risk of mercury release, as preferential corrosion of the η' phase, Cu_6Sn_5, typically occurs.[3] However, release of mercury can still occur even in these materials. Older silver–tin amalgams are based on a silver–tin alloy, while high copper amalgams are based on either a silver–copper–tin alloy or a mixture of silver–tin and silver–copper alloys.[3]

Next-generation metallic biomaterials include porous titanium alloys and porous CoCrMo with elastic moduli that more closely mimic that of human bone; nickel–titanium alloys with shape-memory properties for dental braces and medical staples; rare earth magnets such as the NdFeB family for dental fixatives; and titanium alloys or stainless steel coated with hydroxyapatite for improved bioactivity for bone replacement.[3,5,6] The corrosion resistance, biocompatibility, and mechanical properties of many of these materials still must be optimized; for example, the toxicity and carcinogenic nature of nickel released from NiTi alloys is a concern.[3]

Ceramics and Glasses

Many ceramic materials possess improved biocompatibility as compared to metals, and corrosion is typically not an issue. Ceramics often have high strength but display brittleness, poor crack resistance, and low ductility.[7] Several ceramic materials are bioinert, bioactive (forming bonds with the surrounding tissue such as bone), or bioresorbable (as in the case of some porous ceramics).[7]

Arguably, the most important ceramic biomaterial is hydroxyapatite (HAP), $Ca_{10}(PO_4)_6(OH)_2$, a synthetic analog of bone mineral.[7] Natural bone is a composite comprising small crystalline HAP platelets bonded to collagen.[7] Synthetic hydroxyapatite is known to be bioactive, forming a strong bond with adjacent bone tissue and inducing bone growth along the interface (termed osteoconduction).[6] Hydroxyapatite and its derivatives are used most often in orthopedic and dental applications, for applications such as repair of bone defects and tooth root implants.[7] As mentioned above, HAP powders and coatings are also sometimes used in conjunction with metallic implants to induce adhesion with the surrounding tissue and to promote bioactivity.[6] Solid HAP is very robust in physiological environments and can remain in the body for 5–7 yr, while porous HAP can be resorbed by the body after approximately 1 yr.[7] Over 30 yr of research and clinical practice suggest that HAP and its related compounds are generally nontoxic and produce little or no inflammation or foreign body response. There have been some unfavorable biological reactions reported with porous HAP; these have been attributed to irritation from sharp edges of the implant and micromovement of the implant.[7]

Materials derived from HAP include Mg-HAP, which has been investigated for bone restoration in osteochondrosis; carbonate-HAP, with potential applications in the deposition of calcium and in some root canal fillings; silver-HAP, of interest for infected bone defects; and fluorine-HAP, of interest in the treatment of tooth defects and as coatings for metallic biomaterials.[7] Closely related materials include tricalcium phosphate, $Ca_3(PO_4)_2$, and certain calcium-phosphate glasses (described below), which form microcrystals of HAP on their surfaces in vivo, and thus exhibit similar degrees of bonding to bone tissue as HAP and its derivatives.[7] Additionally, tricalcium phosphate and HAP can both be combined with glasses to optimize the resorption time in vivo or to tune the bioactivity. Many of these so-called "glass-ceramics" typically contain SiO_2 in concentrations from 12% to 50% by mass, and may also contain CaO, Na_2O, or P_2O_5; or less commonly K_2O, MgO, CaF_2, Al_2O_3, and B_2O_3. Glass-ceramic implants with a lower calcium concentration may be resorbed in the body in as little as 10–30 days.[7] Taken together, HAP, tricalcium phosphate, and related materials have been investigated for nearly every conceivable type of bone reconstruction, including skull restoration after surgery or trauma, tooth-root implants, and tooth fillings, maxillofacial reconstruction, joint reconstruction, repair of load-bearing skeletal elements, grafting and stabilizing skull bone, repair of alveolar clefts and augmentation of alveolar ridges, cervical spine fusion, and reconstructive surgery of the middle ear.[7] Additionally, because glass-ceramics containing approximately 45% SiO_2 display strong bonding with soft tissue, these materials are of interest for the restoration of tendons, ligaments, small blood vessels, and nerve fibers; as well as catheters for infected wounds and as microsurgical joints.[7] Finally, HAP is used as a component of composite materials for bone replacement applications, as reviewed recently.[8,9] Examples include collagen–HAP composites and polyethylene–HAP composites.

Another important class of ceramic biomaterials is based on Al_2O_3 (aluminum oxide or corundum). These materials are bioinert and are widely used for the knob of replacement hip joints and other joint endoprostheses.[7] Other uses include tooth-root implants, parts for osteosynthesis, maxillofacial implants, and skull surgery.[7] Aluminum oxide ceramics prepared for use as biomaterials must have a low concentration of impurities such as silicon oxide and alkaline earth oxides, which can interfere with the bioinertness of the implant.[7]

A third class of bioceramics are based on ZrO_2, stabilized by Y_2O_3 or CeO_2. These materials are close to aluminum oxide materials in terms of biocompatibility but exhibit a higher bend strength and crack resistance, though with lower compressive strength.[7] Zirconium dioxide ceramics can be used for many of the same types of biomedical applications as aluminum oxide and is in fact replacing aluminum oxide in many applications.

Glass-based biomaterials have been reviewed recently by Hench, Xynos, and Polak;[6] Stroganova, Mikhailenko, and Moroz;[10] and Knowles.[11] As alluded to above, biocompatible glasses are often used in conjunction with ceramics or alone for applications involving bone and joint reconstruction. Bioactive glasses are more quickly dissolved and resorbed into the body than bioceramics; furthermore, the dissolution time can be tuned using a variety of techniques.[11] Much of the research in this area has involved the ternary Na_2O–CaO–P_2O_5 system, which is of interest for both tissue engineering and antibacterial applications.[11] Calcium phosphate glasses with additional components, such as SiO_2, K_2O, MgO, and Al_2O_3, have also been investigated for bone restoration applications.[10]

Polymers

The literature on polymeric biomaterials is vast. Polymeric materials typically are easier to process than metals or ceramics, and often have lower strength and moduli. Their microstructure and transport properties can often be tuned by varying processing conditions, polymer molecular weight, cross-link density, and environmental conditions (temperature, pH, ionic strength, etc.). For these reasons, they are used more often for soft tissue engineering and drug delivery applications. Polymer-based materials with biomedical applications include hydrogels, temperature, and pH-responsive gels, conventional thermoplastics, block copolymers, polysaccharides, and artificial proteins, to name a few. Several of these are described later in this chapter (see sections Smart Biomaterials and Bioactive Biomaterials below). Here we summarize the major biomedical applications of different classes of polymers. Table 2.2 lists some important classes of polymeric materials along with their main biomaterials applications.[12]

The Next Generation of Biomaterials

Surface-Modified Biomaterials

While traditional biomaterials have saved millions of lives, these materials still suffer from problems with infection, thrombosis, inflammation, and poor healing with resultant fibrous encapsulation of biomaterials. Implanted biomaterials induce an inflammatory foreign body reaction that prevents normal wound healing; the foreign body response may result from nonspecific protein adsorption to biomaterial surfaces. One approach to overcoming these problems is to engineer the tissue-biomaterial interface by modifying the biomaterial surface.[13] Biomaterial surfaces may be engineered to create nonfouling, or stealth, surfaces that inhibit protein adsorption. Poly(ethylene glycol) (PEG) has been attached to biomaterials to create nonfouling materials that repel protein adsorption and cell attachment in vitro; PEG-modified surfaces have been found to resist cell adhesion for up to 2 weeks in vitro. While PEG is effective, its nonfouling properties are dependent on surface chain density, and it is easily damaged by oxidants. Poly(ethylene glycol) oligomers in self-assembled monolayers have been applied to create precision surfaces that are highly protein resistant.[14] Additionally, PEG-like surfaces have been prepared by plasma deposition of tetraethylene glycol dimethylether (tetraglyme) to form a highly nonfouling cross-linked structure; these surfaces are resistant to protein adsorption, as well as adhesion of platelets, monocytes, endothelial cells, and bacteria in vitro. In addition to PEG and PEG-like surfaces, nonfouling surfaces have been constructed using phospholipids, including phosphatidyl choline, and saccharides. In general, a number of different strategies have been successful in reducing nonspecific protein adsorption to biomaterials *in vitro*, but this has not yet translated into success in vivo. For example, tetraglyme-coated implants still induce fibrous capsule formation when implanted subcutaneously in mice, and tetraglyme-treated implants exhibit significantly higher macrophage adhesion than untreated implants in vivo.[13] Better model systems that accurately capture the in vivo environment will be required for the development of nonfouling surfaces for implanted biomaterials.

Smart Biomaterials

In numerous medical device applications, it may be desirable to have biomaterials that can respond to changes in the surrounding environment. Environmental triggers can be used in implanted biomaterials to activate or deactivate drug delivery, cell attachment, or a change in mechanical properties. Smart biomaterials have been designed that are sensitive to changes in pH, temperature, and other physical and chemical stimuli.[15] Hydrogels that exhibit pH-dependent swelling behavior can be created from ionic networks. For

TABLE 2.2

Important Classes of Polymeric Biomaterials

Polymer	Main Biomaterials Applications
Proteins and protein-based polymers	
Collagen	Soft tissue engineering and implants, absorbable sutures, wound dressings, drug delivery
Albumin	Cell encapsulation and drug delivery
Poly(amino acids)	Oligomeric drug carriers, polyelectrolyte complexes for cell encapsulation
Polysaccharides and derivatives	
Carboxymethyl cellulose	Drug delivery, dialysis membranes, polyelectrolyte complexes for cell encapsulation, and cell immobilization
Cellulose sulphate	Complexes for cell encapsulation
Agarose	Supporting material for clinical analysis, cell immobilization
Alginate	Cell encapsulation and immobilization, immobilization of enzymes, controlled release, injectable microcapsules
Carrageenan	Microencapsulation, thermoreversible gelation
Hyaluronic acid	Lubrication applications
Heparin and heparin-like glycosaminoglycanes	Antithrombotic and anticoagulant properties; used in surgery
Dextran and derivatives	Drug delivery
Chitosan and derivatives	Gels, membranes, and microspheres for drug delivery; polyelectrolyte complexes for encapsulation
Aliphatic polymers	
Poly(lactic acid), poly(glycolic acid), and their copolymers	Biodegradable sutures, drug delivery systems, and tissue engineering scaffolds
Poly(hydroxy butyrate), poly(caprolactone), and their copolymers	Biodegradable matrices for controlled release and cell encapsulation
Polyamides	Sutures, dressings, hemofiltration membranes
Polyanhydrides	Biodegradable tissue engineering scaffolds and devices
Poly(ortho esters)	Surface-eroding materials for sustained release and ophthalmology
Poly(cyano acrylates)	Biodegradable surgical adhesives and glues
Polyphosphazenes	Drug delivery, hydrogels, and thin films
Thermoplastic polyurethanes	Permanent implants, prostheses, vascular grafts, catheters, and drug delivery devices
Polyethylene	Sutures, catheters, membranes
Poly(vinyl alcohol)	Gels and membranes for drug delivery and cell encapsulation
Poly(ethylene oxide) and copolymers	Used to render surfaces biocompatible and resistant to protein adhesion; copolymers with poly(propylene oxide) form thermoreversible gels for drug delivery
Poly(hydroxyethyl methacrylate)	Hydrogels for soft contact lenses, drug delivery, skin coatings
Poly(methyl methacrylate)	Dental implants, bone replacement
Polytetrafluoroethylene	Vascular grafts, clips, and sutures
Polydimethylsiloxanes	Implants in plastic surgery and orthopedics blood bags and pacemakers
Poly(vinyl methyl ether)	Temperature-sensitive materials; shape-memory materials
Poly(*N*-alkylacrylamides)	Temperature-sensitive gels

(Adapted from Angelova and Hunkeler.[12])

example, hydrogels prepared from poly(methacrylic acid) grafted with poly(ethylene glycol) (PMAA-g-PEG) shrink at pH 2 because of formation of interpolymer complexes, but swell 3–25 times in size at a physiological pH 7.[16] Ionic strength-dependent and pH-dependent swelling has also been observed in gels of PMAA or poly(acrylic acid) with poly(hydroxyethyl methacrylate).[15] These gels can be loaded with drug and will trap the drug at low pH, then swell to release the drug at a high pH. Such gels may be promising for oral delivery of biopharmaceuticals, as the gel will protect the drug from the acidic pH of the stomach for subsequent successful delivery in the small intestine. Insulin-loaded PMAA-g-PEG gels have been administered orally to diabetic mice, and the glucose levels of the mice decreased following gel administration, suggesting that the gel successfully protected and delivered the protein therapeutic.[17]

In addition to pH-sensitive gels, thermally sensitive biopolymers have been investigated; such materials can respond to changes in temperature from room temperature to body temperature. These hydrogels exhibit a lower critical solution temperature (LCST); the gels expand when cooled below the LCST and contract when heated above the LCST. Hydrogels of poly(N-isopropylacrylamide) (pNIPAAm) have an LCST of 32°C and have been studied extensively for drug release applications.[15] Temperature-modulated drug release from pNIPAAm hydrogels can be achieved via bulk squeezing: drug that is distributed inside the matrix is squeezed out when the hydrogel contracts with heating above the LCST. Another application of pNIPAAm hydrogels is the creation of surfaces for cell culture systems. Coatings of pNIPAAm are hydrophilic below the LCST and hydrophobic above the LCST. Because cells attach on hydrophobic surfaces and detach from hydrophilic surfaces, cells cultured on pNIPAAm surfaces can be nontraumatically lifted as intact sheets from these surfaces, simply by lowering the temperature.

Polymers of elastin-like peptide (ELP) also exhibit temperature-sensitive behavior.[13] The LCST of ELP can be controlled by varying the length of the ELP molecule or its amino acid composition to obtain transition temperatures ranging from 30°C to 90°C. The ELPs have been used for chondrocyte encapsulation and as drug carriers for cancer therapy. An additional thermoresponsive biomaterial is the biodegradable triblock copolymer poly(ethylene glycol)–poly(lactic acid-co-glycolic acid)–poly(ethylene glycol) (PEG–PLGA–PEG). This copolymer exhibits a sol-to-gel transition with increases in temperature.[15] This property may be important for drug delivery applications, as PEG–PLGA–PEG can be injected as a free-flowing solution at room temperature and then becomes a gel upon reaching body temperature. Hydrogels with certain compositions may demonstrate both pH- and temperature sensitivity. For example, hydrogel copolymers of MAA and NIPAAm sense small changes in pH and temperature. Controlled release of antithrombotic agents, including streptokinase and heparin, has been demonstrated with these dual-sensitivity hydrogels.[15]

Other stimuli in addition to pH and temperature have been investigated to create smart biomaterials. These include external physical stimuli such as light, magnetic fields, electric current, and ultrasound. Specific biochemical stimuli also may be used: calcium-responsive hydrogels, antigen-responsive hydrogels, and microbial infection-responsive hydrogels have all been designed for drug delivery applications.[14]

Bioactive Biomaterials

Several advances have been made in rendering materials biologically active.[18] Bioactivity may be used to impart pharmacological activity to a biomaterial, to modify the biocompatibility of a material, or to tune the lifetime and degradation of a biomaterial. One method for creating bioactive biomaterials is to incorporate pharmacological agents into materials.

Polymeric implants that release therapeutic drugs are already in clinical use; examples include progesterone-releasing implants for fertility regulation, and lupron-releasing implants for prostate cancer treatment. Polymer-coated arterial stents that elute sirolimus and tacrolimus have revolutionized the treatment of coronary artery disease, by preventing restenosis following stent placement. Site-specific delivery of chemotherapeutics for brain cancer treatment has been achieved using polyanhydride disks loaded with carmustine (BCNU).[19] Biological growth factors may also be incorporated into materials, either through surface display or through controlled release systems. Polypeptide growth factors may regulate a variety of cellular responses, including cell migration, proliferation, survival, and differentiation. Growth factors have been exploited to create biomaterials that deliver angiogenic growth factors to induce vascular repair; neuronal survival and differentiation factors to treat neurodegenerative disease; transforming growth factor β to induce bone repair; bone morphogenetic protein-4 to enhance bone formation; and tissue growth factors to heal chronic ulcers.

Another approach to creating biological activity in biomaterials is to incorporate adhesion factors, including adhesion-promoting oligopeptides or oligosaccharides.[18] While traditional biomaterials promote cell adhesion via nonspecific adsorption of proteins, a greater degree over cell migration, cell adhesion, and cell-type selectivity can be achieved by incorporating adhesion-promoting factors directly into biomaterials. For example, the RGD tripeptide from fibronectin may be either immobilized on biomaterial surfaces or included directly into the backbone of polymer chains to induce cell adhesion, spreading, focal contact formation, and cytoskeletal organization. The YIGSR domain and the SIKVAV domain from laminin are migration-promoting peptides, and have been incorporated into gels to promote neuronal cell infiltration and nerve regeneration. The tetrapeptide REDV from fibronectin may be employed in vascular grafts to specifically support adhesion and migration of vascular endothelial cells, while also preventing the adhesion of clot-forming platelets. The protein osteopontin has been immobilized on poly(2-hydroxyethyl methacrylate) surfaces to promote endothelial cell adhesion.[13] These approaches to biomaterial modification are clinically relevant, as it has been shown in clinical studies that enhancement of cell adhesion strength improves the performance of endothelial cell-seeded vascular grafts in high-flow regions.[20]

Still another method for engineering bioactive biomaterials is to design materials containing enzymatic recognition sites.[18] Incorporation of enzymatic cleavage sites into a biomaterial allows the degradation rate of the material to be tuned, and also allows the biomaterial to be proteolytically remodeled. For example, gels containing PEG chains with central oligopeptide sites that are substrates for collagenase or plasmin are degradable by cell-associated enzymatic activity.

Tissue Engineering and Regenerative Medicine

The goal of tissue engineering is to replace lost tissues or organs with polymer constructs that contain specific populations of living cells. Traditional strategies for replacing lost tissue include organ transplantation and mechanical device implantation; however, organ transplantation is limited by donor organ shortages and transplant rejection, and mechanical devices have limited durability and limited bioactivity. Tissue engineering has the potential to overcome these limitations by creating functional tissues that have the capacity for growth, remodeling, and self-repair. The general approach to creating engineered tissue is to harvest specific cell populations from the tissue of interest, seed the cells into a biodegradable polymer scaffold, and cultivate the cell/polymer construct in a bioreactor prior to implantation.[21] Upon implantation, the biodegradable polymer will degrade and gradually be replaced by regenerated tissue.

Successful tissue engineering requires appropriate selection of cells, polymers, and bioreactor conditions for the application of interest. Cells used in tissue engineering may be drawn from either primary tissues or cell lines; the bulk of tissue engineering experiments have utilized primary autogenous cells. Scaffold polymers for tissue engineering must be biodegradable, biocompatible, and readily processible into appropriate anatomical shapes. The most commonly used scaffold polymers are poly(glycolic acid) (PGA), poly(lactic acid) (PLA), and their copolymers (PLGA). Polyanhydrides, polycarbonates, and polyurethanes have also been investigated, and hydrogels, particularly algal polysaccharides, have been utilized as cell delivery matrices. Bioreactors for tissue engineering are designed to allow optimal conditioning of the cell/polymer construct to initiate tissue formation prior to implantation. It is desirable to achieve a high and uniform cell density throughout the polymer scaffold. The bioreactor is a dynamic tissue culture environment where gas and nutrient exchange are augmented by constant turnover of tissue culture medium, and where tissue-specific mechanical forces (stretch, pressure, shear forces) are recapitulated.[22]

The tissue engineering approach has been applied to many tissues including skin, cartilage, bone, liver, intestine, urologic tissue, cardiovascular tissue, and neural tissue.[19,22] Tissue engineered skin regeneration systems are already in clinical use to repair burns, wounds, and chronic ulcers. In one approach, neonatal dermal fibroblasts are placed on PLGA scaffolds and grown into sheets to create skin.[19] In another approach, a bilayer system is used, in which a lower dermal layer of human fibroblasts and an upper epidermal layer of human keratinocytes are seeded onto a bovine collagen matrix. Cartilage regeneration has been achieved by delivering chondrocyte suspensions to focal articular cartilage defects; an autologous chondrocyte product has been FDA approved for clinical application.[22] Cartilage tissue has also been engineered in the configuration of the ear, nasal septum, and trachea using PGA and PLGA constructs seeded with chondrocytes; such constructs have also been shown to close full-thickness cranial defects in animals. Clinical trials are under way to use chondrocyte/polymer constructs for cartilage replacement in humans. In the area of bone replacement, PGA meshes seeded with periosteal cells have been shown to generate new bone in animal models. For replacement of liver function, hepatocytes are seeded on PGA sheets; implantation of these sheets into liver enzyme-deficient animals results in partial correction of the enzyme deficiency. Copolymers of PGA and PLA scaffolds fabricated to contain vascular-like channels and seeded with cocultures of hepatocytes and endothelial cells demonstrate remodeling and formation of vascular channels in vitro, suggesting that complex tissue architecture can be achieved with an appropriate cell/polymer construct. For intestinal replacement, mixed enterocyte/stromal cell populations have been seeded onto PGA meshes; such constructs demonstrate formation of structures resembling intestinal villi and crypts when implanted in animals. In the urologic system, functional ureters have been created using tubular PGA constructs seeded with urothelial cells and smooth muscle cells, and functional bladder tissue has been created from PGA sheets seeded with urothelium and smooth muscle cells. For cardiovascular tissue engineering, endothelial cells have been seeded on both tubular polymer constructs to create blood vessels and leaflet-like constructs for heart valve reconstruction; tissue engineered heart valves are functional in large animals.[22,23] For neural regeneration, an electrically conducting polymer, polypyrrole, has been shown to provide a substrate for nerve regrowth in animal models.

Conclusions

Biomaterials have improved the lives of millions by providing material solutions to biomedical problems. Biomaterials have been applied in a variety of clinical disciplines, including cardiovascular medicine, orthopedics, and ophthalmology, and new materials are in use or under development for virtually every organ system in the body. Traditional biomaterials have been designed from polymers, ceramics, and metals. The next generation of biomaterials will incorporate biomolecules, therapeutic drugs, and living cells. Innovative new biomaterials, including surface-modified biomaterials, smart biomaterials, bioactive biomaterials, and tissue-engineered materials, will have improved properties of biocompatibility, tunability, and biological functionality. Successful development of new biomaterials will require an increased understanding of cell–material interactions, as well as better model systems for the biological environment.

References

1. Williams, D.F. Definitions in biomaterials. *Proceedings of a Consensus Conference of the European Society for Biomaterials,* Elsevier: New York, 1987.
2. Ratner, B.D., Hoffman, A.S., Schoen, F.J., Lemons, J.E. *Biomaterials science: an introduction to materials in medicine.* Academic Press: New York, 1996.
3. Blackwood, D.J. Biomaterials: past successes and future problems. *Corros. Rev.* **2003**, *21* (2–3), 97–124.
4. Krug, H.F. Metals in clinical medicine: the induction of apoptosis by metal compounds. *Mat. Wiss. Werkstofftech.* **2002**, *33*, 770–774.
5. Kannan, S.; Balamurugan, A.; Rajeswari, S.; Subbaiyan, M. Metallic implants—an approach for long term applications in bone related defects. *Corros. Rev.* **2002**, *20* (4–5), 339–358.
6. Hench, L.L.; Xynos, I.D.; Polak, J.M. Bioactive glasses for in situ tissue regeneration. *J. Biomater. Sci. Polym. Ed.* **2004**, *15* (4), 543–562.
7. Dubok, V.A. Bioceramics—yesterday, today, tomorrow. *Powder Metall. Met. Ceram.* **2000**, *39* (7–8), 381–394.
8. Mano, J.F.; Sousa, R.A.; Boesel, L.F.; Neves, N.M.; Reis, R.L. Bioinert, biodegradable and injectable polymeric matrix composites for hard tissue replacement: state of the art and recent developments. *Compos. Sci. Technol.* **2004**, *64*, 789–817.
9. Kikuchi, M., Ikoma, T., Itoh, S., Matsumoto, H.N., Koyama, Y., Takakuda, K., Shinomiya, K., Tanaka, J. Biomimetic synthesis of bone-like nanocomposites using the self-organization mechanism of hydroxyapatite and collagen. *Composites Sciences and Technology* **2004**, *64*, 819–825.
10. Stroganova, E.E., Mikhailenko, N.Y., Moroz, O.A. Glass-based biomaterials: present and future (a review). *Glass Ceram.* **2003**, *60* (9–10), 315–319.
11. Knowles, J.C. Phosphate based glasses for biomedical applications. *J. Mater. Chem.* **2003**, *13*, 2395–2401.
12. Angelova, N., Hunkeler, D. Rationalizing the design of polymeric biomaterials. *Trends Biotechnol.* **1999**, *17*, 409–421.
13. Ratner, B.D., Bryant, S.J. Biomaterials: where we have been and where we are going. *Annu. Rev. Biomed. Eng.* **2004**, *6*, 41–75.
14. Ratner, B.D. Reducing capsular thickness and enhancing angiogenesis around implant drug release systems. *J. Controlled Release* **2002**, *78* (1–3), 211–218.

15. Peppas, N.A., Bures, P., Leobandung, W., Ichikawa, H. Hydrogels in pharmaceutical formulations. *Eur. J. Pharm. Biopharm.* **2000**, *50* (1), 27–46.
16. Kim, B., La Flamme, K., Peppas, N.A. Dynamic swelling behavior of pH-sensitive anionic hydrogels used for protein delivery. *J. Appl. Polym. Sci.* **2003**, *89* (6), 1606–1613.
17. Lowman, A.M., Morishita, M., Kajita, M., Nagai, T., Peppas, N.A. Oral delivery of insulin using pH-responsive complexation gels. *J. Pharm. Sci.* **1999**, *88* (9), 933–937.
18. Hubbell, J.A. Bioactive biomaterials. *Curr. Opin. Biotechnol.* **1999**, *10* (2), 123–129.
19. Langer, R. Biomaterials in drug delivery and tissue engineering: one laboratory's experience. *Acc. Chem. Res.* **2000**, *33* (2), 94–101.
20. Meinhart, J., Deutsch, M., Zilla, P. Eight years of clinical endothelial cell transplantation. Closing the gap between prosthetic grafts and vein grafts. *ASAIO J.* **1997**, *43* (5), M515–M521.
21. Langer, R. Selected advances in drug delivery and tissue engineering. *J. Controlled Release* **1999**, *62* (1–2), 7–11.
22. Marler, J.J., Upton, J., Langer, R., Vacanti, J.P. Transplantation of cells in matrices for tissue regeneration. *Adv. Drug Deliv. Rev.* **1998**, *33* (1–2), 165–182.
23. Nugent, H.M., Edelman, E.R. Tissue engineering therapy for cardiovascular disease. *Circ. Res.* **2003**, *92* (10), 1068–1078.

3

Hydrophilic Polymers for Biomedical Applications

Frank Davis and Séamus P. J. Higson

CONTENTS

Introduction

The production of polymeric materials is one of the world's major industries. Polymers are utilized in many applications because of their processability, ease of manufacture, and diverse range of properties. Many of the commonest polymers such as polyethylene, polystyrene, and polytetrafluoroethylene (PTFE) are highly hydrophobic materials rendering them unsuitable for many biomedical applications. For applications that require contact with body fluids such as blood or urine, it is necessary for the materials to be hydrophilic and to be capable of maintaining intimate contact with the fluid in question for prolonged periods of time without significant loss of functional performance.

This chapter begins by describing the nature of hydrophilic polymers and structures, together with properties of some of the most commonly encountered materials. Details are then given of some applications that require the use of polymers and how they can be applied. The chapter finishes with a brief synopsis of the work in this field and possible future research and applications.

Structure and Properties of Hydrophilic Polymers

Polymeric materials made from simple hydrocarbon monomers such as polyethylene are extremely hydrophobic in nature. There is very little in the way of interaction between the polymer backbone and water, which means that these polymers will not adsorb water and their surfaces are not wettable. For a polymer to be hydrophilic, functional groups capable of undergoing interactions with water such as hydrogen bonding are usually required. Such groups include, but are not limited to, alcohols, ethers, esters, carboxylic acids, amines, and amides. These can interact with water via hydrogen bonding or dipolar interactions; such polymers can vary from just being wettable at their surfaces to being permeable to water, and in some cases are capable of adsorbing many times their own weight in water. Some of the most widely used polymers are discussed later. The properties and use of these materials in biomedical applications can only be summarized within the limits of this chapter; however, several other works have been written on this subject.[1–3]

Hydrogels

Many of the polymers mentioned within this work exist as hydrogels. In essence, a hydrogel is a polymer that would normally be soluble in water but has been cross-linked to form a polymer network. The cross-linking process renders the polymers insoluble but does not remove their affinity for water. Therefore, the network can adsorb water with consequent swelling of the polymer. The nature of the polymer and the degree of cross-linking affect the swelling behavior; highly hydrophilic polymers containing few cross-links will tend to adsorb large amounts of water with a high degree of swelling. Less hydrophilic monomers, incorporation of hydrophobic comonomers, or a high degree of cross-linking, all act to reduce water adsorption, usually leading to a more rigid and firmer gel. Hydrogels often show high biocompatibility, usually because they have a high water content, within either a bioinert or a biodegradable polymer network.

Synthetic Hydrophilic Polymers

The structures of a number of synthetic hydrophilic polymers as a representative range of those commonly used are shown in Figure 3.1. Many common hydrophilic polymers are based on methacrylate or acrylate backbones. One of the most common is polyacrylic acid, which forms the basis of many hydrogel materials (Figure 3.1A). Although simple polyacrylic acid is water soluble, it can be easily cross-linked to form an insoluble network polymer that still, however, retains high affinity for water and is capable of adsorbing large amounts of water to form hydrogels; these are discussed later. Polymeth-acrylic acid-based polymers are similar in nature (Figure 3.1B). One of the most commonly used polymers for contact lenses, polyhydroxyethyl methacrylate, is based on a similar backbone (Figure 3.1C).

Polyvinyl alcohol is another widely used material, often as a surface coating because of its high biocompatibility (Figure 3.1D). Usually, it can be obtained by the hydrolysis of polyvinyl acetate, with varying amounts of hydrolysis (Figure 3.1E). Direct synthesis from the monomer is not possible because vinyl alcohol does not exist but tautomerizes to acetaldehyde. The solubility and physical properties of this polymer are highly dependent on its molecular weight and degree of hydrolysis. Cross-linking of the polymer leads to a variety of hydrogels.

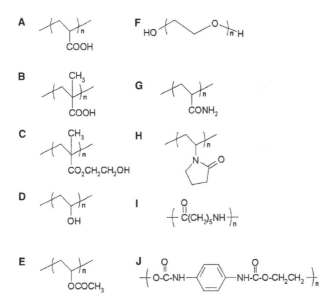

FIGURE 3.1
Structures of some common synthetic hydrophilic polymers: (A) polyacrylic acid, (B) polymethacrylic acid, (C) polyhydroxyethyl methacrylate, (D) polyvinyl alcohol, (E) polyvinyl acetate, (F) PEG/PEO, (G) polyacrylamide, (H) polyvinylpyrrolidinone, (I) Nylon 6, and (J) a simple polyurethane.

Polyethylene glycol (PEG) and the very similar polyethylene oxide (PEO) are used as biocompatible coating agents and hydrogel forming materials, often as block or graft copolymers with other materials (Figure 3.1F). They are often bound to polyurethanes to form hydrophilic foams such as Biopol® (Metabolix Inc.).

Polyacrylamides are also suitable for a variety of biomedical uses; the structure of poly-acrylamide is shown in Figure 3.1G, although the use of acrylamides in copolymers is much more common. Polyvinylpyrrolidinone has also found use as a biocompatible coating material (Figure 3.1H). Polyacrylonitrile, though not suitable in itself, can be hydrolyzed to form some hydrophilic polymers such as the Hypan® (Hymedix Inc.) series of hydrogels.

Condensation polymers have been used although they are often not sufficiently hydrophilic enough to be considered within this chapter. Typical polymers such as Nylon 6 and a typical polyurethane are shown (Figures 3.1I and 3.1J). Some of these have been incorporated into composites with more hydrophilic materials; Biopol, for example, is formed from a copolymer of a polyurethane and PEG.

Natural Hydrophilic Polymers

Besides the synthetic materials, there are a variety of hydrophilic polymers obtained from natural sources. Some examples include collagen, alginate, and carragee, all of which are obtained from natural sources.[1] Collagen is widely used in implants and many other possible uses and applications have been studied for this material.

A series of materials have been made by derivatizing cellulose (Figure 3.2a). This naturally occurring polysaccharide contains reactive hydroxyl groups, which can be easily substituted, normally to form ethers. Typical materials include sodium carboxymethylcellulose, hydroxyethyl, and hydroxypropyl cellulose (Figures 3.2b–3.2d). The properties of

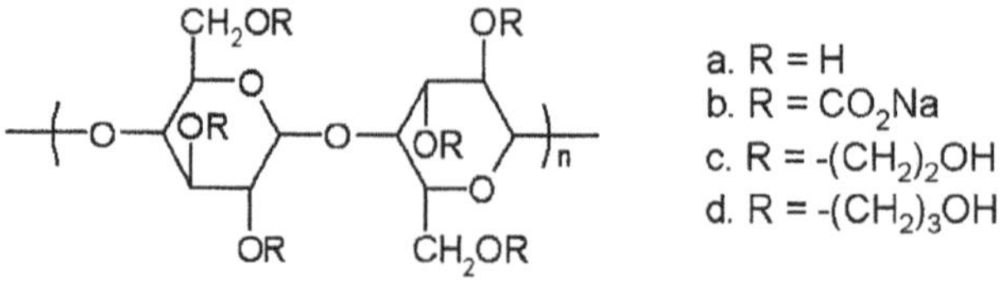

a. R = H
b. R = CO$_2$Na
c. R = -(CH$_2$)$_2$OH
d. R = -(CH$_2$)$_3$OH

FIGURE 3.2
Structures of some common cellulose-based hydrophilic polymers: (a) cellulose, (b) sodium carboxymethyl cellulose, (c) hydroxyethyl cellulose, and (d) hydroxypropyl cellulose.

these polymers depend greatly on the molecular weight and degree of substitution; they have found uses both as gels and as coatings.

Hydrophilic Polymers: Use as Contact Lens Materials

The use of contact lenses, initially made from glass, to correct vision has been known since the 19th century. The first polymeric contact lens was a "hard" contact lens made in 1936 from polymethylmethacrylate (PMMA).[4] However, these lenses had to be taken out at night to prevent eye irritation and much research has been directed to making those that could be worn for much longer periods of time. This led to the development of the so-called "soft" lenses.

To be suitable for long-term wear, a contact lens material has to satisfy several criteria. It must be hydrophilic enough to maintain a stable, continuous tear film on its surface, resist fouling by tear components, not irritate the eye, and be comfortable to wear. It should also be realized that the metabolism of the cornea is highly dependent on dissolution and adsorption of atmospheric oxygen and therefore any lens material must have sufficient oxygen permeability to maintain this, else corneal anoxia will set in.

Hydrogels were the first polymers to be used for this application because of their favorable physical properties and compatibility with the ocular environment. The first commercial product, Polymacon® (Bausch & Lomb), was based on polyhydroxyethyl methacrylate (Figure 3.1C).[5] The materials did suffer somewhat as they did not offer sufficient oxygen mass transport across the lens for prolonged use; hence, much further work has been devoted to developing materials with higher oxygen permeability. Organosilicon-based polymers such as polydimethylsiloxane have high oxygen permeability; however, most of these are very hydrophobic (Figure 3.3A). They have, therefore, been incorporated into lenses along with hydrophilic materials in an attempt to make a material with the required properties. Early studies utilized copolymers containing dimethylsiloxane groups along with hydrophilic monomers such as hydroxyethyl methacrylate or *n*-vinyl pyrrolidinone.[6] Fluoroethers also have high oxygen permeability and a copolymer of a perfluoroether with methyl methacrylate was briefly marketed under the name Fluorofocon A® (3M); however,

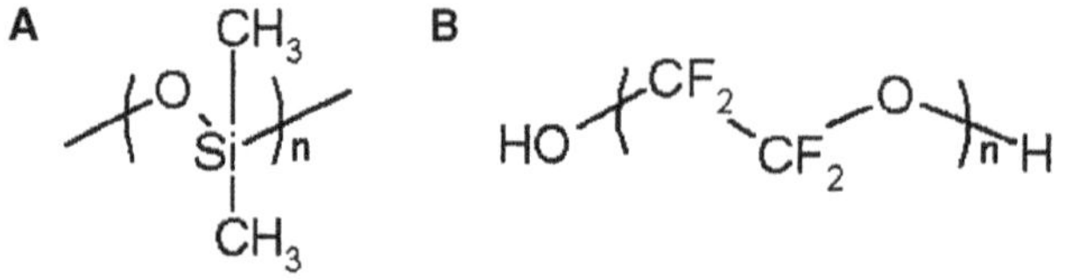

FIGURE 3.3
Structures of some polymers used in contact lens formulations: (A) a typical polysiloxane and (B) a typical polyperfluoroether.

the relatively high cost of the fluorocarbon monomers eventually led to their withdrawal from the market (Figure 3.3B).[7]

Most soft contact lenses commercially available are based on polyhydroxyethyl methacrylate; however, there are a few materials available, such as Lotrafilcon A® (Ciba Vision) and Balafilcon A® (Bausch & Lomb), based on siloxane or fluorosiloxane copolymers, that offer good oxygen permeability and so allow up to 30 days' continuous wear for contact lenses.[4] A survey of the patent literature shows that much interest is still being shown in the field of siloxane-based hydrogels in general and also in improving manufacturing processes for contact lenses formed from such hydrogels.

Besides external contact lenses, there are surgically implanted intraocular contact lenses that are usually used to replace the eye's natural lens after cataract surgery; these lenses are normally based on silicone or PMMA and lie outside the scope of this chapter.

Implantable Membranes

Postsurgical adhesions are a common problem following major abdominal, gynecological, and other forms of surgery. Adhesions are scars that form abnormal connections between tissue surfaces. Postsurgical adhesion formation is a natural consequence of surgery, resulting when tissue repairs itself following incision, cauterization, suturing, or other forms of trauma. This can lead to complications such as bowel obstruction and infertility. Implantable materials are in this case sometimes placed directly between tissue surfaces, for instance, between organs and the abdominal wall, to prevent adhesion formation.

A suitable membrane must be sterile, noninflammatory, and nontoxic, and because these cannot be removed from the body without further surgery, they are often also designed so as to be bioresorbable. A variety of polymers have been studied for this purpose. Typical materials available commercially include Seprafilm® (Genzyme Corp), which is based on a sodium hyaluronate/carboxymethyl cellulose composite, and Oxiplex® (Fziomed Inc.), which is based on a PEO/carboxymethyl cellulose composite.[8] Products designed for similar use but marketed in gel form include Oxiplex-SP®, Spraygel® (Confluent Surgical Inc.), and Resolve® (Life Medical Sciences Inc.). These can be applied directly to the site, rapidly form a solid hydrogel film to prevent adhesions, and are designed to be reabsorbed by the body after approximately 1 week.[9]

Sutures and Implants

Sutures are used to close wounds or incisions made during operations. If the material used for the suturing is bioresorbable, then this eliminates the need for removal of any stitches. Polymers such as polyglycolide [first marketed as Dexon® (Bayer Corp) in the 1960s] and polylactide are commonly used for this purpose, although a wider range of polymers are now available and have been recently reviewed elsewhere (Figures 3.4A and 3.4B).[10]

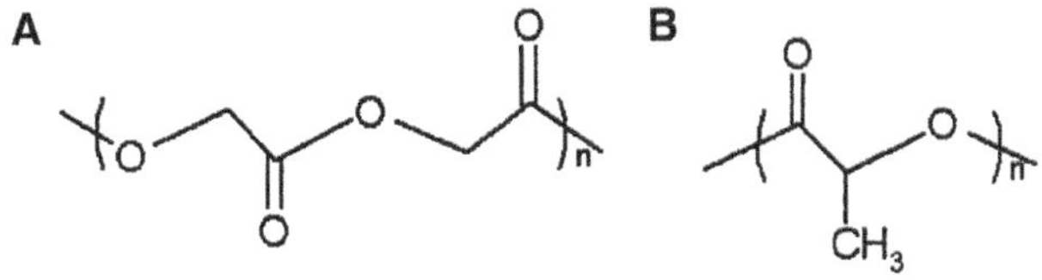

FIGURE 3.4
Structures of some bioresorbable polymers: (A) polyglycolide acid and (B) polylactide.

Besides sutures, other medical devices such as orthopedic fixation devices are of interest because the use of biodegradable polymers means that not only does the device not require removal (a process that can cause refracture of the bone), but because if it degrades it will also slowly transfer stress over time to the damaged area, allowing healing of the tissues. At the time of writing, bioresorbable polymers that offer the necessary strength for use as bone plates for long bones are yet to be developed, although they have found use in the manufacture of rods and pins for fracture fixation.[10]

Natural polymers have also been studied for cosmetic implants such as collagen. Bovine collagen implants have become increasingly popular for use in dermal augmentation of the skin since their introduction in the 1980s. They can be used either to remove unwanted wrinkles or to augment features such as collagen lip injections, which are gaining in popularity.

Burn and Wound Dressings

Many burn dressings are based on keeping a burn site moist while protecting it from the environment as well as helping in pain relief. Because hydrogels contain such a high water content, they have been investigated for use as burn and other wound dressings.[11] They have many desirable properties; for example, imposing the soft, moist texture can alleviate some pain; they also adhere to the wound but not so tightly that they cannot be easily detached from the wound for replacement; and finally, they are also transparent, allowing visual inspection of the healing process. Also, hydrogels prevent dehydration of the wound, have good oxygen permeability, and yet provide a barrier against bacteria. Finally, they can be easily manufactured with coadsorbed drugs for aiding in pain control, promotion of healing, and/or administration of antibiotics.

Materials are based on cross-linked hydrophilic polymers and are manufactured under a variety of trade names such as Spyroflex® (AGS Labs Inc.), Burnfree® (Burnfree Products), and Burnshield® (Levtrade International Ltd). Usually, they contain the hydrogel attached to an impermeable backing sheet and are currently available in up to blanket sizes. Typical polymers used include polyvinyl alcohol, PEG, and polyvinyl pyrrolidinone together with combinations of these materials. Irradiation is often used both to cross-link and to sterilize the hydrogel after it has been formed into the dressing. Collagen hydrogels are also being researched for their ability to allow cell migration and inhibit wound contraction because of their high tensile strength and low extensibility.[12]

Coatings

One major problem with inserting any type of biomedical device into the body that comes into contact with blood or other body fluids is that of biofouling. Both proteins and cells can adhere strongly to many foreign surfaces. In the case of devices such as contact lenses, for example, lipids and tear proteins can adsorb onto the surface of the lens. This can cause clouding of the lens, rendering it unsuitable for use, or lead to irritation. The irreversible adsorption of proteins and cells on synthetic surfaces has adverse effects on a variety of devices. Adverse biological reactions include fibrous encapsulation and blocking of small artificial blood vessels. Other devices, such as heart valves, can give rise to the formation of thromboses. This can cause either arterial blocking or the clot can detach from the surface, leading to the possibility of a stroke or heart attack. The application of hydrophilic polymer coatings has been widely studied to help render many surfaces biocompatible to minimize such risks.

Although a wide variety of polymeric materials have been utilized in the attempted prevention of biofouling as described in the extensive review by Kingschott and Grieser, by far the most popular materials have been ones based on PEG/PEO.[13]

The theoretical considerations of the causes of biofouling and why a PEG/PEO surface should resist the process so well is a topic for a complete review article in itself; however, it is widely reported that the lowest adsorption of proteins occurs for these materials.[13] One factor is thought to be the fact that the PEG/PEO chains are usually highly solvated, meaning that incoming protein molecules experience a surface that is largely composed of water. This surface mimics the typical conditions found within biological systems.

Several different approaches have been used to attach PEG/PEO coatings to a variety of surfaces. Among the simplest is the physical adsorption of Pluronic surfactants, which consist of PEO–polypropylene oxide (PPO)–PEO block terpolymers (Figure 3.4A).[14] The more hydrophobic PPO section of the chain is absorbed onto the substrate being treated with the more hydrophilic PEO blocks stretching out into the aqueous phase. These materials display a variety of biocompatibilities depending on the makeup of the surfactant. Interestingly, increasing the length of the PPO section appeared to have a larger repulsive effect than doing so of the PEO section. This indicates that perhaps only short PEO chains are necessary for effective protein repulsion and the beneficial effect of increasing the PPO section is because of better anchoring of the surfactant and prevention of displacement by protein molecules.[14]

Other methods of attachment have been studied. Alkyl thiols are known to form well-packed chemically bound monolayers on noble metal surfaces and so this method has also been used to attach PEG chains to surfaces. A variety of thiol-substituted PEG oligomers were synthesized and spontaneously assembled onto gold and silver surfaces (Figure 3.5B). [15] Quite short PEG chains were found to be sufficient to induce biocompatibility. It was found that the packing arrangement of the PEG derivatives was important in determining the surface properties with the resultant biocompatibility not simply being a direct function of packing density. Thiols pack more tightly on silver than on gold; yet some treated silver surfaces showed higher levels of fibrinogen adsorption than the corresponding treated gold surface. This was attributed to the PEG chains being in different conformations and showed that simply increasing the surface packing density does not necessarily increase repulsion.

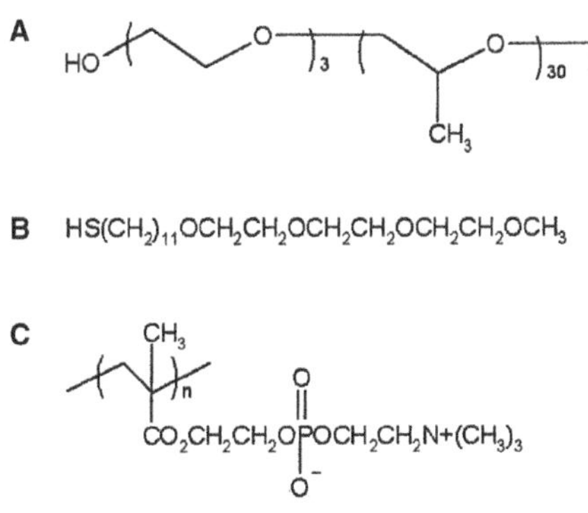

FIGURE 3.5
Structures of some common synthetic polymers used in coatings: (A) a typical Pluronic surfactant, (B) a thiol-substituted PEG derivative, (C) a polymer containing a phospholipid head group analog.

Plasma deposition has also been utilized to deposit PEO-like materials from volatile precursors onto a variety of subjects.[13] This technique involves generating a reactive plasma containing PEO-like monomers, which polymerize and deposit, often with chemical grafting, onto any surface within the plasma.[16] The availability of large-scale vacuum apparatus makes this technique feasible on an industrial scale. The materials deposited by this technique were often shown to contain only short PEO segments; yet greatly reduced protein deposition was observed and the small amounts (ng/cm) that did deposit were easily eluted.[16]

Other materials have also been studied for their ability to reduce protein adsorption onto surfaces. Because many cell membranes are based on phospholipids, polymers containing phospholipid-type head groups have been utilized for this purpose. Poly(2-methacroylethyl phosphoryl choline) could be plasma deposited onto silicone rubber and the adhesion of albumin reduced by factors of up to 80 (Figure 3.5C).[17]

Polysaccharides have also been studied for this purpose, although generally they are not as effective as PEG/PEO systems.[13] Polymers based on dextran substituted with thiol groups could be adsorbed on gold surfaces and have been shown to significantly repel protein deposition, with the degree of repulsion being dependent on both the polymer molecular weight and the amount of thiol substitution.[18]

An additional benefit of coating samples with hydrophilic polymers is their ability to act as a lubricant. Lubrication of a device surface improves device insertion and manipulation; however, traditional treatments that include coatings with low-friction materials such as PTFE or silicone fluid, while improving slip in the body, are difficult for the physician to handle. Incorporation of hydrophilic polymers leads to the adsorption of water molecules in the presence of water or body fluids. This creates a sheath of water at the surface of the device. This watery interface causes the decrease in wet friction, achieving lower friction when wet, while still being easy to handle when dry.

Biosensors

Biosensors are devices that use the unique recognition qualities of biological molecules to selectively detect the presence of desired analytes within complex mixtures such as blood and serum. They generically offer simplified reagent less analyses for a range of biomedical and industrial applications, and for this reason the field has continued to develop into an ever-expanding and multidisciplinary one during the last couple of decades. The world market for biosensors is approximately $5 billion, and at the time of writing, approximately 85% of the world commercial market for biosensors is for blood glucose monitoring.[19] Biosensors can be used as in vivo devices, as in the simple glucose test where a drop of blood is extracted and tested; however, much research is now taking place on implantable, in vivo continuous reading biosensors.[19]

Sensors containing biomolecules such as glucose oxidase suffer from problems such as stability because the enzyme can denature upon storage or use. Also, the testing of complex mixtures infers the danger of unspecific adsorption of interferents or simple blocking of the sensor by protein or cell deposition. Hydrophilic polymers have been utilized in an attempt to solve some of these problems. The use of hydrophilic polymer membranes within biosensors has been recently reviewed.[20] The highly hydrated environment of a swollen hydrogel is a simple analog of conditions found within the biological environment. In this context, the trapping of active ingredients in a hydrogel has been utilized to improve the stability of the biomolecule.[20] Also, a thin hydrophilic polymer coating, such as PEO, can minimize biofouling while also helping to exclude common interferents such as ascorbate ions.[20]

Drug Delivery

The pharmaceutical industry has shown great interest in the development of controlled release systems based on hydrophilic polymers. Many drugs are most efficient when released into the body at a constant rate, rather than in individual doses. For example, a drug could be incorporated into a hydrogel either as it is synthesized or postsynthesis. This can then be ingested by, implanted within, or be simply placed in contact with the patient, e.g., upon the skin. Once in vivo, the drug is released into the environment by diffusion or by the polymer being eroded or dissolved in some way, for example, by digestion. Ideally, this process will occur at a constant rate, thereby continually releasing controlled amounts of the drug. If the polymer is surgically implanted close to an affected site, it means the drug is delivered where it is most needed.

For a polymer to be suitable for this purpose, it must display several properties. First, it must be biocompatible and must biodegrade within a reasonable period of time. Both the polymer itself and its degradation products must be nontoxic and must create neither an allergic nor an inflammatory response. The nature of the drug itself can also affect the method of release; low molecular weight drugs are capable of diffusing out of the polymeric matrix, and if water soluble will be rapidly released. Larger molecules such as proteins tend to be trapped within the matrix until the polymer itself degrades and releases them. The rates of release are dependent on several variables, including the quantity/dosing of drugs within the composite, the rate of degradation of the polymer, the amount of water present (if it is a hydrogel type polymer), and the presence and degree of crosslinking. The application of a wide variety of hydrophilic biodegradable polymers to delivery of proteins has been extensively reviewed.[21] Typical polymers include hydrogels based on polyvinyl alcohol, polyvinylpyrrolidinone, or cellulose, and natural polymers such as alginase or collagen. Besides enabling the controlled release of the active material, the polymer can also be exploited as a stabilizing medium for what are often unstable biological agents.

The polymer-based delivery device can be used to apply the active agent in many ways, including ingestion, suppositories, skin patches, ocular and subcutaneous methods. The treatment of cancer is a field in which these methods are being widely researched, for example, ara-C, a treatment for acute leukemia has the least side effects when introduced subcutaneously. As an alternative to continuous infusion, the drug can be incorporated into cross-linked polyhydroxyethyl acrylate disks, which displayed steady controllable drug release.[22] A polycaprolactone/PEG composite was similarly used as a matrix for the anticonvulsant drug clonazepam and displayed long-term (<45 days) constant release properties.[23] Nitroglycerin is a problematic drug owing to its loss of tablet activity, often by volatilization of the active component.[24] This can be avoided by incorporation of the nitroglycerin into an acrylic-based hydrogel, which is then incorporated into a transdermal patch, as exemplified by products such as Deponit® (Schwarz Pharma), Minitran® (3M Pharma), and Nitrodisc® (G.D. Searle Company).

These products are usually based on simple polymeric systems; however, more complex architectures are known. Insulin can be contained within a "smart" porous membrane containing glucose oxidase encapsulated within a polymethacrylic acid/PEG copolymer. High levels of glucose interact with the glucose oxidase, causing a pH drop and shrinkage of the membrane, which leads to the release of insulin.[25]

Apart from drug delivery, recent work has focused on the use of hydrophilic polymers in gene therapy and has shown that encapsulation of the DNA inside a hydrophilic polymer can increase the transfection efficiency.[3]

Artificial Organs

In the human body, the kidneys remove toxic wastes such as urea from the blood for excretion via the bladder. When they fail to perform this function, it leads to the condition known as uremia, which if not addressed will ultimately lead to death. To prevent this, hemodialysis must be performed, usually by a kidney dialysis machine, which can be thought of as an artificial organ external to the body. A semipermeable membrane based on the hydrophilic polymer cellulose is an integral part of this process. Hemodialysis uses a cellulose membrane tube that is immersed in a large volume of fluid. Blood is first removed from the patient, pumped through this tubing, and then back into the patient intravenously. Cellulose membranes contain pores that allow passage of most solutes in the blood while retaining the proteins and cells.

As blood passes through the membrane, free exchange of the solutes occurs between blood and the external isotonic solution—a salt solution with ionic concentrations near or slightly lower than the desired concentrations in the blood. This means that the two solutions are in dynamic equilibrium, and so the concentration of these species within the blood does not change. Compounds such as urea that are present in the blood in excess pass, however, into the external solution and are thereby removed from the bloodstream.

Hydrophilic polymers are used as major components of artificial internal organs. The physical properties of these materials and, in particular, their high water content, soft texture, and consistency, give them a strong resemblance to actual living soft tissues. A possible application, for example, in the case of organ failure, would be to incorporate living cells from a donor into these hydrogels and then implant them in the patient. The high biocompatibility of the hydrogel would prevent rejection and help to keep the implanted cells viable. For example, hepatocytes could be encased inside highly permeable hydrogel (83% water) tubes with a survival rate in vivo of 85% after 45 days.[26] The tubes maintained viability, prevented rejection, and allowed the passage of albumin from the encapsulated cells. A similar principle using membranes of calcium alginate hydrogel was used to encapsulate hepatocytes. Experiments showed that these units had the ability to replace liver function.[27]

Polymer-based heart valves are widely used as replacements for diseased or damaged human heart valves. Most mechanical heart valves are made from metals, silicone, or polyesters, although some work has gone into incorporating biocompatible coatings such as PEO into these systems.

Tissue Engineering

Biocompatible hydrophilic polymers, both natural and synthetic, can be used to promote tissue repair and regeneration. For example, gelatin scaffolds can be constructed by glutaraldehyde cross-linking, followed by freeze drying of the solution to give a porous matrix. This material can be specifically shaped to provide a support for tissue growth and organization over long periods.[28]

Coronary artery bypass grafting is commonly used to relieve conditions such as angina. Usually, the graft is taken from the patients themselves—a section of vein from the patient's leg being the most common. However, should there be a shortage of supply of this native material, the use of an artificial analog becomes necessary. Synthetic materials have been developed based on Dacron® (Dupont) or PTFE, although these are satisfactory only when used in large arteries; in smaller vessels they tend to cause thrombosis.[29] Current research includes the use of hydrophilic polymers as scaffold materials for the growth of cells with anticoagulatory properties, as recently reviewed elsewhere.[29,30] Both natural polymers such as collagen and synthetics such as biodegradable polyesters have been used.[29,30] The advantages of the polyesters are that not only are they stronger than many of the natural materials, but that over a period of time they are also reabsorbed by the body to be replaced with endogenous endothelial cells.[30]

Conclusions

The primary focus of this chapter has been to provide a basic understanding of the field of hydrophilic polymers and their uses. These materials have a range of applications both in the bulk and as thin films. For example, hydrogel membranes are used as inhibitors of postsurgical adhesion formation and also as coatings. Dispensing these polymers onto the surface of medical devices can confer desirable surface properties that the substrate material lacks. Hydrogel coatings especially can result in vast improvements in biocompatibility or reduction in surface friction. In the bulk, their predictable and controllable diffusivity can lead to their use as drug delivery agents. Both natural and synthetic polymers have been utilized for these purposes.

The present polymer technology has resulted in improvements in several different medical fields, including drug therapy and surgical procedure. It is always difficult to predict future developments, but it appears likely that hydrophilic polymers will help further the development of more viable localized drug delivery, implantable sensors, and artificial organ substitutes than currently exist. Much research is being carried out worldwide in these fields, making this both an interesting and a fast moving field.

References

1. LaPorte, R.J. *Hydrophilic Polymer Coatings for Biomedical Devices*; Technomic: Lancaster, 1997.
2. Ottenbrite, R.M. *Frontiers in Biomedical Polymer Applications*; Technomic: Lancaster, 1998; Vol. 1.
3. Ottenbrite, R.M. *Frontiers in Biomedical Polymer Applications*; Technomic: Lancaster, 1999; Vol. 2.
4. Nicolson, P.C., Vogt, J. Soft contact lens polymers: an evolution. *Biomaterials* **2001**, *22*, 3273–3283.
5. Lui, Y., Wilson, A., Zantos, S. Contact lens. In *Kirk-Othmer Encyclopedia of Chemical Technology*, 4th Ed.; Wiley: New York, 1993; Vol. 7, 191–228.
6. Mueller, K., Kleiner, E. Polysiloxane Hydrogels. US Patent 4,136,250, Jan 3, 1979.
7. Rice, D., Ihlenfeld, J. Contact lens containing a fluorinated telechelic polyether. US Patent 4,440,918, Apr 3, 1984.
8. Becker, J.M., Dayton, M.T., Fazio, V.M., Beck, D.E., Stryker, S.J., Wexner, S.D., Wolff, B.G., Roberts, P.L., Smith, L.E., Sweeney, S.A., Moore, M. Prevention of postoperative abdominal adhesions by a sodium hyaluronate-based bioresorbable membrane: a prospective, randomized, double-blind multicenter study. *J. Am. Coll. Surg.* **1996**, *183*, 297–306.
9. Mettler, L., Audebert, A., Lehmann-Willenbrock, E., Jacobs, V.R., Schive, K. Prospective clinical trial of spraygel as a barrier to adhesion formation: an interim analysis. *J. Am. Assoc. Gynecol. Laparosc.* **2003**, *10* (3), 339–344.
10. Middleton, J.C., Tipton, A.J. Synthetic biodegradable polymers as medical devices. *Med. Plast. Biomater. Mag.* **1998**, *30*, 30–40.

11. Rosiak, J. US Patent, 487, 1,490, 1989.
12. Doillon, C.J., Whyne, C.F., Brandwein, S., Silver, F.H. Collagen-based wound dressings—control of the pore structure and morphology. *J. Biomed. Mater. Res.* **1986,** *20,* 1219–1228.
13. Kingschott, P., Grieser, H.J. Surfaces that resist bioadhesion. *Curr. Opin. Solid State Mater. Sci.* **1999,** *4,* 403–412.
14. Green, R.J., Davies, M.C., Roberts, C.J., Tendler, S.J.B. A surface plasmon resonance study of albumin adsorption to PEO-PPO triblock copolymers. *J. Biomed. Mater. Res.* **1998,** *42,* 165–171.
15. Harder, P., Grunze, M., Dahint, R., Whiteside, G.M., Laibinis, P.E. Molecular conformation in oligo(ethylene glycol)-terminated self assembled monolayers on gold or silver surfaces determines their ability to resist protein adsorption. *J. Phys. Chem. B* **1998,** *102,* 426–436.
16. Shen, M.C., Martinson, L., Wagner, M.S., Castner, D.G., Ratner, B.D., Horbett, T.A. PEO-like plasma polymerized tetraglyme surface interactions with leukocytes and proteins: in vitro and in vivo studies. *J. Biomater. Sci. Polym. Ed.* **2002,** *13* (4) 367–390.
17. Hsuie, G.H., Lee, S.D., Chang, P.C., Kao, C.Y. Surface characterization and biological properties study of silicone rubber material grafted with phospholipid as biomaterial via plasma induced graft copolymerization. *J. Biomed. Mater. Res.* **1998,** *42,* 134–147.
18. Frazier, R.A., Mattjis, G., Davis, M.C., Roberts, C.J., Schacht, E., Tendler, S.J.B. Characterization of protein-resistant dextran monolayers. *Biomaterials* **2000,** *21,* 957–966.
19. Newman, J.D., Tigwell, L.J., Turner, A.P.F., Warner, P.J. *Biosensors: A Clearer View, Biosensors 2004—The 8th World Congress on Biosensors;* Elsevier: New York, 2004.
20. Nagels, L.J., Staes, E. Polymer (bio)materials design for amperometric detection in LC and FIA. *Trends Anal. Chem.* **2001,** *20* (4) 178–185.
21. Gombotz, W.R., Pettit, D. Biodegradable polymers for protein and peptide drug-delivery. *Bioconjugate Chem.* **1995,** *6,* 332–351.
22. Teijon, J.M., Trigo, R.M., Garcia, O., Blanco, M.D. Cytarabine trapping in poly (2-hydroxyethyl methacrylate) hydrogels: drug delivery studies. *Biomaterials* **1997,** *18,* 383–388.
23. Cho, C.S., Han, S.Y., Ha, J.H., Kim, S.H., Lim, D.Y. Clonazepam release from bioerodible hydrogels based on semi-interpenetrating polymer networks composed of poly(epsilon-caprolactone) and poly(ethylene glycol) macromer. *Int. J. Pharm.* **1999,** *181,* 235–242.
24. Markovich, R.J., Taylor, A.K., Rosen, J. Drug migration from the adhesive matrix to the polymer film laminate facestock in a transdermal nitroglycerin system. *J. Pharm. Biomed. Anal.* **1997,** *16* (4) 651–660.
25. Gander, B., Meinel, L., Walter, E., Merkle, H.P. Polymers as a platform for drug delivery: reviewing our current portfolio on poly(lactide-co-glycolide) (PLGA) microspheres. *Chimia* **2001,** *55,* 212–217.
26. Hongiger, H., Balladur, P., Mariani, P., Calmus, Y., Vaubourdolle, M., Delelo, R., Capeau, J., Nordlinger, B. Permeability and biocompatibility of a new hydrogel used for encapsulation of hepatocytes. *Biomaterials* **1995,** *16,* 753–759.
27. Yanagi, K., Ookawa, K., Mizuno, S., Ohshima, N. Performance of a new hybrid artificial liver support system using hepatocytes entrapped within a hydrogel. *ASAIO Trans.* **1989,** *35* (3) 570–572.
28. Kang, H.W., Tabata, Y., Ikada, Y. Fabrication of porous gelatin scaffolds for tissue engineering. *Biomaterials* **1999,** *20,* 1339–1344.
29. Thomas, C.T., Campbell, G.R., Campbell, J.H. Advances in vascular tissue engineering. *Cardiovasc. Pathol.* **2003,** *12,* 271–276.
30. Berglund, J.D., Galis, Z.S. Designer blood vessels and therapeutic revascularisation. *Brit. J. Pharmacol.* **2003,** *140,* 627–636.

4

Hydrogels

Jae Hyung Park, Kang Moo Huh, Mingli Ye, and Kinam Park

CONTENTS

Introduction

A hydrogel is a three-dimensional polymer network made of a hydrophilic polymer or a mixture of polymers. In general, at least 10–20% of the total weight of a hydrogel is water. When a hydrogel is dried, it is called xerogel, or simply a dried hydrogel. When a dried hydrogel is placed in an aqueous environment, it can absorb a large amount of water and swell isotropically to maintain its original shape. Swollen hydrogels maintain their shape without dissolving even in abundant water because of the presence of chemical or physical cross-linking of polymer chains. The extent of swelling depends inversely on the cross-linking density. When more than 95% of the total weight is water, the hydrogel is also called superabsorbent. It is not unusual to see hydrogels with more than 99% of water.

Because of the presence of high water content and the rubbery property, hydrogels have been frequently compared with the natural tissues. The similarity to natural tissues renders hydrogels useful in biomedical and pharmaceutical applications. Further, depending on the chemical structures of the constituting polymers, hydrogels can be tailored to respond to external stimuli, such as temperature, pH, solvent composition, electric field, light, and specific biomolecules. Those hydrogels can undergo change in swelling/ deswelling, shape change, and sol–gel transformation upon stimulation by external factors, and they are often called "smart hydrogels." Such an interesting nature of the smart hydrogels has allowed their use in controlled drug delivery, biomechanical devices, and separation systems.

The advent of nanotechnology has provided new avenues for engineering materials in nano- and microscales. In recent years, there have been extensive studies on potential applications of hydrogels in nanotechnology. Most of the studies have tried to exploit the unique properties of hydrogels, such as the hydrophilic nature of the surface, soft physical properties, and environmental sensitivity. Of particular interest has been nanoscale fabrication and manipulation of hydrogel-based materials that may lead to scientific and technological advances. Here, we review the current technologies on the preparation and potential applications of hydrogel-based nanomaterials, including hydrogel nanoparticles, hydrogel-coated nano/micro devices, inorganic (or organic) nanoparticle-entrapped hydrogels, and molecularly imprinted hydrogels.

Hydrogels in Nanotechnology

The hydrophilic polymer molecules of a hydrogel are interconnected by cross-linking, and this structure prevents dissolution of the polymer chains in an aqueous solution despite absorption of a large amount of water by the hydrogel. Hydrogels are generally classified into chemical and physical gels, according to the type of cross-links. Chemical gels are produced by cross-linking of hydrophilic polymers via covalent bonding. In an aqueous solution, they absorb water until they reach equilibrium swelling, which depends on the cross-linking density. On the other hand, physical gels are generated by noncovalent bonding, such as molecular entanglements, electrostatic interactions, hydrogen bonding, and hydrophobic interactions. These interactions, in contrast to covalent bonding, are reversible and can be disrupted by changes in physical conditions, such as temperature, pH, ionic strength, and stress.

Figure 4.1 illustrates representative mechanisms of hydrogel formation. There are a number of different macromolecular structures that form physical or chemical hydrogels. Hydrogels can be designed to undergo biodegradation in a physiological solution,[1–3] exhibit rapid response to physical stimuli,[4–6] or reach the equilibrium swelling level within a few minutes.[7–9]

Furthermore, there are differences in preparation methods and properties between chemical and physical gels, and each gel type has its own advantages and disadvantages for the design of specific materials or devices involving nanotechnology. Physical gels have been primarily used for biomedical applications, especially in the form of polymer micelles or self-aggregates for controlled drug delivery to the specific sites of action. They are spherical in shape and have the mean diameter ranging from nanometers to micrometers. Because nanoparticles are frequently administrated via the systemic route, it is desirable to construct them with biodegradable polymers so that they can degrade into low molecular weight entities eligible for renal excretion. Chemical gels that are covalently cross-linked provide a variety of potential applications because of their high stability in harsh environments, such as high temperature, acidic/basic solutions, and high stresses. In addition to drug delivery, chemical hydrogels have been considered as the constituents of diagnostic, electronic, and photonic devices.[10–12]

Nanoparticle-Bearing Hydrogels

Incorporation of functional nanoparticles into a hydrogel matrix produces unique properties, which cannot be found in other conventional organic/inorganic materials. Nanoparticles can be entrapped into a hydrogel matrix by chemical bonding or physical interactions with the polymer backbone of the hydrogel that have rubber elasticity and the stimuli-sensitive swelling/shrinking behavior. The entrapment of polymer nanoparticles,

(A) Chemical Hydrogel

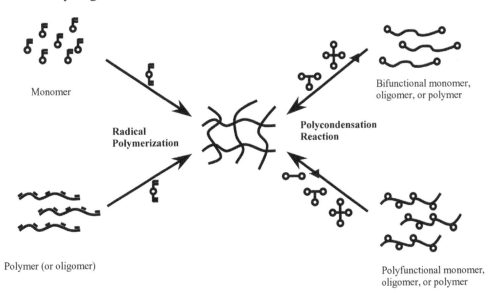

Monomer

Radical Polymerization

Polymer (or oligomer)

Polycondensation Reaction

Bifunctional monomer, oligomer, or polymer

Polyfunctional monomer, oligomer, or polymer

(B) Physical Hydrogel

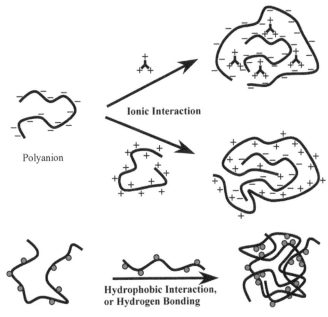

Polyanion

Ionic Interaction

Hydrophobic Interaction, or Hydrogen Bonding

Amphiphilic polymer

FIGURE 4.1

Representative methods of hydrogel formation. (A) Chemically cross-linked hydrogels are prepared from monomers, oligomers, or polymers in the presence of cross-linking agents. The chemical cross-linking proceeds via radical polymerization or polycondensation reaction. (B) Physically cross-linked hydrogels can be formed by ionic interactions, hydrophobic interaction, or hydrogen bonding.

forming a crystalline colloidal array (CCA), into a hydrogel matrix has displayed different colors without adding coloring agents, responding to external stimuli.[10,13] By incorporating 10–12 nm magnetite (Fe_3O_4) or maghemite (γ-Fe_2O_3) into a hydrogel matrix, the shape of hydrogels can be modulated by applying magnetic field.[14,15]

Monodispersed colloidal particles are reported to form CCAs via slow particle sedimentation, centrifugation, and spin coating.[10,13,16] The CCA diffracts visible and near-infrared light at wavelengths dependent on the lattice spacing, which produces an intense color. By combining the characteristic of the CCA with stimuli-sensitive hydrogels, novel functional hydrogels have recently been prepared. Of various hydrophilic polymers, poly(N-isopropylacrylamide) (PNIPAM) and its copolymers have been widely used to prepare chemical gels containing CCAs, because they show fast stimuli-sensitive volume phase transition and are readily prepared by the free radical polymerization of monomers in the presence of a difunctional cross-linking agent.[10–12,14] Weissman et al.[10] prepared PNIPAM hydrogels by photopolymerization in the presence of polystyrene nanoparticles (99 nm in diameter) as a CCA component and N,N-methylene-*bis*-acrylamide as a cross-linking agent. This chemically cross-linked CCA hydrogels exhibited various colors according to the temperature that changes the hydrogel volume affecting the array lattice constant. Polyacrylamide-based hydrogels have also been investigated to entrap the CCA lattice.[13] Although polyacrylamide does not provide thermo-sensitive volume transition like PNIPAM, it allows incorporating molecular-recognition groups (e.g., crown ethers for metal ions and glucose oxidase for glucose) during its polymerization or by simple chemical modification. The recognition events may make the gel swell because of changes in environments within the hydrogel matrix, such as an osmotic pressure and pH, which increases the mean separation between the colloidal spheres, and thus changes the diffracted light to longer wavelengths. More recently, the use of monodispersed nanoparticles as a hydrogel matrix has been developed.[11,16,17] This technique involves preparation of monodispersed hydrogel nanoparticles by radial polymerization in the presence of the cross-linker and surfactant, and formation of nanoparticle networks (colloidal crystal gels) by centrifugation[16] or chemical cross-linking.[11,17] The resultant hydrogel matrices displayed a bright iridescence in the visible region of the spectrum and underwent a reversible color change in response to changes in the temperature or pH. For chemically cross-linked nanoparticle networks, the color was changed depending on the concentration and size of the nanoparticles.

Nanoparticles have also been incorporated into a hydrogel matrix to develop novel materials sensitive to stimuli, such as light[18] and magnetic fields.[14] The gold nanoshell, composed of a thin layer of gold surrounding a dielectric core (e.g., gold sulfide), is one of the representative nanoparticles that have been incorporated into chemical gels, resulting in unique optical properties.[19–21] The diameters of both the core and shell are known to be responsible for the optical properties of nanoshells.[19,20] As the core size and shell thickness can be readily manipulated during the fabrication process, the optical extinction profiles of the nanoshells can be adjusted to observe light at desired wavelengths. Recently, by using nanoshells which absorb near-infrared light, hydrogel/nanoshell composites have been prepared for photothermally modulated drug delivery.[22] Embedding the nanoshells in PNIPAM hydrogel produced unique properties by which the composite hydrogel exhibited volume transition upon the irradiation of near-infrared light, i.e., the nanoshells in the hydrogel matrix converted light to heat, raising the temperature of the composite above the low critical solution temperature of PNIPAM. Because near-infrared light is capable of being transmitted through tissue, such hydrogels bearing gold nanoshells may have promising potential as an injectable drug delivery system for low molecular weight drugs, peptides, proteins, and genes.[18,19,21,22]

Molecularly Imprinted Hydrogels

Design of a precise macromolecular architecture that can selectively recognize target molecules has gained significant attention because of its potential applications for separation processes, immunoassays, biosensors, and catalysis.[23,24] Molecular imprinting technology has been developed as a response to the need to create such architectures. In general, molecular imprinting within polymers involves formation of prepolymerization complex between the template molecule and functional monomers, polymerization in the presence of a cross-linking agent and an appropriate solvent, and removal of the template.[25] Once the template is removed, the polymer network may have specific recognition elements for the target (or template) molecules. This nanoimprinting process is of great importance because it can create three-dimensional binding cavities for specific target molecules.

Because most recognition processes are associated with three-dimensional structure of the recognition site, it is preferable to limit the movement of the polymer chain that may affect affinity or selectivity to target molecules. Therefore, conventional methods to prepare molecularly imprinted polymers have used high ratios of the cross-linking agent to functional monomers, which leads to formation of rigid polymer matrix with low average molecular mass between cross-links.[26] On the contrary, imprinting within hydrogels requires different methods because they undergo changes in three-dimensional structure upon coming in contact with water. To maintain imprinting structure in an aqueous environment, hydrogels have been prepared by spatially varying cross-linking density.[25,27] As density fluctuations in the polymer network include microregions of localized higher cross-linking, hydrogels could retain an effective imprinting structure as well as proper rigidity to produce adequate specificity. Another promising strategy for imprinting within hydrogels is to match polymerization and rebinding solvents in terms of dielectric constant, polarity, and protic nature. This may reduce differences in swelling behavior, resulting in high binding affinity to target molecules. Also, in designing the network architecture for hydrogels, it is important to choose the length of the functional monomer and the molecular mass of the cross-linking monomer to endow with specificity to target molecules.[25,28]

Molecular imprinting provides shape-specific cavities (or nanovacuoles) that match the template molecule or chemical groups capable of specifically interacting with the template molecule. Of various polymers, ethylene glycol dimethyacrylate and methacrylic acid have been most widely used for the formation of imprinted polymers, where template molecules can be antibiotics,[29,30] carbohydrates,[19] peptides,[31] and enzymes.[32] Alvarez-Lorenzo et al.[33,34] and Hiratani et al.[35] have developed molecularly imprinted hydrogels by polymerizing in the presence of template molecule, functional monomer, and thermosensitive monomer. These hydrogels showed high affinity to target molecules as well as stimuli-sensitive recognition, by which the imprinted sites disappeared upon gel swelling and reformed upon shrinking.

The imprinted hydrogels, sensitive to analyte, have been the focus of many investigations for controlled drug release. A few examples are shown in Figure 4.2. Figure 4.2A shows that enzymes may be included in the hydrogel to invoke local pH changes by binding to analyte, and thus initiate the hydrogel swelling, modulating the drug release rate;[36,37] in Figure 4.2B, cross-links can lose their function by free analytes and this leads to hydrogel swelling to change the drug release rate;[38,39] and in Figure 4.2C, binding of

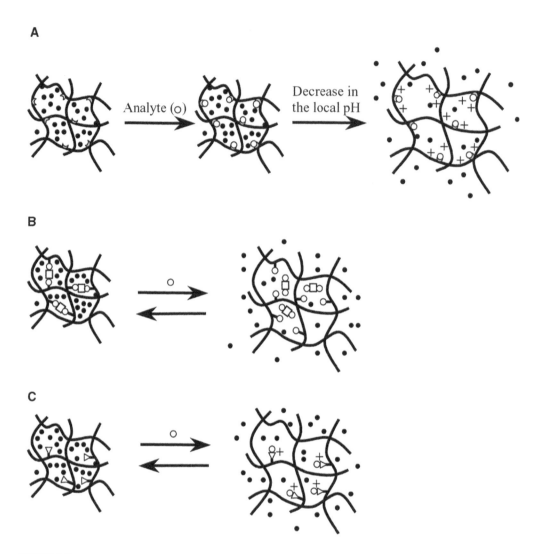

FIGURE 4.2
Molecularly imprinted hydrogels for drug delivery. (A) Binding of analytes (O) to enzyme (©) induces changes in the local pH. For cationic hydrogels, the acidic local pH results in ionization and swelling of hydrogel, resulting in faster release of drug (●). (B) Cross-linking agents (□) bind to analytes (O) anchored to the polymer backbone to maintain the hydrogel structure, which can swell and release the incorporated drugs as the concentration of free analyte increases, replacing the polymer-bound analytes. (C) Binding of analyte (O) of specific site or functional groups (∇) on the polymer backbone increases hydrophilicity of hydrogel, which induces swelling and drug release. (Modified from Byrne, Park, and Peppas.[26])

analytes to specific sites or functional groups on the polymer backbone may change hydrophilicity (or hydrophobicity) of hydrogels, inducing swelling (or shrinking) in an aqueous environment.[40] It should be emphasized, however, that the currently available imprinting techniques involve nonspecific binding sites that decrease the specificity to target molecules. The precise control of the network structure of hydrogels via nanotechnology will contribute to a number of applications, including microfluidic devices, biomimetic sensors, drug delivery system, and membrane separation technology.

Hydrogel Nanoparticles

Because of their biocompatibility and soft rubbery nature, hydrogels have been extensively studied in biomedical and pharmaceutical fields. Macroscopic hydrogels have been studied for sustained drug delivery owing to their slow swelling kinetics. Microparticulate hydrogels for drug delivery have been examined more widely because of their unique properties resulting from small size. Initially, hydrophobic nano- and microparticles have been used extensively for systemic drug delivery, but it was soon found out that they were readily taken up by the reticuloendothelial system, and thus exhibited short residence time in blood. Thus, in an attempt to improve hydrophilicity for prolonged circulation time, the surfaces of the hydrophobic nanoparticles have been modified by conjugating, blending, and coating with hydrophilic polymers, such as polyethylene glycol (PEG) and PEG-containing block copolymers.[41,42]

Hydrogel nanoparticles have a special role in drug delivery in the sense that they have all the advantages of both nanoparticles and hydrogels with regard to the particle size and hydrophilicity. Hydrogel nanoparticles can swell rapidly because of large surface area and short diffusion path length for water. They can be modified to have reactive groups on the surface and are useful to introduce functional moieties, such as targeting and other bioactive moieties. Hydrogel nanoparticles consisting of stimuli-responsive polymers may exhibit corresponding responsive properties, which are often found to become much faster than bulk hydrogels. Studies of hydrogel nanoparticle have intensified during the past decade because of enormous potentials in the development and implementation of new stimuli-responsive or smart materials, biomimetics, biosensors, artificial muscles, drug delivery systems, and chemical separation systems.

Chemically cross-linked hydrogel nanoparticles have been prepared in the presence of hydrophilic monomers, cross-linking agents, and emulsifiers. Advances in technology have enabled precise control of the core–shell structure of hydrogel nanoparticles.[43] The core–shell nanoparticles have been synthesized to modify surface properties of core particles or to provide stimuli-sensitivity for nonresponsive particles. Recently, multiresponsive core–shell hydrogel nanoparticles have also been developed by Jones and Lyon[43] and Gan and Lyon.[44] They synthesized temperature- and pH-responsive hydrogel nanoparticles with core–shell morphologies, where core particles composed of PNIPAM were prepared via aqueous free radical polymerization, and then used as nuclei for subsequent polymerization of acrylic acid copolymers. Their swelling/deswelling thermodynamics were easily controlled by chemical manipulation of the core and shell structures, thus displaying both temperature and pH dependence. Without chemical cross-linking, core–shell hydrogel nanoparticles can also be prepared on the basis of the electrostatic interaction between water-soluble polymers.[45] Prokop et al.[46] have demonstrated that by atomizing the aqueous solution containing core polymer with negative charge, the nanosize droplets are encapsulated with cationic polymer solution by the electrostatic interaction (Figure 4.3). This methodology showed promising potential as a protein delivery system.

Self-assembled hydrogel nanoparticles based on hydrophobically modified polysaccharides have been extensively studied as a drug carrier because of their excellent biocompatibility and ease of preparation. It is well known that polymeric amphiphiles, upon contact with an aqueous environment, spontaneously form micelles or micelle-like self-aggregates via undergoing intra- or intermolecular associations between hydrophobic moieties, primarily to minimize interfacial free energy. Hydrophobically modified polysaccharides are also known to self-assemble in aqueous media to form a unique core–shell structure that consists of hydrophobic segments and hydrophilic segments, respectively

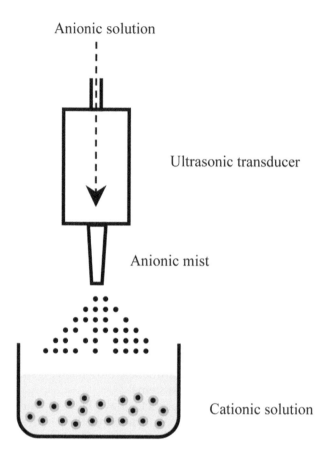

Anionic solution

Ultrasonic transducer

Anionic mist

Cationic solution

FIGURE 4.3

Formation of the core–shell type hydrogel nanoparticles by electrostatic interactions. The anionic solution which contains core polymer is introduced as a mist into a cationic solution of shell polymer. (Modified from Prokop, Holland, Kozlov, Moore, and Tanner.[45])

(Figure 4.4). This type of hydrogels have multiple inner cores, which physically crosslink the hydrophilic polymer chains.[47] A number of polysaccharides have been investigated to create self-assembling systems, including dextran,[48] glycol chitosan,[49,50] pullulan,[51,52] and curdlan.[53] These polysaccharides are natural water-soluble polymers that are inherently biocompatible and biodegradable. The core–shell structure of self-assembled hydrogels can be employed as a potential delivery system that can effectively deliver hydrophobic drugs. It has been recently demonstrated that hydrophobically modified polysaccharides capable of forming nano-sized self-aggregates can imbibe hydrophobic drugs and release them in a sustained manner.[49] Hydrophobic moieties, conjugated to polysaccharides, can either be small molecules (e.g., cholesterol, alkyl chains, and bile acids)[47,51,53–55] or oligomers.[56] The conjugation of stimuli-sensitive hydrophobic moieties to polysaccharides may produce hydrogel nanoparticles, responsive to corresponding stimuli.[57] For example, Na et al.[53,56,57] have recently developed pH-sensitive hydrogel nanoparticles as an anticancer drug carrier. The extracellular pH of most solid tumors and inflammatory regions in the body is known to be lower than that in the normal tissues and blood (pH 7.4). To target the extracellular matrix of such disease sites, they prepared pullulan acetate-based nanoparticles bearing sulfonamide moieties, which show the hydrophobic nature at the low pH.

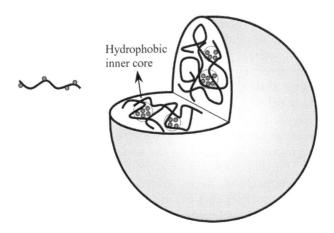

Hydrophobic
inner core

FIGURE 4.4
Self-assembled hydrogel nanoparticles of hydrophobically modified polysaccharides. Note that nanoparticles have multiple inner cores which physically cross-link the hydrophilic polysaccharide chain. (Modified from Akiyoshi, Deguchi, Tajima, Nishikawa, and Sunamoto.[47])

The resulting nanoparticles rapidly released the anticancer drug (doxorubicin) at pH < 7.0, whereas the drug release rate was substantially reduced at normal tissue pH (7.4).

Hydrogel Coating on the Surfaces

Surfaces of hydrophobic substrates have been frequently modified with hydrophilic polymers to achieve desirable properties for in vivo applications. Surface modification with hydrophilic polymers is known to minimize nonspecific interactions with blood proteins, cells, and tissues. The hydrophilic polymers commonly used for surface coating include PEG, polysaccharides, and poly(vinyl alcohol). To physically coat hydrophilic polymers on the nanoparticles, the solvent extraction/evaporation method has been used.[58,59] After the oil-in-water emulsion is prepared by adding the organic solution containing a hydrophobic polymer into the aqueous solution of a hydrophilic homopolymer or a block copolymer, the organic solvent is removed by evaporation or extraction, thus forming the hydrogel layer on the formed nanoparticles. The outer layer of a hydrophilic polymer in the nanoparticle is anchored by various interactions with the core polymer chains, such as physical entanglement, hydrophobic interaction, and hydrogen bonding. This approach has been frequently used for surface modification of biodegradable nanoparticles, such as poly(D,L-lactide) (PLA),[58] poly(lactide-co-glycolide),[59] and polyphosphazene.[60]

The hydrogel coating on the nanoparticles has also been achieved by radical polymerization. For poly(isobutyl cyanoacrylate) (PIBCA), the monomer was emulsified in an aqueous solution containing PEG that acts as a nucleophile initiator of polymerization through its hydroxyl terminal groups.[61] Once the aqueous pH is adjusted to 1, polymerization is initiated, thus forming PEG-coated PIBCA nanoparticles.[62,63] PNIPAM has been coated on the nanoparticles by radical polymerization in the presence of hydrophobic nanoparticles to be coated, N,N'-methylenebisacrylamide (cross-linker), ammonium persulfate (initiator), and sodium dodecyl sulfate (emulsifier). The thickness of the outer hydrogel layer on the nanoparticles was readily controlled by varying the concentrations of the monomer and emulsifier.[64]

Hydrogels that coat metal and semiconductor nanoparticles are of considerable interest because of their unique size-dependent physicochemical properties.[65–67] Precise control of the structure and surface properties of nanoparticle would make them more attractive for use in biomedical applications. Inorganic nanoparticles have been conjugated to biomolecules such as sugars, peptides, proteins, and DNA. Such conjugates showed many advantages as fluorescent biological labels,[66–68] primarily appearing from inorganic nanoparticles, including high quantum efficiencies, optical activity over biocompatible wavelengths, and chemical or photochemical stability. It should be emphasized that in spite of numerous potential applications, inorganic nanoparticles have suffered from their aggregation and lack of biocompatibility. Hydrogel coating on such nanoparticles may not only prevent their aggregation by changing the surface hydrophilicity, but also improve their biocompatibility. Furthermore, use of stimuli-responsive hydrogels may provide unique properties for nanoparticles. Recent efforts have led to the development of hydrogel-coated inorganic nanoparticles that exhibit structural changes responsive to stimuli such as light. For example, hydrogel-coated gold nanoparticles have been prepared using surfactant-free emulsion polymerization method, as shown in Figure 4.5.[69] The hydrogel layer was constructed with a mixture of *N*-isopropylacrylamide and acrylic acid, and its thickness could be varied by adjusting the amount of monomer and initiator, as well as the reaction time. The results revealed that the hydrogel can be thermally activated by exposure to light via the strong plasmon absorption of the gold nanoparticle core.

As mentioned earlier, the surface coating with hydrogel can improve the biocompatibility as well as provide specific functions. One of the promising strategies to improve the surface characteristics is to attach polymeric micelles onto the surfaces, thus forming the polymeric micelle-entrapped hydrogel layer. This approach is useful to maximize the number of tethered hydrophilic chains because the polymeric micelle has a high density of hydrophilic polymer on the surface, resulting in an effective nonfouling property. Further, as the polymeric micelles contain hydrophobic inner core as a reservoir of hydrophobic drugs, the surface coating with polymeric micelles may allow developing of biocompatible devices that can release the drug in a sustained manner. The structure of the polymeric micelles, however, is readily disrupted upon attachment to the surface, leading to the formation of a loosely packed layer structure.[70,71] In an attempt to stabilize the polymeric micelles, Ijima et al.[72] prepared heterobifunctional block copolymer of PEG–PLA, in which PEG had a reactive aldehyde group at the chain end, whereas PLA possessed a methacryloyl group that can be polymerized in the presence of the initiator. This amphiphilic copolymer was then exposed to an aqueous solution, which enabled it to form the polymeric micelle, followed by the polymerization of the hydrophobic inner core. The resulting micelles showed high stability in harsh environments.[72] The aldehyde groups at the end of PEG chain was

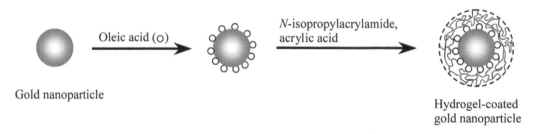

FIGURE 4.5
Hydrogel-coated gold nanoparticles prepared by surfactant-free emulsion polymerization. After coating oleic acids on the gold nanoparticle, polymerization was carried out in the presence of *N*-isopropylacrylamide, acrylic acid, and ammonium persulfate (initiator). The size of resulting nanoparticles was in the range of 100–230 nm. (Modified from Kim and Lee.[69])

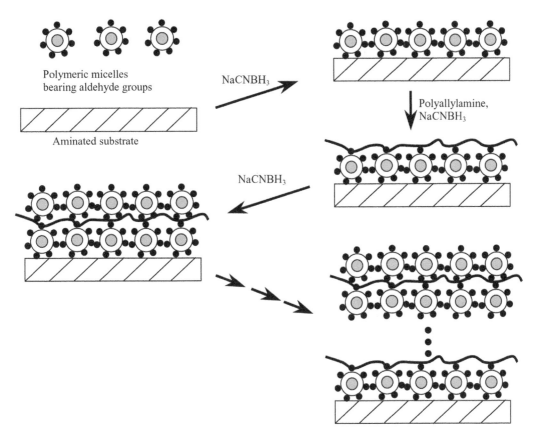

FIGURE 4.6
Schematic illustration of the multilayered micellar coating on the surface. Polymeric micelles to be coated were first stabilized by polymerization of the hydrophobic inner core. The stable micelles were then immobilized on the aminated substrate by the reaction with aldehyde groups on the surface of polymeric micelle. (Modified from Emoto, Iijima, Nagasaki, and Kataoka.[74])

used to chemically attach to the surfaces bearing the amino groups so that a single layer of polymer micelle is formed on the surface.[73] By introducing amino groups on the top of the micellar layer through tethering polyallylamine, multilayered highly organized micellar hydrogel can be coated on the surfaces.[74] Figure 4.6 shows the schematic illustration of the formation of the multilayered micellar coating on the surface. The resulting surface with micellar hydrogel layers exhibited excellent resistance to protein adsorption. In addition, the incorporation of the hydrophobic drug into the micellar hydrophobic core (~10 nm in diameter) made it possible to release the drug in a controlled manner, depending on the number of coated layers.[75]

Conclusions

Recent advances in nanotechnology have enabled us to extend potential applications of micro- and nano-particulate hydrogels. Combination of hydrogel and nanotechnology may afford a powerful means for manipulating the properties of surfaces and interfaces. Fabrication of nanostructures using hydrogels involves hydrophilic nanoparticles, molecular imprinting, nanoparticle-entrapped hydrogel, and nanoengineering for surface modification. These technologies will accelerate the development of various drug delivery and biomedical devices, as well as other electronic and photonic devices.

Acknowledgments

This study was supported in part by the National Institute of Health through GM67044 and GM65284.

References

1. Jeong, B., Bae, Y.H., Lee, D.S., Kim, S.W. Biodegradable block copolymers as injectable drug-delivery systems. *Nature* **1997**, *388*, 860–862.
2. Kissel, T., Li, Y., Unger, F. ABA–triblock copolymers from biodegradable polyester A-blocks and hydrophilic poly(ethylene oxide) B-blocks as a candidate for in situ forming hydrogel delivery systems for proteins. *Adv. Drug Deliv. Rev.* **2002**, *54*, 99–134.
3. Holland, T.A., Tessmar, J.K., Tabata, Y., Mikos, A.G. Transforming growth factor-beta 1 release from oligo(poly(ethylene glycol) fumarate) hydrogels in conditions that model the cartilage wound healing environment. *J. Control. Release* **2004**, *94*, 101–114.
4. Wang, C., Stewart, R.J., Kopecek, J. Hybrid hydrogels assembled from synthetic polymers and coiled-coil protein domains. *Nature* **1999**, *397*, 417–420.
5. Qiu, Y., Park, K. Environment-sensitive hydrogels for drug delivery. *Adv. Drug Deliv. Rev.* **2001**, *53*, 321–339.
6. Kopecek, J. Smart and genetically engineered biomaterials and drug delivery systems. *Eur. J. Pharm. Sci.* **2003**, *20*, 1–16.
7. Chen, J., Park, H., Park, K. Synthesis of superporous hydrogels: hydrogels with fast swelling and superabsorbent properties. *J. Biomed. Mater. Res.* **1999**, *44*, 53–62.
8. Chen, J., Park, K. Synthesis and characterization of superporous hydrogel composites. *J. Control. Release* **2000**, *65*, 73–82.
9. Gemeinhart, R.A., Park, H., Park, K. Effect of compression on fast swelling of poly(acrylamide-co-acrylic acid) superporous hydrogels. *J. Biomed. Mater. Res.* **2001**, *55*, 54–62.
10. Weissman, J.M., Sunkara, H.B., Tse, A.S., Asher, S.A. Thermally switchable periodicities and diffraction from mesoscopically ordered materials. *Science* **1996**, *274*, 959–960.
11. Hu, Z., Lu, X., Gao, J., Wang, C. Polymer gel nanoparticle networks. *Adv. Mater.* **2000**, *12*, 1173–1176.
12. Jones, C.D., Lyon, L.A. Photothermal patterning of microgel/gold nanoparticle composite colloidal crystals. *J. Am. Chem. Soc.* **2003**, *125*, 460–465.
13. Holtz, J.H., Asher, S.A. Polymerized colloidal crystal hydrogel films as intelligent chemical sensing materials. *Nature* **1997**, *389*, 829–832.
14. Xulu, P.M., Filipcsei, G., Zrinyi, M. Preparation and responsive properties of magnetically soft poly(*N*-isopropylacrylamide) gels. *Macromolecules* **2000**, *33*, 1716–1719.
15. Shipway, A.N., Willner, I. Nanoparticles as structural and functional units in surface-confined architectures. *Chem. Commun. (Camb.)* **2001**, 2035–2045.

16. Debord, J.D., Lyon, L.A. Thermoresponsive photonic crystals. *J. Phys. Chem. B* **2000**, *104*, 6327–6331.
17. Hu, Z., Lu, X., Gao, J. Hydrogel opals. *Adv. Mater.* **2001**, *13*, 1708–1712.
18. Sershen, S., West, J. Implantable, polymeric systems for modulated drug delivery. *Adv. Drug Deliv. Rev.* **2002**, *54*, 1225–1235.
19. Averitt, R.D., Sarkar, D., Halas, N.J. Plasmon resonance shifts of Au coated Au_2S nanoshells: insight into multicomponent nanoparticle growth. *Phys. Rev. Lett.* **1997**, *78*, 4217–4220.
20. Averitt, R.D., Westcott, S.L., Halas, N.J. Linear optical properties of gold nanoshells. *J. Opt. Soc. Am. B* **1999**, *16*, 1824–1832.
21. Sershen, S.R., Westcott, S.L., Halas, N.J., West, J.L. An optomechanical nanoshell-polymer composite. *Appl. Phys. B* **2001**, *73*, 1–3.
22. Sershen, S.R., Westcott, S.L., Halas, N.J., West, J.L. Temperature-sensitive polymer-nanoshell composites for photothermally modulated drug delivery. *J. Biomed. Mater. Res.* **2000**, *51*, 293–298.
23. Takeuchi, T., Haginaka, J. Separation and sensing based on molecular recognition using molecularly imprinted polymers. *J. Chromatogr. B Biomed. Sci. Appl.* **1999**, *728*, 1–20.
24. Piletsky, S.A., Alcock, S., Turner, A.P. Molecular imprinting: at the edge of the third millennium. *Trends Biotechnol.* **2001**, *19*, 9–12.
25. Langer, R., Peppas, N.A. Advances in biomaterials, drug delivery, and bionanotechnolgy. *AIChE J.* **2003**, *49*, 2990–3006.
26. Byrne, M.E., Park, K., Peppas, N.A. Molecular imprinting within hydrogels. *Adv. Drug Deliv. Rev.* **2002**, *54*, 149–161.
27. Bures, P., Huang, Y., Oral, E., Peppas, N.A. Surface modifications and molecular imprinting of polymers in medical and pharmaceutical applications. *J. Control. Release* **2001**, *72*, 25–33.
28. Hilt, J.Z., Byrne, M.E. Configurational biomimesis in drug delivery: molecular imprinting of biologically significant molecules. *Adv. Drug Deliv. Rev.* **2004**, *56*, 1599–1620.
29. Lubke, C., Lubke, M., Whitcombe, M.J., Vulfson, E.N. Imprinted polymers prepared with stoichiometric template-monomer complexes: efficient binding of ampicillin from aqueous solutions. *Macromolecules* **2000**, *33*, 5098–5105.
30. Lai, E.P.C., Wu, S.G. Molecularly imprinted solid phase extraction for rapid screening of cephalexin in human plasma and serum. *Anal. Chim. Acta* **2003**, *481*, 165–174.
31. Kempe, M. Antibody-mimicking polymers as chiral stationary phases in HPLC. *Anal. Chem.* **1996**, *68*, 1948–1953.
32. Piletsky, S.A., Piletsky, E.V., Elgersma, A.V., Yano, K., Karube, I. Atrazine sensing by molecularly imprinted membranes. *Biosens. Bioelectron.* **1995**, *10*, 959–964.
33. Alvarez-Lorenzo, C., Guney, O., Oya, T., Sakiyama, T., Takeoka, Y., Ito, K., Wang, G., Annaka, M., Hara, K., Du, R., Chuang, J., Wasserman, K., Grosberg, A.Y., Masamune, S., Tanaka, T. Polymer gels that memorize elements of molecular conformation. *Macromolecules* **2000**, *33*, 8693–8697.
34. Alvarez–Lorenzo, C., Hiratani, H., Tanaka, K., Stancil, K., Grosberg, A.Y., Tanaka, T. Simultaneous multiple-point adsorption of aluminum ions and charged molecules by a polyampholyte thermosensitive gel: controlling frustrations in a heteropolymer gel. *Langmuir* **2001**, *17*, 3616–3622.
35. Hiratani, H., Alvarez–Lorenzo, C., Chuang, J., Guney, O., Grosberg, A.Y., Tanaka, K. Effect of reversible cross–linker, *N,N*-bis(acryloyl)cystamine, on calcium ion adsorption by imprinted gels. *Langmuir* **2001**, *17*, 4431–4436.
36. Podual, K., Doyle, F.J., III, Peppas, N.A. Dynamic behavior of glucose oxidase-containing microparticles of poly(ethylene glycol)-grafted cationic hydrogels in an environment of changing pH. *Biomaterials* **2000**, *21*, 1439–1450.
37. Podual, K., Doyle, F.J., III, Peppas, N.A. Glucose-sensitivity of glucose oxidase-containing cationic copolymer hydrogels having poly(ethylene glycol) grafts. *J. Control. Release* **2000**, *67*, 9–17.
38. Obaidat, A.A., Park, K. Characterization of glucose dependent gel–sol phase transition of the polymeric glucose-concanavalin A hydrogel system. *Pharm. Res.* **1996**, *13*, 989–995.

39. Obaidat, A.A., Park, K. Characterization of protein release through glucose-sensitive hydrogel membranes. *Biomaterials* **1997**, *18*, 801–806.
40. Kataoka, K., Miyazaki, H., Bunya, M., Okano, T., Sakurai, Y. Totally synthetic polymer gels responding to external glucose concentration: their preparation and application to on-off regulation of insulin release. *J. Am. Chem. Soc.* **1998**, *120*, 12,694–12,695.
41. Storm, G., Belliot, S.O., Daemen, T., Lasic, D.D. Surface modification of nanoparticles to oppose uptake by the mononuclear phagocyte system. *Adv. Drug Deliv. Rev.* **1995**, *17*, 31–48.
42. Otsuka, H., Nagasaki, Y., Kataoka, K. PEGylated nanoparticles for biological and pharmaceutical applications. *Adv. Drug Deliv. Rev.* **2003**, *55*, 403–419.
43. Jones, C.D., Lyon, L.A. Synthesis and characterization of multiresponsive core–shell microgels. *Macromolecules* **2000**, *33*, 8301–8306.
44. Gan, D., Lyon, L.A. Tunable swelling kinetics in core–shell hydrogel nanoparticles. *J. Am. Chem. Soc.* **2001**, *123*, 7511–7517.
45. Prokop, A., Holland, C.A., Kozlov, E., Moore, B., Tanner, R.D. Water-based nanoparticulate polymeric system for protein delivery. *Biotechnol. Bioeng.* **2001**, *75*, 228–232.
46. Prokop, A., Kozlov, E., Newman, G.W., Newman, M.J. Water-based nanoparticulate polymeric system for protein delivery: permeability control and vaccine application. *Biotechnol. Bioeng.* **2002**, *78*, 459–466.
47. Akiyoshi, K., Deguchi, S., Tajima, H., Nishikawa, T., Sunamoto, J. Microscopic structure and thermoresponsiveness of a hydrogel nanoparticle by self-assembly of a hydrophobized polysaccharide. *Macromolecules* **1997**, *30*, 857–861.
48. Kim, I.S., Jeong, Y.I., Kim, S.H. Self-assembled hydrogel nanoparticles composed of dextran and poly(ethylene glycol) macromer. *Int. J. Pharm.* **2000**, *205*, 109–116.
49. Son, Y.J., Jang, J.-S., Cho, Y.W., Chung, H., Park, R.-W., Kwon, I.C., Kim, I.-S., Park, J.Y., Seo, S.B., Park, C.R., Jeong, S.Y. Biodistribution and anti-tumor efficacy of doxorubicin loaded glycolchitosan nanoaggregates by EPR effect. *J. Control. Release* **2003**, *91*, 135–145.
50. Park, J.H., Kwon, S., Nam, J.-O., Park, R.-W., Chung, H., Seo, S.B., Kim, I.-S., Kwon, I.C., Jeong, S.Y. Self-assembled nanoparticles based on glycol chitosan bearing 5beta-cholanic acid for RGD peptide delivery. *J. Control. Release* **2004**, *95*, 579–588.
51. Akiyoshi, K., Kobayashi, S., Shichibe, S., Mix, D., Baudys, M., Wan Kim, S., Sunamoto, J. Self–assembled hydrogel nanoparticle of cholesterol-bearing pullulan as a carrier of protein drugs: Complexation and stabilization of insulin. *J. Control. Release* **1998**, *54*, 313–320.
52. Gupta, M., Gupta, A.K. Hydrogel pullulan nanoparticles encapsulating pBUDLacZ plasmid as an efficient gene delivery carrier. *J. Control. Release* **2004**, *99*, 157–166.
53. Na, K., Park, K.-H., Kim, S.W., Bae, Y.H. Self-assembled hydrogel nanoparticles from curdlan derivatives: characterization, anti-cancer drug release and interaction with a hepatoma cell line (HepG2). *J. Control. Release* **2000**, *69*, 225–236.
54. Kwon, S., Park, J.H., Chung, H., Kwon, I.C., Jeong, S.Y., Kim, I.S. Physicochemcial characteristics of self-assembled nanoparticles based on glycol chitosan bearing 5beta-cholanic acid. *Langmuir* **2003**, *19*, 10,188–10,193.
55. Park, J.H., Kwon, S., Nam, J.O., Park, R.W., Chung, H., Seo, S.B., Kim, I.S., Kwon, I.C., Jeong, S.Y. Self-assembled nanoparticles based on glycol chitosan bearing 5beta-cholanic acid for RGD peptide delivery. *J. Control. Release* **2004**, *95*, 579–588.
56. Na, K., Lee, K.H., Bae, Y.H. pH-sensitivity and pH-dependent interior structural change of self-assembled hydrogel nanoparticles of pullulan acetate/oligo-sulfonamide conjugate. *J. Control. Release* **2004**, *97*, 513–525.
57. Na, K., Bae, Y.H. Self-assembled hydrogel nanoparticles responsive to tumor extracellular pH from pullulan derivative/sulfonamide conjugate: characterization, aggregation, and adriamycin release in vitro. *Pharm. Res.* **2002**, *19*, 681–688.
58. Landry, F., Bazile, D., Spenlehauer, G., Veillard, M., Kreuter, J. Release of the fluorescent marker prodan from poly(D,L-lactic acid) nanoparticles coated with albumin or polyvinyl alcohol in model digestive fluids (USP XXII). *J. Control. Release* **1997**, *44*, 227–236.

59. Win, K.Y., Feng, S.S. Effects of particle size and surface coating on cellular uptake of polymeric nanoparticles for oral delivery of anticancer drugs. *Biomaterials* **2005**, *26*, 2713–2722.

60. Vandorpe, J., Schacht, E., Dunn, S., Hawley, A., Stolnik, S., Davis, S.S., Garnett, M.C., Davies, M.C., Illum, L. Long circulating biodegradable poly(phosphazene) nanoparticles surface modified with poly(phosphazene)–poly(ethylene oxide) copolymer. *Biomaterials* **1997**, *18*, 1147–1152.

61. Peracchia, M.T., Vauthier, C., Puisieux, F., Couvreur, P. Development of sterically stabilized poly(isobutyl 2-cyanoacrylate) nanoparticles by chemical coupling of poly(ethylene glycol). *J. Biomed. Mater. Res.* **1997**, *34*, 317–326.

62. Peracchia, M.T., Vauthier, C., Passirani, C., Couvreur, P., Labarre, D. Complement consumption by poly(ethylene glycol) in different conformations chemically coupled to poly(isobutyl 2-cyanoacrylate) nanoparticles. *Life Sci.* **1997**, *61*, 749–761.

63. Peracchia, M.T., Vauthier, C., Desmaele, D., Gulik, A., Dedieu, J.C., Demoy, M., d'Angelo, J., Couvreur, P. Pegylated nanoparticles from a novel methoxypolyethylene glycol cyanoacrylate–hexadecyl cyanoacrylate amphiphilic copolymer. *Pharm. Res.* **1998**, *15*, 550–556.

64. Gan, D., Lyon, L.A. Fluorescence nonradiative energy transfer analysis of crosslinker heterogeneity in core–shell hydrogel nanoparticles. *Anal. Chim. Acta* **2003**, *496*, 53–63.

65. Schmid, G. Large clusters and colloids: metals in the embryonic state. *Chem. Rev.* **1992**, *92*, 1709–1727.

66. Bruchez, M., Jr., Moronne, M., Gin, P., Weiss, S., Alivisatos, A.P. Semiconductor nanocrystals as fluorescent biological labels. *Science* **1998**, *281*, 2013–2016.

67. Jaiswal, J.K., Mattoussi, H., Mauro, J.M., Simon, S.M. Long-term multiple color imaging of live cells using quantum dot bioconjugates. *Nat. Biotechnol.* **2003**, *21*, 47–51.

68. Cao, Y.W., Jin, R., Mirkin, C.A. DNA-modified core–shell Ag/Au nanoparticles. *J. Am. Chem. Soc.* **2001**, *123*, 7961–7962.

69. Kim, J.H., Lee, T.R. Thermo- and pH-responsive hydrogel-coated gold nanoparticles. *Chem. Mater.* **2004**, *16*, 3647–3651.

70. Frinha, J.P.S., d'Oliveira, J.M.R., Martinoho, J.M., Xu, R., Winnik, M.A. Structure in tethered chains: polymeric micelles and chains anchored on polystyrene latex spheres. *Langmuir* **1998**, *14*, 2291–2296.

71. Bijsterbosch, H.D., Stuart, M.A.C., Fleer, G.J. Adsorption kinetics of diblock copolymers from a micellar solution on silica and titania. *Macromolecules* **1998**, *31*, 9281–9294.

72. Iijima, M., Nagasaki, Y., Okada, T., Kato, M., Kataoka, K. Core-polymerized reactive micellesm from heterotelechelic amphiphilic block copolymers. *Macromolecules* **1999**, *32*, 1140–1146.

73. Emoto, K., Nagasaki, Y., Kataoka, K. A core–shell structured hydrogel thin layer on surfaces by lamination of a poly(ethylene glycol)-b-(D,L-lactic acid) micelle and polyallylamine. *Langmuir* **2000**, *16*, 5738–5742.

74. Emoto, K., Iijima, M., Nagasaki, Y., Kataoka, K. Functionality of polymeric micelle hydrogels with organized three-dimensional architecture on surfaces. *J. Am. Chem. Soc.* **2000**, *122*, 2653–2654.

75. Otsuka, H., Nagasaki, Y., Kataoka, K. PEGylated nanoparticles for biological and pharmaceutical applications. *Adv. Drug Deliv. Rev.* **2003**, *55*, 403–419.

5

Polyanhydrides

Maria P. Torres, Amy S. Determan, Surya K. Mallapragada, and Balaji Narasimhan

CONTENTS

Introduction

Polyanhydrides are a class of bioerodible polymers that have shown excellent characteristics as drug delivery carriers. The properties of these biomaterials can be tailored to obtain desirable controlled release characteristics. Extensive research in this promising area of biomaterials is the focus of this chapter. In the first part of the chapter, the chemical structures and synthesis methods of various polyanhydrides are discussed. This is followed by a discussion of the physical, chemical, and thermal properties of polyanhydrides and their effect on the degradation mechanism of these materials. Finally, a description of drug release applications from polyanhydride systems is presented, highlighting their potential in biomedical applications.

Background

The need for suitable materials for the delivery of drugs in a safe and controlled manner has led to the development of numerous biodegradable polymers. Controlled release of a variety of therapeutic agents has been achieved with the use of biodegradable polymeric devices. Research has focused on poly(α-hydroxy acids), poly(orthoesters), and poly(anhydrides). Poly(α-hydroxy acids), e.g., poly(lactic acid) (PLA) and poly(glycolic acid) (PGA), and copolymers undergo bulk erosion and the drug release kinetics from these carriers is not well defined. On the other hand, poly(orthoesters) and polyanhydrides undergo surface erosion with predictable kinetics. In the case of polyanhydrides, the degradation rates can be tailored to suit specific applications by changing the chemistry.

Polyanhydrides comprise monomer units connected by water-labile anhydride bonds. In the presence of water, the polymer is cleaved across the anhydride bond into two carboxylic acid groups (Figure 5.1). It is precisely this hydrolytic instability that precluded their use in the textile industry in the 1950s and led researchers to suggest their potential as drug delivery carriers in the 1980s. Since then, polyanhydrides have been synthesized with a wide range of chemistries for a variety of biomedical applications.

The promising characteristics of polyanhydrides for biomedical applications rely on the surface erosion mechanism that translates into well-controlled release kinetics, where the drug release rate coincides with the degradation rate of the polymer. In an aqueous environment, the macromolecules at the surface break into smaller chains before water penetrates into the device. Thus, the drug is released as the polymer degrades. In contrast, bulk eroding polymers degrade slowly and water penetrates into the system much faster, having, in consequence, less predictable kinetics as the drug is released from the entire matrix. A comparison of surface and bulk erosion mechanisms is shown in Figure 5.2.

Polyanhydride-based drug delivery devices (Gliadel®) have been approved by the Food and Drug Administration (FDA) for the treatment of brain tumors. This device is a polyanhydride wafer composed of sebacic acid and 1,3-bis(*p*-carboxyphenoxy) propane [1,3-bis(*p*-carboxyphenoxy) propane : poly(sebacic acid) (CPP : SA) copolymer in 20 : 80 molar ratio] loaded with the chemotherapeutic agent, carmustine, 1,3-bis[2-chloroethyl]-1-nitro-sourea (BCNU). Other potential applications of CPP : SA copolymers include the release of bethanechol for the treatment of Alzheimer's disease and the controlled release of insulin.[1] The treatment of osteomyelitis, which is a bone infection difficult to treat by conventional methods, has been carried out with 20 : 80 CPP : SA copolymer loaded with

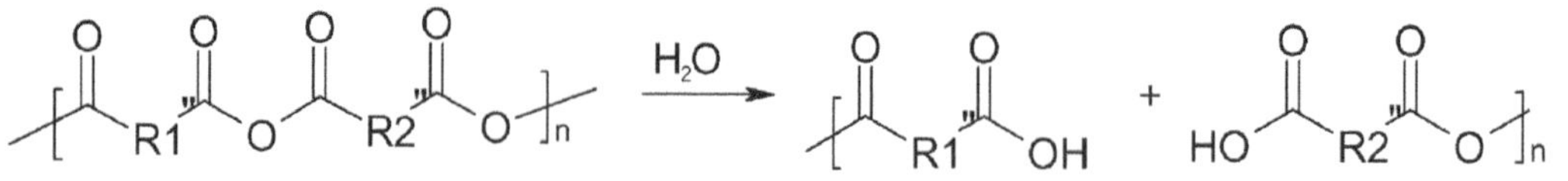

FIGURE 5.1
Hydrolysis of polyanhydrides.

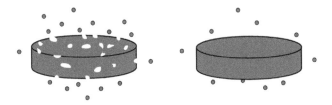

FIGURE 5.2
Mechanisms of polymer erosion: bulk (left) and surface (right).

gentamicin sulfate.[1] Several chemotherapeutic drugs, local anesthetics, anticoagulants, neuroactive drugs, and vaccines have been delivered using polyanhydrides.

Classification

There are three major classes of polyanhydrides: aliphatic, unsaturated, and aromatic. The chemical structures are shown in Table 5.1.

Aliphatic Polyanhydrides

The first aliphatic polyanhydride was synthesized from the monomer adipic acid (AA), which is thermally unstable and forms cyclic dimers and polymeric rings when heated at high temperatures. Poly(sebacic acid) (SA), the aliphatic polyanhydride most widely used in drug delivery applications at present, was synthesized for the first time in the 1930s.[2] Typical properties of aliphatic polyanhydrides include crystallinity, a melting temperature range of 50–90°C (increasing with monomer chain length), and solubility in chlorinated hydrocarbons. These degrade and are eliminated from the body within weeks. When copolymerized with aromatic polyanhydrides, the degradation time can be extended to several months as the aromatic composition increases. The most widely studied aliphatic–aromatic copolymer system is based on SA and 1,3-bis(*p*-carboxyphenoxy) propane (CPP).[3]

Unsaturated Polyanhydrides

The development of unsaturated polyanhydrides responded to the necessity of improving the mechanical properties of the polymers in applications such as the temporary replacement of bone.[4] Unsaturated polyanhydrides, prepared by melt or solution polymerization, include homopolymers of fumaric acid (FA), acetylenedicarboxylic acid (ACDA), and 4,4'-stilbenzenedicarboxylic acid (STDA). The chemical structures of poly(FA) and poly(ACDA) are shown in Table 5.1. These polymers are highly crystalline and insoluble in common organic solvents. The double bonds of these monomers make them suitable for further crosslinking to improve mechanical properties of polyanhydrides. When copolymerized with aliphatic diacids, less crystalline polymers with enhanced solubility in chlorinated solvents result.

Aromatic Polyanhydrides

The first aromatic polyanhydrides synthesized were poly(isophthalic acid) (IPA) and poly(terephthalic acid) (TA).[5] A few common aromatic polyanhydrides are shown in Table

TABLE 5.1

Typical Polyanhydrides Used for Drug Delivery Applications

Classification	R group	Examples
Aliphatic polyanhydrides	$\left[\!\!\left[\text{C}_{\text{H}_2}\right]_x\!\!\right]$	$x = 4$, Adipic anhydride (AA) $x = 8$, Sebacic anhydride (SA)
Unsaturated polyanhydrides	$\left[\text{C}_{\text{H}}\!=\!\text{C}_{\text{H}}\right]$	Fumaric anhydride (FA)
	$\left[\text{C}\!\equiv\!\text{C}\right]$	Acetylenediacarboxylic anhydride (ACDA)
Aromatic polyanhydrides		meta: isophthalic anhydride (IPA) para: terephthalic anhydride (TA)
		$x = 1$, bis(p-Carboxyphenoxy) methane (CPM) $x = 3$, 1,3-bis(p-Carboxyphenoxy) propane (CPP) $x = 6$, 1,6-bis(p-Carboxyphenoxy) hexane (CPH)
Novel polyanhydrides		Trimellitylimido glycine (TMAgly)
		1,8-bis(p-Carboxyphenoxy)-3,6-dioxaoctane (CPTEG)

5.1. Homopolymers of aromatic diacids are crystalline, insoluble in common organic solvents, and have melting points greater than 100°C. Their hydrophobicity results in a slow degradation rate that can last over a year in some cases. Thus aromatic polyanhydrides are not suitable for drug delivery when used as homopolymers. To overcome their slow degradation rates, they have been copolymerized with aliphatic diacids, i.e., CPP : SA copolymers, and with other aromatic monomers. The copolymers of the aromatic monomers TA and IPA are amorphous and soluble, with a faster degradation and a melting point below 120°C.[6]

Other Polyanhydride Chemistries

Although it is impossible to discuss in detail all the polyanhydrides that have been synthesized, some distinguishable classes are discussed here. Typical examples of novel classes of polyanhydrides include those derived from amino acids and fatty acids, and those modified by copolymerization with esters and ethers. The polyanhydrides derived from amino acids, including trimellitylimido glycine (shown in Table 5.1), pyromellitylimido alanine, and trimellitylimido l-tyrosine, have been copolymerized with aliphatic (SA) and aromatic (CPP and CPH) monomers to obtain enhanced degradation and improved mechanical strength because of the presence of the imide bond.[7] These polymers have been studied as vaccine carriers.[8] Some polyanhydrides have been synthesized from dimer and trimer unsaturated fatty acids, and form nonlinear hydrophobic fatty acid esters such as ricinoleic and maleic acid. Other classes of polyanhydrides include ones copolymerized with esters and ethers, which have been suggested as potential drug carriers in the last decade.[9,10] Uhrich and coworkers recently synthesized novel poly(anhydride-co-ester)s containing salicylic acid in the backbone.[11–14] The in vitro/in vivo release of salicylic acid (the active form of aspirin) was studied for the treatment of Crohn's disease and tuberculosis. Copolymers of aliphatic polyanhydrides with ε-caprolactone, trimethylene carbonate, ethylene glycol,[15] and lactic acid have been synthesized. Several modifications of anhydride monomers have been carried out to obtain desired characteristics for particular applications.[16–18] An example is the incorporation of triethylene glycol (TEG) into an aromatic monomer (CPH) to enhance the hydrophilicity of the monomer, resulting in a faster degradation rate.[19] The resulting polymer (Table 5.1) is poly(1,8-bis(*p*-carboxyphenoxy)-3,6-dioxaoctane) (CPTEG).

Synthesis

The most widely used method to synthesize polyanhydrides is melt condensation polymerization, which results in high molecular weight polymers.[20] Other methods include Schotten–Baumann condensation, dehydrative coupling, and ring opening polymerization.

Melt Polycondensation

The general process for melt polycondensation of polyanhydrides is shown in Figure 5.3. It consists of reacting dicarboxylic acid monomers with an excess of acetic anhydride to form oligomers that are polymerized at high temperature under vacuum. The degree of polymerization is influenced by the monomer purity, the strength of the vacuum, the reaction temperature, and the reaction time. It has been found that for most polyanhydrides, the optimal polymerization temperature is in the range of 170–190°C.[21] In general, the condensation reaction is conducted for 2–3 hr, as significant depolymerization can occur after heating for longer periods.[22] With optimal conditions, molecular weights greater than 100,000 can be produced.

The polyanhydrides synthesized by melt condensation have fiber-forming properties in the molten state. They hydrolyze when exposed to air and this degradation is mainly controlled by the composition of the polymer. Homopolymers of aromatic monomers, such as CPH, degrade at a rate that is several orders of magnitude lower than that of homopolymers of aliphatic monomers.[22]

Several variations have been made to the melt condensation process. In the case of polymerization with propionic anhydride and butyric anhydride, harsh conditions can be used for the removal of unreacted anhydride owing to the high boiling point of both chemicals.[6] A variety of catalysts have been used to polymerize polyanhydrides within 20–60 min; but the main disadvantage for biomedical applications is the potential toxicity from catalysts such as cadmium acetate, earth metal oxides, and $ZnEt_2$-H_2O.[6]

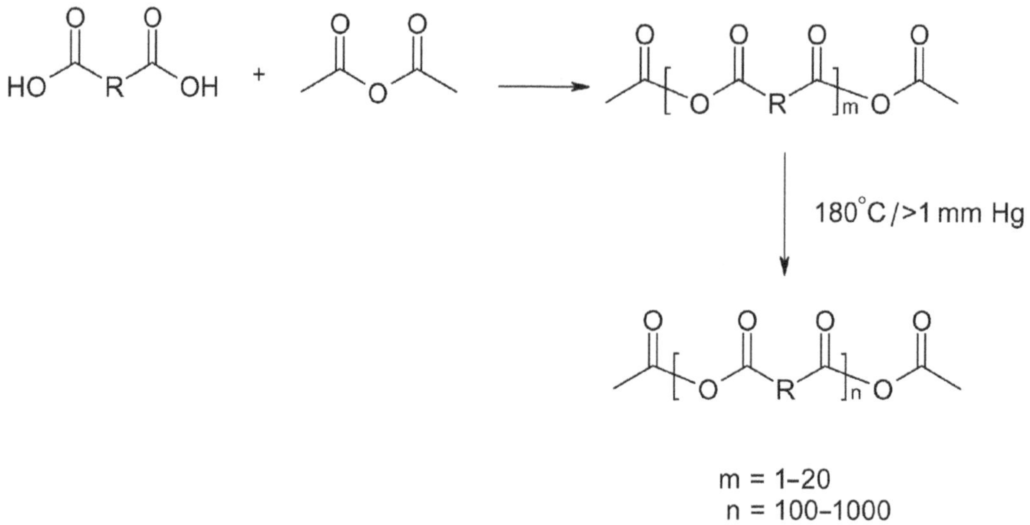

$$180°C/>1\ mm\ Hg$$

m = 1–20
n = 100–1000

FIGURE 5.3
Melt condensation polymerization of polyanhydrides.

Schotten–Bauman Condensation

The Schotten–Bauman condensation produces polyanhydrides with moderate molecular weights by a dehydrochlorination reaction between a diacid chloride and a dicarboxylic acid.[21] The polymerization takes place by reacting the monomers for 1 hr at room temperature, and it can be conducted via solution or interfacial methods. Solvents that are used in solution polymerization include dichloromethane, chloroform, benzene, and ethyl ether. The degree of polymerization obtained with this method is approximately 20–30. Lower molecular weight products are obtained for less reactive monomers such as isophthaloyl chloride.

Polymerization conducted in aqueous interfacial systems suffers from hydrolytic decomposition. The decomposition reaction can be minimized when contact with water is avoided. In the case of polymerization in nonaqueous interfacial environments, products with number average molecular weights up to 5000 can be obtained.[22] Various aromatic polymers were prepared from the reaction of equimolar amounts of the acid dissolved in an aqueous base and the corresponding diacid chloride dissolved in an organic solvent. Reaction occurred between dibasic acid in one phase and an acid chloride in the other. Polar solvents for this reaction include dimethylformamide and 1,4-dicyanobutane.

Dehydrative Coupling

Another method to synthesize polyanhydrides is by dehydrative coupling of two carboxyl groups. Even though this method produces lower molecular weight products (mostly oligomers) compared with the methods described above, it is a single step polymerization where a dicarboxylic acid monomer can be directly converted into the polymer. Moreover, it can be conducted at low temperatures suitable for monomers that cannot resist harsh reaction conditions.

A number of dehydrative agents have been effective in coupling the carboxyl groups. The most effective agents are bis[2-oxo-3-oxazolidinyl]phosphinic chloride, N-phenylphosphoroamidochloridate, diphenyl chlorophosphate, diethyl phosphorobromidate, dicyclohexylcarbodiimide, chlorosulfonylisocyanate, and 1,4-phenylene diisocyanate.[22] In general, higher molecular weight polymers were obtained with polar solvents such as dichloromethane and chloroform. The major disadvantage of this method is the problematic isolation and purification of the final products while preventing hydrolytic decomposition.

Ring Opening Polymerization

Low molecular weight linear polymers undergo transformations between linear and cyclic forms. When a mixture of low and high molecular weight polymers is subjected to molecular distillation, cyclic monomers and dimers are distilled off and a high molecular weight polymer remains behind. The cyclic molecules are transformed to a polymer that contains large ring structures.[6]

Another variation of this process is the preparation of adipic acid from cyclic adipic anhydride (oxepane-2,7-dione). The monomer is prepared by the reaction of adipic acid with acetic anhydride followed by catalytic depolymerization under vacuum. Factors that affect this reaction include temperature, reaction time, and the concentration of catalyst, if used. When catalyzed, reaction at 180°C for 30 min produced polymers with molecular weights up to 300,000. Uncatalyzed reactions that were carried out for more than 2 hr at 180°C yielded low molecular weight polymers.[6]

Characterization

To understand the properties that make polyanhydrides suitable drug carriers, their chemical, physical, and thermal behavior need to be characterized. This section discusses the methods to determine the chemical structure and composition, the molecular weight, the thermal properties, the phase behavior, the stability, and the erosion mechanism of polyanhydrides.

Chemical Structure and Composition

The technique most widely used for determining the chemical structure and composition of polyanhydrides is ^{1}H NMR spectroscopy. The chemical structure is assigned in accord with the chemical shifts characteristic of aliphatic and aromatic protons. The protons close to electronegative groups, i.e., aromatic groups, absorb at lower frequencies (6.5–8.5 ppm), while aliphatic protons absorb at higher frequencies (1–2 ppm).[23] ^{1}H NMR has also been used to determine the degree of randomness in polyanhydride copolymers.[24] By integration of NMR peaks it is possible to determine if a copolymer has a random or block-like structure. Other useful information obtainable from ^{1}H NMR spectra include details of the conversion of polymerization reactions, the actual composition of the polymer, the polymer molecular weight, and degradation rate.

Fourier transform infrared (FTIR) spectroscopy and Raman spectroscopy have also been used to authenticate polyanhydride structures. Aliphatic polymers absorb at 1740 and 1810 cm^{-1}, while aromatic polymers absorb at 1720 and 1780 cm^{-1}.[24] All the polyanhydrides show methylene bands because of deformation, stretching, rocking, and twisting. Aside from being used to ascertain polyanhydride structures, these techniques can be used to determine degradation progress, by monitoring the area of carboxylic acid peak (1770–1675 cm^{-1}) with respect to the characteristic anhydride peaks over time.

Molecular Weight

The molecular weight of polyanhydrides can be determined by gel permeation chromatography (GPC), viscosity measurements, and ^{1}H NMR spectra. Vapor pressure osmometry (VPO) cannot be used for molecular weight determination, as depolymerization occurs during the experiment. The weight average molecular weight (M_w) of polyanhydrides ranges from 5000 to 300,000. Typical polydispersity indexes are in the range of 2–15, which increases with molecular weight. Gel permeation chromatography determines the molecular weight relative to polystyrene standards. The intrinsic viscosity (η) is proportional to M_w, as shown by the Mark Houwink relationship for CPP : SA copolymer [Eq. (1)]. This relationship was calculated from viscosity experiments and M_w values from GPC.

$$[\eta]_{\mathrm{CHCL_3}}^{23°C} = 3.88 \times 10^{-7} \times M_{\mathrm{W}}^{0.658} \tag{1}$$

An alternative way to estimate the molecular weight of polyanhydrides is by end group analysis from ^{1}H NMR spectra. The degree of polymerization can be calculated from the ratio of the area of the inner chain protons to the area of terminating groups. The number average degree of polymerization (DP) of CPP : SA copolymers is represented in Eq. (2), where (CPP) and (SA) depict the area of scaled inner chain protons, (Ac) represent the acetylated end group, and (SA*) and (CPP*) designate the carboxylic terminated polymer chain.[25]

$$DP = \frac{2[(CPP)+(SA)]}{[(Ac)+(SA^*)+(CPP^*)]} \tag{2}$$

Thermal Properties

The thermal transitions of polyanhydrides have been determined from differential scanning calorimetry (DSC). Differential scanning calorimetry thermal scans provide properties such as glass transition temperature (T_g), melting temperature (T_m), and heat of fusion (ΔH). It is important to know the values of T_g and T_m in the fabrication of drug delivery devices such as tablets and microspheres. While T_g determines the minimum temperature required for compression molding, T_m determines the minimum temperature necessary for injection molding or melt compression. A general decreasing trend in T_g's has been observed as methylene groups are added into the main chain of an anhydride monomer. As mentioned earlier, aliphatic polyanhydrides melt at temperatures below 100°C and aromatic polyanhydrides have melting points greater than 100°C.

It has been shown that the crystallinity of polymers affects erosion and drug release rates, because crystalline regions erode slower than amorphous ones.[26] Moreover, highly crystalline polyanhydrides affect the device morphology as it creates irregular external surfaces. The crystallinity of polyanhydrides has been determined using x-ray diffraction, DSC, ^{1}H NMR spectroscopy, and small-angle x-ray scattering (SAXS). It has been demonstrated that homopolymers of aromatic and aliphatic diacids are crystalline. When copolymerized, polyanhydrides exhibited a decrease in crystallinity in copolymers of equimolar compositions, i.e., CPP : SA, CPH : SA, and FA : SA copolymers.[26] The ΔH from

DSC thermographs exhibited a decrease as the copolymers approached equimolar compositions. This decrease in crystallinity is representative of the random behavior of the polymer chain, as determined by ^{1}H NMR spectra. In general, the copolymers rich in one monomer had higher crystallinity.

Phase Behavior

Polymer blends, which display distinct physical and chemical properties, are used for the design of materials for diverse applications. This variation in properties may lead to microphase separation, which in turn affects the drug release because of a thermodynamical partition of drugs between the phases, depending on their compatibility with the phase.[27] Research has shown that aliphatic, aromatic, and copolymers of anhydride monomers are miscible and the blends had a single melting temperature that was lower than that of the starting polymers.[6] On the other hand, polyanhydrides that are partially miscible with poly(orthoesters), poly(hydroxybutyric acids), and low molecular weight poly(esters) and have two melting temperatures are clearly indicative of the phase separation. Blends of polyanhydrides with poly(caprolactone) [poly(CL)] are completely immiscible. Degradation studies in blends of poly(CL) with poly(dodecanedioic anhydride) [poly(DD)] indicated that the anhydride component degraded rapidly and was released from the blend, without affecting the poly(CL) degradation.[24] Other studies include the characterization of microphase-separated copolymers of poly(SA) with poly(CPH) or poly(ethylene glycol).[28,29] The phase diagram for the poly(CPH)/poly(SA) blend system has been determined using SAXS, optical microscopy, and molecular simulations, while the blends of poly(SA) and PEG were characterized by DSC and infrared (IR) spectra. The poly(CPH)/poly(SA) system exhibits an upper critical solution temperature behavior.

Stability

The stability of polyanhydrides has been studied in solid state and in dry chloroform. Aromatic polyanhydrides such as poly(CPP), poly(CPH), and poly(CPM) maintained their original molecular weight for at least one year in solid state upon storage under dry argon or vacuum at 21°C. In contrast, aliphatic polyanhydrides such as poly(SA) have a high-degradation rate at the same storage conditions.[1] Studies performed with GPC revealed that M_w of polyanhydrides tends to undergo a rapid decrease initially, and later a constant stabilized decrease in molecular weight is observed. The decrease in molecular weight was explained by an internal anhydride interchange mechanism resulting in ring formation, as revealed by ^{1}H NMR. This mechanism was supported by the fact that the decrease in molecular weight was reversible and heating of the depolymerized polymer at 180°C for 20 min yielded the original high molecular weight polymer.[24] It is important to mention that polyanhydrides experienced significant weight loss when stored at ambient conditions in which water attacks the anhydride bonds.

The stability of polyanhydrides in a solution was studied using chloroform under dry nitrogen atmosphere at 37°C.[1] The aromatic polyanhydrides remained stable under these conditions during a three-day period, while copolymers with aliphatic SA had a significant molecular weight loss during the same time period. Therefore, polyanhydrides can be processed in a solution environment as long as the time is not extended more than this period.

γ-Irradiation methods have been utilized for the sterilization of polyanhydrides. In this technique, aliphatic and aromatic homo- and copolymers were irradiated at 2.5 Mrad, and the chemical structure as well as the physical properties were found to be the same before and after irradiation.[24] The studies showed that saturated polyanhydrides are stable during γ-irradiation, as a slight increase in molecular weight was observed. Electron paramagnetic resonance (EPR) spectroscopy was used to characterize free radicals in γ-sterilized polyanhydrides.[30] Polymers with high melting temperatures produced the highest yields of room temperature radicals, which in turn transform into less conjugated polyanhydrides that leads to lower molecular weight polymers.

Degradation and Erosion

Polymer erosion (i.e., mass loss) is a complex process that is determined by numerous factors, including the molecular weight loss (degradation), swelling, dissolution and diffusion of oligomers and monomers, and morphological changes.[31] Polyanhydrides undergo degradation prior to erosion, as a consequence of the chemical instability of the anhydride bond. Thus, degradation and erosion are limited to the surface, as water does not penetrate into the device.[32]

Erosion kinetics is complicated when the anhydride monomers of a copolymer system exhibit microphase separation that leads to the erosion of different phases at different rates. The erosion of a fast eroding phase may leave the slow eroding phase intact.[33] At this point, the monomer solubility plays a major role in polyanhydride erosion kinetics, as monomers are accumulated in eroding zones of the matrix and its dissolution will depend on the pH of the microenvironment.[34] It is known that the saturation concentration of the monomers CPH, SA, and CPTEG is a function of pH and that, at a particular pH, the order of solubility of the monomers is CPTEG > SA > CPH. This provides valuable information when describing the drug release from polymers containing any of these monomers.[19,33]

Polyanhydride-Based Drug Delivery Systems

Biocompatibility

The biocompatibility of implantable polyanhydride disks was studied in the brain of rats, rabbits, monkeys, and eventually in human clinical trials.[35] Wafers of poly(CPP : SA) and poly(FAD : SA) were implanted in the frontal lobes of rats, rabbits, and monkeys. In all these studies, the animals receiving the implants showed no behavioral changes or neurological deficits, indicating that the polymers were not invoking a systemic or local toxicity. To determine how the body metabolized the poly(CPP : SA), radio-labeled copolymers were implanted in the brains of rats.[21] Seven days after the implantation, 40% of the ^{14}C SA-labeled polymer had been excreted as CO_2, 10% was excreted along with the urine, 2% with the feces, and 10% still in the implanted device. In the same period only 4% of the ^{14}C CPP-labeled polymer was excreted along with the urine and feces.

The biocompatibility of poly(CPP), poly(TA), and copolymers of CPP : SA and CPP : TA implanted in the corneas of rabbits was studied.[36,37] Six weeks after implantation, the cornea remained clear and showed no evidence of corneal edema or neovascularization, indicating biocompatibility of the polymer matrix implant.

Subcutaneous implants of 20 : 80 CPP : SA copolymer were administered in rats at doses of 40 and 120 times the size that is to be used in humans. The purpose of these experiments was to test the systemic toxicity of the polymers. Eight weeks after implant-ing the disks, the rats were sacrificed and their organs underwent histopathologi-cal evaluations. In general, there was little to no difference between the organs of the experimental group (receiving the implant) and those of the control group. Again, in all the cases the polymers underwent degradation and were found to cause minimal inflammation at the site of implantation. Thus, polyanhydrides are inert and suitable for in vivo drug delivery.[38]

Drug/Polymer Interactions

When selecting polymers as drug delivery carriers, it is necessary to establish whether the polymer will react with the incorporated or the released drug. Three factors need to be considered: the reactivity of the drug, the hydrophobicity of the drug, and the fabrication method. The reactivity of CPP : SA copolymer with the para substituted anilines: *p*-nitroan-iline (PNA), *p*-bromoaniline, *p*-anisidine, and *p*-phenylenediamine was examined.[37] The model drugs were incorporated into the polymer matrix using injection and compression molding. When injection molding was used to encapsulate the drugs at 120°C, the more reactive drugs (*p*-bromoaniline, *p*-anisidine, and *p*-phenylenediamine) reacted with the polymer forming amides. However, when the drugs were incorporated into the polymer matrix using compression molding at room temperature, the drugs did not react with the polymer during the fabrication process.

The hydrophobicity of the drug can also influence the interactions between the drug and the polymer. When hydrophilic dyes (acid orange and brilliant blue) were encapsulated in polyanhydrides, the T_m's of the polymers were unchanged. When hydrophobic dyes (PNA and methyl red) encapsulated, the T_m's of the polymers changed thereby indicating an interaction between the polymer and the drug.[39]

Device Fabrication

Polyanhydride drug delivery devices have been fabricated as implantable[40] and injectable devices. Implantable devices are fabricated by compression molding, melt compression, or solvent casting. The first step of compression molding is to obtain a fine powder of the drug and the polymer. The powders are physically mixed and placed in a piston mold. The wafer is formed by applying a pressure (typically 30 kpsi) and by heating the sample to a temperature 5–10°C above the T_g of the polymer.[1,41] One drawback of this method is the uneven distribution of the drug in the polymer, leading to poor reproducibility. The Gliadel system is a compression-molded wafer. To overcome the problem of uneven drug distribution, the drug and polymer are spray dried together to form microspheres. The microspheres are then compression molded to form the wafer.

An alternative to compression molding is melt compression. This procedure requires the polymer and drug to be heated 10°C above the T_m of the polymer, forming a viscous solution.[1,41] The solution can then be either placed in a conventional mold under low pres-sure or injection molded. This fabrication method results in an even drug distribution. However, the elevated temperatures needed to melt the polymer could cause adverse reac-tions in temperature sensitive drugs such as proteins.

Solvent casting is done by codissolving or suspending the drug in the polymer solution. The solution is then poured into a flat open mold and cooled on dry ice. The resulting film is often fragile. If the drug is not soluble in the polymer, it will settle on the bottom of the film, leading to an uneven drug distribution.[27,30]

To form an injectable drug delivery device, the drug is loaded into polymer microspheres. Drug-loaded polyanhydride microspheres have been fabricated using different methods. The most common method is the solvent extraction method, which includes water/oil/ water (w/o/w), water/oil/oil (w/o/o), or solid/oil/oil (s/o/o) (dependent on whether the drug is soluble in the polymer solvent). In the w/o/w method, the drug (typically proteins) is dissolved in an aqueous phase and then emulsified with a larger volume of polymer that is dissolved in an organic solvent, typically methylene chloride. The inner emulsion is then added to a larger volume of water that contains a surfactant, usually PVA, and allowed to stir for several hours to extract the solvents. In the case of w/o/o or s/o/o the outer aqueous PVA phase is replaced with an immiscible organic solvent, i.e., silicon oil. The spheres are typically collected by either centrifugation or filtration.

Microspheres can also be fabricated by the hot-melt procedure; however, this method is not ideal for encapsulating temperature sensitive drugs such as proteins.[1] Spray drying or atomizing the polymer and drug together can also be used to fabricate microspheres. This method requires the use of either a spray dryer or atomization.[1,42] In the case of a spray dryer, the polymer/drug suspension is pumped into the spray drier. As the suspension is sprayed, a stream of air causes the polymer spheres to harden. Microspheres are fabricated by atomization by passing the drug/polymer suspension through an atomizing nozzle. As the polymer/drug spheres leave the nozzle, they are collected in a bath of liquid nitrogen sitting on top of a frozen layer of ethanol.[42] The liquid nitrogen/ethanol bath is then stored at − 80°C for three days. During this time the ethanol slowly thaws and the frozen microspheres fall into it. As the microspheres sit in the ethanol, the organic solvent (methylene chloride) slowly diffuses out, leaving solid spheres that could be collected by filtration.

In Vitro Release

The rate at which an encapsulated drug will be released from a polyanhydride device, either a wafer or a microsphere, is strongly dependent on polymer composition and drug distribution. Other factors that contribute to the release rate of drugs includes: fabrication technique, size/shape of the device, and pH of the surrounding media.

The hydrophobicity of a drug influences its distribution within the polymeric device. [27] *p*-Nitroaniline and disperse yellow have higher affinities for poly(CPH) and poly(SA), respectively. The two drugs were encapsulated in tablets of poly(CPH), poly(SA), and copolymers of the two to determine if the drugs would partition into the more favorable polymer microdomain. When the dominant polymer had a low affinity for the drug, a burst effect was seen. In the case of 50 : 50 CPH : SA copolymer, each drug followed the release of the monomers, indicating that the drug was partitioning into the more favorable domain.

The size of the device may also influence drug distribution and release rate.[43] Monodisperse microspheres of differing average diameters were studied to determine the influence of size of the device on the delivery. Smaller diameter microspheres showed a more prolonged release rate of drug than did microspheres that had a large diameter. As the diameter increased, the time it took for the microsphere to form by precipitation increased; thus, increasing the time for the drug to segregate toward the surface of the microspheres.

As the anhydride bonds in the polymer backbone are hydrolyzed, carboxylic acid is formed. The formation of the acidic degradation products reduces the local pH of the eroding device. The diffusion of the acidic degradation products away from the device is expedited when the device is in a basic solution. However, when the device is in an acidic solution, the erosion process is slowed significantly.[1,44]

Polyanhydrides have also been investigated as protein carriers.[45] Poly(SA) and 20 : 80 CPH : SA copolymer microspheres were found to conserve both the primary structure of the released protein [bovine serum albumin (BSA)] and the secondary structure of the encapsulated and released protein, and showed a sustained delivery for approximately 15 and 30 days, respectively. As the CPH content in the copolymer increased, the secondary structure of BSA was not conserved, as indicated by the steep decrease in the α-helix content.

In Vivo Delivery

The 20 : 80 CPP : SA copolymer was the first polyanhydride to be clinically tested in humans. The copolymer was used to encapsulate BCNU, a chemotherapeutic drug used to treat a fatal form of brain cancer known as glioblastoma multiforme. Along with the polymer, BCNU was codissolved, and the disks were fabricated by compression molding. The preclinical trials in rat, rabbit, dog, and monkey brains demonstrated the effectiveness of the polymer in delivering an active drug that remained localized, minimizing systemic reaction to the drug.[46] The wafer, once implanted into the brain of the glioblastoma patients, releases the BCNU for approximately three weeks.[47] Scanning electron microscopy was used to monitor the erosion of the wafer both in vitro and in vivo.[48] It was found that the erosion of the wafer was controlled by diffusion of BCNU and erosion of the polymer. The delivery device was approved in 1996 by the USFDA for use in conjunction with surgery for patients suffering from recurrent glioblastoma. In 2003, the USFDA approved the use of the device in newly diagnosed advanced cases of malignant gliomas, in conjunction with surgery and radiation.

The use of 20 : 80 CPP : SA and 18 : 82 FAD : SA copolymers disks as drug delivery devices for carboplatin, a treatment for glioma, was also investigated in rodents.[49] The majority of the drug was released in seven days from the CPP : SA copolymer disk, and 65% of the drug was released from the FAD : SA copolymer disk in seven days. This method of delivery was more effective than systemic therapy and did not cause systemic toxicity.

A separate polyanhydride system has also been investigated for the treatment of osteomyelitis, a bone infection typically caused by bacteria.[50–52] Copolymer implants (50 : 50 FAD : SA) containing gentamicin were tested in the backs of rats, in the infected tarsocrural joints of horses, and in humans with infected prosthetic hips or knees.[51] In all the cases, the local delivery of gentamicin was successful and the systemic exposure to the drug was avoided.

Conclusions

Polyanhydrides are promising as biomaterials because they possess a unique combination of properties that include hydrolytically labile backbone, hydrophobic bulk, and chemistry that can be easily combined with other functional groups to design novel materials. These materials are primarily surface-erodible and offer the potential to stabilize protein drugs and sustain release from days to months. The microstructure characteristics of copolymer systems can be exploited to tailor drug release profiles. The versatility of polyanhydride chemistry promises a new class of drug release systems for specific applications.

References

1. Tamada, J., Langer, R. The development of polyanhydrides for drug delivery applications. *J. Biomater. Sci. Polym. Ed.* **1992**, *3* (4), 315–353.
2. Hill, J., Carothers, W. Studies of polymerization and ring formation. XIV. A linear superpolyanhydride and a cyclic dimeric anhydride from sebacic acid. *J. Am. Chem. Soc.* **1932**, *54*, 1569–1579.
3. Domb, A.J., Ron, E., Langer, R. Poly(anhydrides) based on aliphatic–aromatic diacids. *Macromolecules* **1989**, *22*, 3200–3204.
4. Domb, A.J., Mathiowitz, E., Ron, E., Giannos, S., Langer, R. Polyanhydrides. IV. Unsaturated and crosslinked polyanhydrides. *J. Polym. Sci. A.* **1991**, *29*, 571–579.
5. Bucher, J., Slade, W.C. Anhydrides isophthalic and terephthalic acids. *J. Am. Chem. Soc.* **1909**, *31*, 1319–1321.
6. Domb, A.J., Amselem, S., Shah, J., Maniar, M. Polyanhydrides: synthesis and characterization. *Adv. Polym. Sci.* **1993**, *107*, 94–141.
7. Staubli, A., Ron, E., Langer, R. Hydrolytically degradable amino acid containing polymers. *J. Am. Chem. Soc.* **1990**, *112*, 4419–4424.
8. Hanes, J., Chiba, M., Langer, R. Degradation of porous poly(anhydride-co-imide) microspheres and implications for controlled macromolecule delivery. *Biomaterials* **1998**, *19* (1–3), 163–172.
9. Jiang, H.L., Zhu, K.J. Preparation, characterization and degradation characteristics of polyanhydrides containing poly(ethylene glycol). *Polym. Int.* **1999**, *48*, 47–52.
10. Weinberg, J.M., Gitto, S.P., Wooley, K.L. Synthesis and characterization of degradable poly(silyl ester). *Macromolecules* **1998**, *31*, 15–21.
11. Erdmann, L., Macedo, B., Uhrich, K.E. Degradable poly(anhydride ester) implants: effects of localized salicylic acid release on bone. *Biomaterials* **2000**, *21* (24), 2507–2512.
12. Erdmann, L., Uhrich, K.E. Synthesis and degradation characteristics of salicylic acid-derived poly(anhydride-esters). *Biomaterials* **2000**, *21* (19), 1941–1946.
13. Schmeltzer, R., Anastasiou, T., Uhrich, K. Optimized synthesis of salicylate-based poly (anhydride-esters). *Polym. Bull.* **2003**, *49* (6), 441–448.
14. Anastasiou, T., Uhrich, K. Aminosalicylate-based biodegradable polymers: synthesis and in vitro characterization of poly(anhydride-esters) and poly(anhydride-amides). *J. Polym. Sci. A: Polym. Chem.* **2003**, *41*, 3667–3679.
15. Jiang, H., Zhu, K. Pulsatile protein release from a laminated device comprising polyanhydrides and pH-sensitive complexes. *Int. J. Pharm.* **2000**, *194* (1), 51–60.
16. Campo, C.J., Anastasiou, T., Uhrich, K.E. Polyanhydrides. The effects of ring substitution changes on polymer properties. *Polym. Bull.* (Berlin) **1999**, *42* (1), 61–68.
17. Anastasiou, T.J., Uhrich, K.E. Novel polyanhydrides with enhanced thermal and solubility properties. *Macromolecules* **2000**, *33* (17), 6217–6221.
18. Anseth, K.S., Shastri, V.R., Langer, R. Photopolymerizable degradable polyanhydrides with osteocompatibility. *Nature Biotech.* **1999**, *17* (2), 156–159.
19. Torres, M.P., Vogel, B.M., Narasimhan, B., Mallapragada, S.K. Synthesis and characterization of novel polyanhydrides with tailored erosion mechanisms. *J. Biomed. Mater. Res.* Submitted.
20. Domb, A.J., Ehrenfreund, T., Golenser, J., Langer, R., Israel, Z. Biodegradable polyanhydrides: synthesis and drug delivery applications. *Biodegrad. Polym.* **2003**, *2*, 121–151.
21. Domb, A.J., Amselem, S., Langer, R., Maniar, M. Polyanhydrides as carriers of drugs. In *Biomedical Polymers Designed-to-Degrade Systems*; Shalaby, S.W., Ed.; Hanser Publishers: New York, 1994; 69–96.
22. Leong, K.W., Simonte, V., Langer, R. Synthesis of polyanhydrides: melt-polycondensation, dehydrochlorination, and dehydrative coupling. *Macromolecules* **1987**, *20* (4), 705–712.
23. Narasimhan, B., Kipper, M.J. Surface-erodible biomaterials for drug delivery. *Adv. Chem. Eng.* **2004**, *29*, 169–218.

24. Kumar, N., Langer, R.S., Domb, A.J. Polyanhydrides: an overview. *Adv. Drug Delivery Rev.* **2002**, *54* (7), 889–910.

25. McCann, D.L., Heatley, F., D'Emanuele, A. Characterization of chemical structure and morphology of eroding polyanhydride copolymers by liquid-state and solid-state 1H n.m.r. *Polymer* **1999**, *40* (8), 2151–2162.

26. Mathiowitz, E., Ron, E., Mathiowitz, G., Amato, C., Langer, R. Morphological characterization of bioerodible polymers. 1. Crystallinity of polyanhydride copolymers. *Macromolecules* **1990**, *23*, 3212–3218.

27. Shen, E., Kipper, M.J., Dziadul, B., Lim, M.-K., Narasimhan, B. Mechanistic relationships between polymer microstructure and drug release kinetics in bioerodible polyanhydrides. *J. Control. Release.* **2002**, *82* (1), 115–125.

28. Chan, C.-K., Chu, I.-M. Phase behavior and miscibility in blends of poly(sebacic anhydride)/poly(ethylene glycol). *Biomaterials* **2002**, *23* (11), 2353–2358.

29. Kipper, M.J., Seifert, S., Thiyagarajan, P., Narasimhan, B. Understanding polyanhydride blend phase behavior using scattering, microscopy, and molecular simulations. *Polymer* **2004**, *45* (10), 3329–3340.

30. Domb, A.J., Elmalak, O., Shastri, V.R., Ta-Shma, Z., Masters, D.M., Ringel, I., Teomim, D., Langer, R. 8. Polyanhydrides. *Drug Target. Delivery* **1997**, *7*, 135–159.

31. Gopferich, A., Tessmar, J. Polyanhydride degradation and erosion. *Adv. Drug Delivery Rev.* **2002**, *54* (7), 911–931.

32. Kumar, N., Ravikumar, M.N.V., Slivniak, R., Krasko, M.Y., Domb, A.J. Biodegradation of polyanhydrides. In *Biopolymers*; Matsumura, S., Steinbuechel, A., Eds.; Wiley-VCH Verlag GmbH: Weinheim, Germany, 2003; Vol. 9, 423–456.

33. Kipper, M., Narasimhan, B. Molecular description of erosion phenomena in biodegradable polymers. Macromolecules **2005**, *38* (5), 1989–1999.

34. Goepferich, A., Langer, R. The influence of microstructure and monomer properties on the erosion mechanism of a class of polyanhydrides. *J. Polym. Sci. A: Polym. Chem.* **1993**, *31*, 2445–2458.

35. Katti, D.S., Lakshmi, S., Langer, R., Laurencin, C.T. Toxicity, biodegradation and elimination of polyanhydrides. Adv. *Drug Delivery Rev.* **2002**, *54*, 933–961.

36. Brem, H., Kader, A., Epstein, J.I., Tamargo, R.J., Domb, A., Langer, R., Leong, K.W. Biocompatibility of a biodegradable, controlled-release polymer in the rabbit brain. *Selec. Cancer Ther.* **1989**, *5* (2), 55–65.

37. Leong, K.W., D'Amore, P., Marletta, M., Langer, R. Bioerodible polyanhydrides as drug-carrier matrices. II. Biocompatibility and chemical reactivity. *J. Biomed. Mater. Res.* **1986**, *20*, 51–64.

38. Kumar, M.N., Ravi, V., Domb, A.J. Drug delivery, controlled. In *Encyclopedia of Biomaterials and Biomedical Engineering*; Wnek, G.E., Bowlin, G.L., Eds.; Marcel Dekker, Inc.: New York, 2004; 1, 467–477.

39. Shen, E., Pizsczek, R., Dziadul, B., Narasimhan, B. Microphase separation in bioerodible copolymers for drug delivery. *Biomaterials* **2001**, *22* (3), 201–210.

40. Jain, J.P., Mode, S., Domb, A.J., Kumar, N. Role of polyanhydrides as localized drug carriers. *J. Control. Rel.* **2005**, *103* (3), 541–563.

41. Chasing, M., Lewis, D., Langer, R. Polyanhydrides for controlled drug delivery. *Biopharm. Manuf.* **1988**, *2*, 33–39.

42. Gombotz, W.R., Healy, M.S., Brown, L.R. *A Very Low Temperature Casting of Controlled Release Microspheres*; Enzytech, Inc.: U.S.A., 1991.

43. Berkland, C., Kipper, M.J., Narasimhan, B., Kim, K.K., Pack, D.W. Microsphere size, precipitation kinetics and drug distribution control drug release from biodegradable polyanhydride microspheres. *J. Control. Release.* **2004**, *94* (1), 129–141.

44. Chasin, M., Domb, A., Ron, E., Mathiowitz, E., Langer, R., Leong, K., Laurencin, C., Brem, H., Grossman, S. Polyanhydrides as drug delivery systems. *Drugs Pharm. Sci.* **1990**, *45*, 43–70.

45. Determan, A.S., Trewyn, B.G., Lin, V.S.-Y., Nilsen-Hamilton, M., Narasimhan, B. Encapsulation, stabilization, and release of BSA-FITC from polyanhydride microspheres. *J. Control. Release.* **2004**, *100* (1), 97–109.
46. Brem, H. Polymers to treat brain tumours. *Biomaterials* **1990**, *11*, 699–701.
47. Brem, H., Mahaley, S., Vick, N.A., Black, K.L., Schold, S.C., Burger, P.C., Friedman, A.H., Ciric, I.S., Eller, T.W., Cozzens, J.W., Kenealy, J.N. Interstitial chemotherapy with drug polymer implants for the treatment of recurrent gliomas. *J. Neurosurg.* **1991**, *74*, 441–446.
48. Dang, W., Daviau, T., Brem, H. Morphological characterization of polyanhydride biodegradable implant Gliadel during in vitro and in vivo erosion using scanning electron microscopy. *Pharm. Res.* **1996**, *13* (5), 683–691.
49. Olivi, A., Ewend, M.G., Utsuki, T., Tyler, B., Domb, A.J., Brat, D.J., Brem, H. Interstitial delivery of carboplatin via biodegradable polymers is effective against experimental glioma in the rat. Cancer Chemother. *Pharmacol.* **1996**, *39*, 90–96.
50. Perez, C., Castellanos, I.J., Costantino, H.R., Al-Azzam, W., Griebenow, K. Recent trends in stabilizing protein structure upon encasulation and release from bioerodible polymers. *J. Pharm. Pharmacol.* **2002**, *54*, 301–313.
51. Stephens, D., Li, L., Robinson, D., Chen, S., Chang, H.-C., Liu, R.M., Tian, Y., Ginsburg, E.J., Gao, X., Stultz, T. Investigation of the in vitro release of gentamicin from a polyanhydride matrix. *J. Control. Release* **2000**, *63*, 305–317.
52. Li, L.C., Deng, J., Stephens, D. Polyanhydride implant for antibiotic delivery-from bench to the clinic. *Adv. Drug Delivery Rev.* **2002**, *54*, 963–986.

6

Functional Biomaterials

Chun Wang

CONTENTS

Introduction

The term "functional biomaterials" is a broad definition of biomaterials that carry biologically relevant function(s). These may include all the contemporary biomaterials used in medical implants and devices and are classified traditionally as metals, ceramics, composites, and polymers. As the development of biomaterials relies more and more heavily on the understanding and adaptation of principles in biology, functional biomaterials often refer to materials that combine biological molecules, such as proteins, peptides, and nucleic acids, forming systems that actively interact with biological entities (such as cells) and modulate biological processes. Functional biomaterials are also responsive materials that are able to recognize signals in biological environment, change their structures, and carry out their functions accordingly.

    The layout of this chapter is structured around "biologically relevant functions," which specifically refer to the ability of recognizing cells and regulating cellular activities and the ability of delivering biologically active substances precisely. Topics discussed in detail include material surfaces functionalized with biomolecules that bind to cells and support cell growth in 2-D, material matrices that encapsulate and support cell and tissue growth in 3-D, "intelligent" materials and material surfaces that respond to biological signals, and materials that control specific delivery of genetic therapeutics to cells in vivo. Preparation, processing, and functionalization strategies are described for synthetic polymeric biomaterials, synthetic systems combined with biomolecules, and biopolymers that are derived naturally or synthesized de novo.

Functional Biomaterials that Promote Cell Adhesion

Inside the human body, cell adhesion to foreign biomaterial surfaces is mediated by a layer of proteins found in the blood or serum. Biomaterials that are able to control the adsorption of blood proteins will be able to control selectively the adhesion of cells. This function underlies the so-called "biocompatibility" of biomaterials and is ultimately responsible for the success or failure of medical implants and devices in vivo.[1]

Minimizing Nonspecific Cell Adhesion

Functional biomaterial surfaces that absorb proteins minimally are desirable in prolonging the lifetime of medical implants and providing a clean background for introducing specific cell adhesion functionalities.[1] Nonspecific protein adsorption occurs in various degrees to all surfaces, but more readily to hydrophobic and positively charged surfaces. To date, the most effective way to minimize nonspecific protein and cell adhesion is to use surfaces comprised of chains of polyethylene oxide (PEO; also named polyethylene glycol, or PEG).[1]

PEG chains are extremely flexible and hydrophilic. Protein adsorption to a PEG-surface restricts the freedom of the PEG chains and is therefore thermodynamically unfavorable. The effectiveness of reducing protein adsorption depends on the length of PEG—longer chains are more effective than short ones at any given surface coating density. PEG chains can be chemically coupled at one end to many biomaterial surfaces bearing reactive groups such as amines and carboxylates. Block copolymers of PEG, such as di- or tri-block copolymers of PEG and polypropylene oxide (PPO), known as Pluronics® or Poloxamers®, and PEG graft copolymers, can be used to physically coat hydrophobic material surfaces such as polystyrene beads or dishes. PEG-based polymer networks, or hydrogels, are well known to absorb proteins minimally and are often used to coat otherwise protein-absorbing surfaces. Gold surfaces can be similarly passivated through forming self-assembled monolayers (SAMs) of alkanethiols terminated with oligoethylene glycol chains. Recently, a new approach to more stable surface coating was developed[2] combining thiol-gold interaction with PEG block copolymers (PEG-polypropylene sulfide-PEG). The free-ends of surface-bound PEG chains can then be used to attach cell adhesion motifs to facilitate specific interactions with cells on top of a nonadhesive "low-noise" background.

In addition to PEG, certain polysaccharides, such as hyaluronic acid (HA) and dextran, have been used as low protein-adsorption, low cell adhesion surface coatings. Synthetic polymer surfactants consisting of poly(vinyl amine) with dextran and alkanoyl side chains, which mimicks the glycocalyx—negatively charged sugar layer outside cell membrane, are also shown to reduce protein adsorption to hydrophobic graphite surface.[3]

Surface Immobilization of Cell Adhesion Motifs

Early attempts to functionalize biomaterial surfaces with biological molecules[1] were focused on improving blood compatibility of cardiovascular devices, such as the artificial heart and synthetic blood vessels, by immobilizing heparin or albumin on polyurethane or Dacron®. To enhance cell adhesion to biomaterial surfaces, entire extracellular matrix (ECM) proteins, such as fibronectin and laminin, have been used directly as coatings. However, because of the nonspecific manner of whole protein adsorption, most of the cell binding capability is often lost. Using a molecular templating technique, it may be possible to select which protein(s) to absorb on biomaterial surfaces.[4]

It was found that ECM proteins contain short stretches of amino acids (peptides) that bind specifically to a group of receptors on most cell surface, the integrin receptors, which triggers intracellular events that lead to various kinds of cell behavior. These cell adhesion peptides have been immobilized on synthetic biomaterial surfaces to promote specific attachment of selected cell types, which is desirable in many tissue engineering applications. Comparing with ECM protein adsorption, immobilizing cell adhesion peptides can be done in a much better controlled fashion with higher cell binding functionality.[5]

A common method to graft cell adhesion peptides on biomaterial surface is chemical end-coupling, attaching reactive amino acid residues in the peptides to reactive groups on the surface. If the biomaterial surface does not naturally contain reactive groups, they can be generated either chemically or by radiation (UV, plasma gas discharge, etc.). A linker or "arm" is often bridging the peptide and the surface to enhance the freedom and activity of the adhesion motif.[1]

One of the most prevalent cell adhesion peptides is the tripeptide arginine-glycine-aspartate (RGD) from fibronectin.[5] This tripeptide has been used extensively to promote adhesion and spreading of many cell types. For example, polyurethane surface can be activated and coupled to RGD via the carboxy terminus. Surfaces containing photo-activable groups, such as benzophenone or aryl azide, can be modified easily with RGD by UV radiation. This tripeptide and other similar adhesion peptides have also been grafted to activated PEG surface, which supported long-term highly specific cell adhesion and spreading. Through copolymerized lysines, RGD has been immobilized on biodegradable polymers of the poly(glycolic acid) and poly(lactic acid) families. Glass surface can be silylated by trialkoxysilanes to couple with adhesion motifs bearing amines, hydroxyls, or carboxyls. In addition to peptide motifs, monosaccharides, such as glucose and lactose have been used to functionalize polystyrene surfaces to enhance hepatocyte binding.[5]

Immobilization of cell adhesion motifs using physical interactions is more straightforward than chemical attachment.[5] One of the earlier reports on the practical application of RGD involved peptide adsorption to many polymer and inorganic surfaces via a hydrophobic tail of oligoleucine. Polymeric surfactants such as the PEG block copolymers can be modified to display RGD on the end of the hydrophilic block and coat hydrophobic substrates. Alkanethiols bearing cell adhesion peptides form SAMs on gold surfaces. Peptide amphiphiles containing RGD and other sequences on one end, and long alkyl ester lipid tails on the other, have been shown to self-assemble on hydrophobic surfaces and used to probe specific cell adhesion.[6]

The strength and the specificity of cell adhesion by surface-functionalized biomaterials are dependent on the density, composition, and 2-D distribution of the immobilized adhesion motif.[5] It is generally accepted that there is a trade-off between cell affinity and cell mobility, with the optimal cell mobility achieved at an intermediate ligand density. The explanation is that low ligand density does not provide enough adhesion sites and strength, whereas high density creates too much adhesion, limiting mobility. Cell adhesion motifs with different chemical composition mediate adhesion and spreading of different cell types. For example, a peptide derived from laminin, isoleucine-lysine-valine-alanine-valine (IKVAV), mediates neurite extension, whereas the RGD peptide attaches to most cell types. The combination of cell adhesive surfaces with nonadhesive surfaces makes it possible to generate 2-D patterns that lead to cell attachment to selected and predetermined areas on biomaterials. Micropatterning techniques such as photolithography and microcontact printing (or soft lithography) have been applied[7] either to create patterns of large populations of one or multiple cell types or to control the shape of cells, which adopt the shape of the "adhesive islands."[8] The ability of organizing cells spatially is of interest in tissue engineering applications, and cell shape control serves an excellent model for elucidating the biology of mechanotransduction.

It has long been realized that the surface topography of biomaterials plays a very important role in addition to the cell adhesive motifs effecting cell attachment and spreading. Texture of biomaterial surfaces, such as micrometer-size grooves, influences cell behavior through contact guidance.[5] Recently it became clear[9] that nanometer-size topographic features are more important than micrometer-size features; however, the exact mechanism of contact guidance remains unclear. The design of next generation functional biomaterial surfaces for controlling cell behavior must take into account both the chemical signals (the cell adhesion motifs) and the topographic cues.

Besides solid material surfaces, functional water-soluble polymers have also been created to promote cell-cell adhesion in aqueous media. For example, cell adhesion peptides of RGD or tyrosine-isoleucine-glycine-serine-arginine (YIGSR) have been attached to the ends of biofunctional PEG and shown to induce neural cell aggregation in culture media.[10]

Responsive Material Surfaces

Functional biomaterial surfaces have been created to change between being hydrophobic and hydrophilic, in response to external signals, such as differences in temperature, solvent environment, light, or electrical current.[11] Temperature-sensitive poly(N-isopropylacrylamide) (PNIPAm)-coated substrate is hydrophobic at a cell culture temperature of 37°C that favors cell adhesion and becomes hydrophilic at a lower temperature of 20°C, causing the detachment of cell sheets.[12] These reversible responsive surfaces have been used to culture and harvest layers of endothelial, epithelial, lung, liver, cardiac, and kidney cells, and could eventually enable assembly of complex tissues and organs.

Functional Biomaterials that Promote Cell and Tissue Growth in 3-D

Functional biomaterials that support growth of cells and tissues in 3-D are divided into two categories: polymeric scaffolds and hydrogels. These structures not only provide mechanical support of cells, but also provide necessary chemical and biological signals to allow cell attachment, migration, proliferation, and differentiation.

Material Processing for Cell Macroencapsulation

The inclusion of cells into functional biomaterial scaffolds carries several critical requirements. During and after the macroencapsulation process, cells must remain viable and physiologically functional. It is desirable to encapsulate as many cells as possible and have them distributed evenly throughout the scaffolds, because the mass and the structure of the tissue product depend on the initial cell number and distribution. For in vivo applications, it is often necessary to perform cell encapsulation quickly inside body cavities with irregular shapes and sizes. These requirements demand special processes for material preparation and scaffold formation.

There are generally two approaches for cell macroencapsulation. One is to fabricate the scaffold first, followed by cell seeding. The other approach is to combine cells with a mixture of precursors in a liquid state and form the scaffold structure around cells, or in situ. The first cell seeding approach requires scaffolds with high porosity, so that large numbers of cells could be introduced into the scaffold interior with ease. One way of generating porous polymeric scaffolds is by solvent casting in the presence of porogens such as PEG or salt crystals.[13] The polymer is first dissolved in an organic solvent and cast into a mold. The solvent is then removed, followed by the removal of the porogens, leaving pores in the polymer bulk. Other pore-generating techniques for water-insoluble polymer scaffolds such as poly(lactic acid) (PLA) have also been developed, using gas-foaming and supercritical carbon dioxide.[13] Polymer fibers can also be woven into porous structures as cell scaffolds. The in situ encapsulation approach usually requires the chemistry of scaffold formation to be triggerable externally, such as photo-initiated polymerization and gelation of precursor molecules, temperature-controlled reversible sol-gel transformation, or mixing-controlled chemical coupling of two activated polymer precursors. Scaffold formation and cell encapsulation can also be driven by self-assembly of molecules such as block copolymers, peptides, proteins, and polysaccharides, into supra-molecular structures such as fibers and gels.

Synthetic Biodegradable Polymers

Biological functionalization of synthetic biodegradable polymers as 3-D scaffold for cells involves grafting ECM proteins or cell adhesion motifs such as the RGD peptide. One example is introducing reactive groups such as amines into the polymer backbone through copolymerization and using these groups later for peptide conjugation, as in the case of poly(lactic acid-*co*-lysine).[14] Other scaffolds of biodegradable polymers may be treated to reveal reactive groups by plasma, by surface-restricted controlled hydrolysis (by, e.g., sodium hydroxide), or may be blended with another functional polymer near the surface for peptide conjugation. Surface treatment involving breaking polymer chains should be done carefully, because of the risk of compromising the mechanical strength of the scaffold as a result of excessive polymer chain scission.[1]

Most of the commonly used degradable polymer scaffolds are mechanically strong, but for certain applications such as engineering muscles and tendons, which require considerable elasticity, these polymers are not optimal. Novel biodegradable polyesters have been developed with superior elasticity and strength that resemble vulcanized rubber and are hence termed as "biorubber."[15] Scaffolds made with these mechanically functional materials may be useful especially in engineering elastic tissue such as muscular-skeletal tissues and blood vessels.

Synthetic Polymer Hydrogels

Because of the excellent biocompatibility and hydrophilicity of PEG and well-known chemical derivatizations of PEG end groups, PEG and PEG-based copolymer hydrogels are used extensively to encapsulate cells in tissue engineering.[16] PEG containing terminal acrylates and α-hydroxy acid can be photopolymerized to form hydrogels. Cross-linkers containing peptide sequences have been introduced into these gels, which are susceptible to degradation by specific enzymes secreted by cells. These systems have been explored to mimic the cell-initiated remodeling of natural ECM, allowing cell migration, degradation, secretion of natural ECM, and releasing growth factors on demand.[17] They have also been used to form protective layer of endothelium to block restenosis, the unwanted clogging of blood vessels after balloon angioplasty.[5] Another synthetic pathway of PEG hydrogels uses Michael-type addition reaction between a nucleophile such as a thiol group and an acrylate or acrylamide. Peptides containing terminal cysteines can crosslink branched or multiarmed (or star-shaped) PEG upon simple mixing and form hydrogels in minutes with encapsulated cells under room temperature.[18] In addition to peptide-cross-linkers allowing biodegradation, PEG hydrogels can also be modified with sugar residues to enhance adhesion to certain cell types. Extensive work has been reported on tailoring the bioadhesive, mechanical, and degradation properties of PEG hydrogels to achieve better tissue engineering outcomes.[19]

PEO-PPO-PEO triblock copolymers (Pluronics or Poloxamers) form reversible physically cross-linked hydrogels under certain concentration range and temperature. The use of this system in tissue engineering is scarce because of its inability to degrade. Di- or tri-block copolymers of PEG with PLA have been developed to overcome this problem. Multiple blocks of PEG and PLA, synthesized by condensation reaction of L-lactic acid in the presence of succinic acid, can also form reversible hydrogels near body temperature. Cells may be combined with the block copolymers in the liquid state at a lower temperature and can be subsequently injected into the body where gels are formed. Biodegradation occurs later through the PLA segments.[19]

Poly(vinyl alcohol) (PVA) has tunable hydrophilicity and water-solubility through controlling the extent of hydrolysis of its precursor, poly(vinyl acetate), and its molecular weight. It can be chemically cross-linked into gels by glutaraldehyde or epichlorohydrin, which are highly toxic small molecules. Alternative methods of gelation have been developed using a repeated freeze/thaw method or using an electron beam. Because of its low degradability in vivo, PVA has been primarily used as a long-term or permanent scaffold material or as blends with other degradable polymers.[19]

Polyphosphazene is a class of organometallic polymer that contains alternating phosphorus and nitrogen atoms with two side groups attached to each phosphorus atom. It is synthesized by a substitution reaction between poly(dichlorophosphazene) and alcohols or amines. The backbone of the polymer is hydrophilic. The two side groups on the phosphorus atom can take on various structures and properties, which potentially can be modified to create biofunctionalities. Cross-linked polyphosphazene gels have been investigated for skeletal tissue regeneration.[19]

Poly(2-hydroxyethylmethacrylate) (HEMA) hydrogels were developed initially as soft contact lenses. Macroporous polyHEMA gels have been prepared by freeze/thaw method or salt leaching technique and used for nondegradable cartilage replacement. The modification of polyHEMA gels by dextran enables enzymatic degradation of the scaffolds. A novel cross-linking chemistry of polyHEMA with grafted enantiomeric oligo(L-lactide) and oligo(D-lactide) has been developed, allowing in situ formation of polyHEMA hydrogels, which are potentially useful in cell encapsulation.[19]

Owing to their temperature-sensitive sol-gel transition behavior, PNIPAAm and copolymers with acrylic acid or acrylamide can be used to encapsulate cells. Chondrocytes have been successfully encapsulated without using any organic solvent or toxic compound. The application of these systems as injectable bioadhesive (RGD) scaffold for cartilage tissue engineering is being explored. Degradability can be introduced into the otherwise nondegradable polymer system by peptide- or polysaccharide-based cross-linkers.[19]

Poly(propylene fumarate-*co*-ethylene glycol) (PF-*co*-EG) is a hydrophilic block copolymer that can be cross-linked chemically or by UV light. When used as an injectable cell scaffold in bone and vascular tissue engineering, this block copolymer degrades through the ester bonds in the PF blocks.[20]

To afford biological functionality to the otherwise inert synthetic polymer hydrogels, cell adhesion peptides are either incorporated as cross-links or attached to pendant polymer chains in the bulk or on the surface. Biologically active growth factors, such as TGF-β, bone morphogenetic proteins (BMPs), and VEGF, have been incorporated in hydrogels through covalent tethering, physical tethering, or entrapment, as stimuli for cell growth and differentiation.[20]

Natural Polymer Hydrogels

Collagen is a major component of the ECM in many tissues, such as bone, cartilage, and skin. Collagen forms thermal reversible gels physically and gels cross-linked by dialdehyde or diazide chemically. Collagen contains numerous cell adhesion motifs and sequences that are degradable by specific cell proteases. Collagen gels are often modified to include other biomolecules such as fibronectin, chondroitin sulfate, or hyaluronic acid. Although there are large variations of collagen extracted from natural sources, it serves as an excellent functional biomaterial matrix for engineering tissues and organs such as liver, skin, blood vessel, and small intestine. Single-stranded partial degradation product of collagen is gelatin, which also forms gels and is used in various tissue engineering applications.[19]

Silkworm silk has a long history of medical use as degradable sutures with good biocompatibility. Recently, the mechanism of natural processing to create high strength silk fibers was elucidated. Water-soluble silk fibroin protein extracted from silkworm silk has been processed into hydrogels with or without PEG by varying temperature, pH, and salt concentration. Because of its long degradation time, silk protein may be especially useful in reconstructing tissues requiring relatively stable scaffolds.[21]

Hyaluronic acid (or hyaluronate) is a giant molecule of glycosaminoglycan found in natural ECM and is especially important in wound healing. Hydrogels of HA can be prepared by cross-linking HA chains with hydrazide derivatives or by polymerization of glycidyl methacrylate-containing HA macromers, and HA gels are degraded by a natural enzyme in the body, hyaluronidase. Applications of HA gels include artificial skin, wound dressing, and soft tissue augmentation. Blending with synthetic polymers is often necessary to enhance the relatively weak mechanical properties of pure HA gels.[19]

Fibrin is the major protein component of blood clots and is formed by enzymatic cleavage and polymerization of fibrinogen. Because of its important role in natural wound healing process, fibrin is an attractive functional material for tissue engineering. Additional cell binding peptides, such as RGD, have been incorporated into fibrin using transglutaminase factor XIIIa activity. The resulting fibrin gels have been shown to promote neurite extension. Other applications of fibrin gels include engineering skeletal muscles, smooth muscles, and cartilage.[19]

Alginate is a natural polysaccharide obtained from brown algae. Divalent ions such as Ca^{2+} cross-link alginate into physical gels, which are used for microencapsulation of chondrocytes, hepatocytes, and islet cells to treat diabetes. Ionically cross-linked alginate gels degrade in vivo by losing the ions, and the degradation is difficult to control. Covalent cross-link of alginate has been achieved using adipic dihydrazide or bifunctional PEGs. It is possible to control alginate-PEG hydrogel formation, swelling properties, mechanical properties, and degradation. To incorporate cell binding ability, alginate gels have been modified with sugar-binding lectin and RGD and have been used to culture skeletal muscle cells.[19]

Agarose is another polysaccharide extracted from algae, and it forms thermally reversible gels. The concentration of agarose determines the pore size and stiffness of agarose gels and has been shown to influence cell migration. Agarose gels covalently modified with chitosan have been used to promote neurite outgrowth. Certain cell adhesion peptides have also been coupled to agarose gels.[19] Recently a method was developed to enable 3-D spatially controlled coupling of cell adhesive RGD within agarose gels, and directed growth of neurites has been demonstrated.[22]

Chitosan is a cationic polysaccharide derived from chitin, which is most abundant in crab shells. Because of its insolubility in neutral buffers and most organic solvents, chitosan has to be chemically modified to increase its solubility. Chitosan gels can be prepared by ionic crosslinking or chemical crosslinking by glutaraldehyde. Azide derivatives of chitosan can also be gelled by UV light. Chitosan gels have been modified with sugar moieties to interact with hepatocytes and with natural proteins such as collagen, gelatin, and albumin for neural regeneration.[19]

Genetically Engineered Protein Polymers

Recombinant DNA technology and bacterial fermentation techniques have enabled the design of artificial genes and the production of artificial protein materials. Typically a peptide repeating sequence is designed and translated into a DNA sequence. The DNA monomer is chemically synthesized and enzymatically polymerized and cloned into a bacterial vector or a plasmid, which is later transferred into a bacterial host to express the coded protein polymer. Because the protein products are genetically coded in the DNA sequences, it is possible to produce protein-based biomaterials with precisely defined molecular weight, composition, biologically functional domains, and molecular assembly properties.[5,14]

Natural silk proteins constitute long repeated sequences of alanine-glycine (AG), which form aligned, stacked, antiparallel β-sheets. Hydrophilic amino acids such as glutamic acid have been placed with regular distances from one another, creating β-turns. It has been shown that the thickness of the β-sheet crystals can be precisely controlled and that non-natural amino acid residues can be placed precisely at one face of the crystals. Cell adhesion peptides, such as RGD and YIGSR, have been incorporated into these structures to be exposed at a particular crystal surface. These materials have been used to coat polystyrene substrates, exposing the cell adhesion motifs as support for cell culture.[5]

Elastin is an amorphous natural protein that affords elasticity to the ECM and connective tissues. It contains a five-residue repeat of glycine-valine-glycine-valine-proline (GVGVP), which has been polymerized genetically to yield elastin-analogs. These protein materials display distinct temperature-dependent phase transition, and the transition temperature is determined by the sequence of the elastin repeating unit and the substitution of the proline residue. Elastin polymer gels have been prepared by gamma-radiation or chemical cross-linking at precisely positioned lysine residues. Cell adhesion peptides such as RGD and others have been incorporated genetically into the elastin polymer at predetermined places. Elastin hydrogels are being evaluated for engineering vascular graft and cartilage tissue. Hydrogels based on elastin polymer grafted with PEG can be formed by photopolymerization, which, along with the temperature-triggered sol-gel transition behavior, is being exploited to encapsulate cells in situ.[5,14]

Block protein polymers containing hydrophilic segments and self-associating α-helical segments form reversible hydrogels spontaneously under certain temperature and pH conditions.[23] Natural and synthetic α-helical coiled coil protein segments have been combined with a synthetic copolymer of *N*-hydroxypropylmethacrylamide (HPMA) to create hybrid gels with unique responsive properties.[24] These gel systems have the potential of acquiring biofunctionalities that are relevant to tissue engineering, and they are being developed to control cell behavior.

Self-Assembling Peptides

Short peptide sequences containing alternating positively and negatively charged amino acid residues have been discovered serendipitously from a natural DNA-binding protein and found to self-assemble in salt-containing water into nanofibers, membranes, and gels.[25] Because of the high hydrophilicity of the sequences and the nanofiber-woven structure, these peptide hydrogels contain extremely high water-content (close to 99.5%). Salt-triggered gelation enables in situ encapsulation of cells into hydrogels, which have been shown to support growth of several types of cells and tissues including cartilage, liver cells, and neuronal cells.[25] Similar high water-content hydrogels have been created from chemically synthesized amphiphilic block copolypeptides. Toxicity of such materials to cells depends on the amino acid sequences of the polypeptides.[26]

Peptide amphiphiles containing an amino-terminal nonpolar hydrocarbon tail and a hydrophilic peptide head self-assemble into similar nanofibers, which subsequently form 3-D hydrogels. By incorporating a phosphorylated serine residue in the peptide region of the amphiphiles, calcium deposition and mineralization of hydroxyapatite can be initiated, forming bonelike structures. More recently, by incorporating cell adhesion peptide IKVAV into the amphiphile, neural progenitor cells cultured within these hydrogels differentiated into neurons, instead of astrocytes, owing to the specific binding and signaling of the IKVAV peptide.[27] This initial demonstration of how biomaterial matrix alone can be tailored to steer the fate of progenitor cells suggests the possibility of using these materials in vivo to promote regeneration of damaged tissues.

Functional Biomaterials in Gene Delivery

Functional Requirements for Biomaterials in Gene Delivery

Gene therapy is broadly defined as supplying the patient's cells with genetic materials, which prevent or correct a disease directly, or do so through their protein products. These genetic therapeutics include DNA (both large and small) and RNA. Genetic materials such as DNA plasmids can also be used as vaccines, if the plasmids encode antigenic proteins. The ability to deliver genetic therapeutics and vaccines specifically to target tissues and cells is crucial for the success of gene therapy, which promises to revolutionize modern medicine.[28]

There are two classes of gene delivery vehicles, or vectors, that are based on viruses or nonviral biomaterials. The viral delivery approach is limited by potential toxicity, immunogenecity, lack of targeting, production and packaging problems, and high cost. Nonviral vectors, on the other hand, have the potential to overcome these limitations of viral vectors.[28] Unlike delivery of small molecular drugs, peptide, or protein drugs, gene delivery presents unique challenges and demands biomaterials assisting delivery to be equipped with specific biological functions.

Barriers to gene delivery are categorized into two levels: systemic and cellular. The systemic barrier is the reticular endothelial system (RES) of the body that includes liver, spleen, and phagocytic cells, which capture and clear any foreign material such as an exogenous DNA. Unwanted delivery to nontarget tissues and cells is also undesirable. At the cellular level, there are multiple barriers for DNA uptake, escaping endosomes following endocytosis, nuclear targeting/entry, and DNA unpackaging to allow transcription.[28] Functional biomaterials are expected to overcome these barriers, in addition to having excellent biocompatibility.

Targeted Gene Delivery to Cells

One way to localize gene delivery to a particular site in the body is using controlled release materials, mainly biodegradable polymers.[28] Microparticles of polymers based on poly(lactic-co-glycolic acid) (PLGA) have been extensively used to deliver DNA in vivo after local injections. By tuning particle size to the micrometer range, it is possible to passively target phagocytic cells such as macrophages and dendritic cells, which could be advantageous in DNA vaccine delivery. Recently, microparticles of novel poly(ortho ester) polymers have been developed[29] to accelerate degradation and DNA release in response to intracellular pH. DNA can also be tethered to or entrapped in a solid substrate, such as a tissue engineering scaffold, and be released to specific cells upon cell-initiated substrate degradation.

Prolonged systemic delivery of DNA can be achieved by coating the delivery vehicles with PEG, which avoids nonspecific cell adhesion by being "stealth." Various biologically specific targeting ligands have also been coupled to delivery vectors.[30] These include transferrin, monoclonal antibody, mannose, galactose, lactose, folic acid, low-density lipoproteins, and RGD peptides. By binding to cell surface receptors, these targeting strategies enable DNA to enter target cells through receptor-mediated endocytosis. Recently, it is discovered that certain cationic peptides, such as a segment of the HIV Tat protein, when conjugated to a gene delivery vector, can enter cells directly and rapidly without relying on endocytosis. The benefit of direct cell entry by a gene delivery vector is that it bypasses the barrier of endosomal membrane.

Gene Condensation

Genomic DNA in the cell nucleus is highly condensed by cationic proteins such as histone, so that it can be packaged within a compacted organelle. Gene delivery also requires condensation of the exogenous DNA, because for the highly negatively charged hydrophilic DNA molecules it will otherwise be extremely difficult to cross the cell membrane. This is often accomplished by cationic vectors that neutralize the negative charges on the DNA, resulting in small complexes less than 200 nm in size.

Since the early 1970s, one of the first gene delivery systems useful in vivo was developed based on lipids of cationic *N*[1-(2,3-dioleyloxy)propyl]-*N,N,N*-trimethylammonium chloride) (DOTMA) and dioleoylphosphatidylethanolamine (DOPE), known as Lipofectin. Other commonly used cationic lipids include 2,3-dioleyloxy-*N*-[2-(spermine-carboxamido) ethyl]-*N,N*-dimethyl-1-propanaminium trifluoroacetate (DOSPA), dioctamido-decylamidoglycylspermine (DOGS), 1,2-dimyristyloxypropyl-3-dimethylhydroxyethyl ammonium bromide (DMRIE), 1,2-bis(oleoyloxy)-3-(trimethylammonio) propane (DOTAP), and 3β-(*N,N*-dimethylaminoethane) carbamoyl cholesterol (DC-Chol).[30] DNA is compacted and encapsulated in these cationic lipid vesicles, or liposomes, which enter cells by endocytosis.

Cationic polymers, such as poly(L-lysine) (PLL), polyethylenimine (PEI), chitosan, polyamidoamine (PAMAM) dendrimers, poly(2-dimethylamino) ethyl methacrylate, and polyphosphoesters, condense DNA to form compacted polyplexes.[30] The size and the stability of polyplexes depend on the ratio of cations vs. anions, temperature, ionic strength, and the solvent. Stability of polyplexes can be enhanced by conjugating PEG to the polycations or by using PEG-containing block or graft polymers that form micelles. Small cationic peptides are also able to condense DNA, however, six-consecutive-cations is the minimal requirement to achieve this effectively.

Subcellular Transport Processes

Endocytosed DNA/vector complexes end up in the endosomal compartments, which are filled with digestive enzymes in a slightly acidic microenvironment. Therefore, it is crucial for the DNA to be removed from that environment, which can result in DNA degradation and loss of activity. Polymers such as PEI and polyhistidine are thought to mediate the endosomal escape process by a hypothesized "proton sponge" mechanism—tertiary amine groups in the polymers buffer the acidic pH in endosome, causing it to burst or leak owing to osmotic effect.[30] Small "fusogenic" peptide sequences derived from virus such as influenza are known to change their conformation at endosomal pH and fuse with endosomal membrane. Similar peptides have been conjugated to cationic polymers and shown to enhance transfection efficiency of the exogenous DNA, presumably because of enhanced endosomal escape capability.[29] Synthetic anionic polymers such as poly(2-alkylacrylic acid) behave similarly to the fusogenic peptides in mediating pH-triggered release of DNA from endosome into the cytoplasm.

Once in the cytoplasm, the DNA or DNA/vector complex has to be transported to and enter the cell nucleus for gene transcription. This process is thought to be facilitated by specific peptide sequences termed nuclear localization signals (NLS).[29] In fact, it was shown that one NLS peptide appears sufficient for transporting a DNA plasmid to the nucleus. NLS peptides are frequently coupled to cationic vectors to aid in nuclear transport of DNA/vector complexes. Interestingly, it is reported recently that PEI/DNA complexes can be localized readily to the cell nucleus without using any NLS. Further elucidation of the transport mechanism inside cells should lead to better strategies of designing functional gene delivery materials.

Gene Unpackaging

To be transcriptionally active, the DNA must recover from its condensed form and be separated from its cationic carrier.[29] Disintegration of DNA/vector complex can be engineered by using disulfide crosslinks that are reduced in the cytoplasm by glutathione activity. The cationic polymer carriers (such as poly[α-(4-aminobutyl)-L-glycolic acid], and poly(4-hydroxy-1-proline ester)) can be programmed to degrade overtime by hydrolysis, releasing uncomplexed DNA for transcription.[30]

Conclusions

This chapter focuses on biomaterials that carry biologically relevant functions. Methods are introduced for modifying biomaterial surface to minimize nonspecific cell adhesion and enhance specific cell adhesion and spreading by incorporating biomolecules. Preparation and processing of biomaterials that form 3-D scaffolds for supporting cell and tissue growth and differentiation are described. Biodegradable (synthetic and natural) polymers, hydrogels, and responsive biomaterials play pivotal roles in performing these biological functions. Finally, the development of functional biomaterials to facilitate gene delivery is discussed in the context of overcoming various biological barriers inside and outside cells.

References

1. Ratner, B.D., Hoffman, A.S., Schoen, F.J., Lemons, J.E., Lemons, J.E. *Biomaterials Science: An Introduction to Materials in Medicine*; Academic Press: San Diego, 1996.
2. Bearinger, J.P., Terrettaz, S., Michel, R., Tirelli, N., Vogel, H., Textor, M., Hubbell, J.A. Chemisorbed poly(propylene sulphide)-based copolymers resist biomolecular interactions. *Nat. Mater.* **2003**, *2*, 259–264.
3. Holland, N.B., Qiu, Y., Ruegsegger, M., Marchant, R.E. Biomimetic engineering of non-adhesive glycocalyx-like surfaces using oligosaccharide surfactant polymers. *Nature* **1998**, *392* (6678), 799–801.
4. Shi, H., Tsai, W.B., Garrison, M.D., Ferrari, S., Ratner, B.D. Template-imprinted nanostructured surfaces for protein recognition. *Nature* **1999**, *398* (6728), 593–597.
5. Hubbell, J.A. Biomaterials in tissue engineering. *Biotechnology* **1995**, *13*, 565–576.
6. Tirrell, M., Kokkoli, E., Biesalski, M. The role of surface science in bioengineered materials. *Surf. Sci.* **2002**, *500*, 61–83.
7. Whitesides, G.M., Ostuni, E., Takayama, S., Jiang, X., Ingber, D.E. Soft lithography in biology and biochemistry. *Annu. Rev. Biomed. Eng.* **2001**, *3*, 335–373.
8. Chen, C.S., Mrksich, M., Huang, S., Whitesides, G.M., Ingber, D.E. Geometric control of cell life and death. *Science* **1997**, *276* (5317), 1425–1428.
9. Curtis, A.S., Wilkinson, C.D. Reactions of cells to topography. *J. Biomater. Sci. Polym. Ed.* **1998**, *9* (12), 1313–1329.
10. Dai, W., Belt, J., Saltzman, W.M. Cell-binding peptides conjugated to poly(ethylene glycol) promote neural cell aggregation. *Biotechnology* **1994**, *12* (8), 797–801.
11. Russell, T.P. Surface-responsive materials. *Science* **2002**, *297*, 964–967.
12. Shimizu, T., Yamato, M., Kikuchi, A., Okano, T. Cell sheet engineering for myocardial tissue reconstruction. *Biomaterials* **2003**, *24* (13), 2309–2316.
13. Lanza, R.P., Langer, R., Chick, W.L. Eds.; *Principles of Tissue Engineering*; Academic Press: San Diego, 1997.

14. Langer, R., Tirrell, D.A. Designing materials for biology and medicine. *Nature* **2004**, *428*, 487–492.
15. Wang, Y., Ameer, G.A., Sheppard, B.J., Langer, R. A tough biodegradable elastomer. *Nat. Biotechnol.* **2002**, *20* (6), 602–606.
16. Hoffman, A.S. Hydrogels for biomedical applications. *Adv. Drug Delivery Rev.* **2002**, *54* (1), 3–12.
17. Nguyen, K.T., West, J.L. Photopolymerizable hydrogels for tissue engineering applications. *Biomaterials* **2002**, *23* (22), 4307–4314.
18. Lutolf, M.P., Hubbell, J.A. Synthesis and physicochemical characterization of end-linked poly(ethylene glycol)-co-peptide hydrogels formed by Michael-type addition. *Biomacromolecules* **2003**, *4* (3), 713–722.
19. Lee, K.Y., Mooney, D.J. Hydrogels for tissue engineering. *Chem. Rev.* **2001**, *101* (7), 1869–1879.
20. Drury, J.L., Mooney, D.J. Hydrogels for tissue engineering: scaffold design variables and applications. *Biomaterials* **2003**, *24*, 4337–4351.
21. Altman, G.H., Diaz, F., Jakuba, C., Calabro, T., Horan, R.L., Chen, J., Lu, H., Richmond, J., Kaplan, D.L. Silk-based biomaterials. *Biomaterials* **2003**, *24* (3), 401–416.
22. Luo, Y., Shoichet, M.S. A photolabile hydrogel for guided three-dimensional cell growth and migration. *Nat. Mater.* **2004**, *3*, 249–253.
23. Petka, W.A., Harden, J.L., McGrath, K.P., Wirtz, D., Tirrell, D.A. Reversible hydrogels from self-assembling artificial proteins. *Science* **1998**, *281*, 389–392.
24. Wang, C., Stewart, R.J., Kopecek, J. Hybrid hydrogels assembled from synthetic polymers and coiled-coil protein domains. *Nature* **1999**, *397*, 417–420.
25. Zhang, S. Fabrication of novel biomaterials through molecular self-assembly. *Nat. Biotechnol.* **2003**, *21* (10), 1171–1178.
26. Pakstis, L.M., Ozbas, B., Hales, K.D., Nowak, A.P., Deming, T.J., Pochan, D. Effect of chemistry and morphology on the biofunctionality of self-assembling diblock copolypeptide hydrogels. *Biomacromolecules* **2004**, *5* (2), 312–318.
27. Silva, G.A., Czeisler, C., Niece, K.L., Beniash, E., Harrington, D.A., Kessler, J.A., Stupp, S.I. Selective differentiation of neural progenitor cells by high-epitope density nanofibers. *Science* **2004**, *303* (5662), 1352–1355.
28. Luo, D., Saltzman, W.M. Synthetic DNA delivery systems. *Nat. Biotechnol.* **2000**, *18*, 33–37.
29. Wang, C., Ge, Q., Ting, D., Nguyen, D., Shen, H.-R., Chen, J., Eisen, H.N., Heller, J., Langer, R., Putnam, D. Molecularly engineered poly(ortho ester) microspheres for enhanced delivery of DNA vaccines. *Nat. Mater.* **2004**, *3*, 190–196.
30. Han, S., Mahato, R.I., Sung, Y.K., Kim, S.W. Development of biomaterials for gene therapy. *Mol. Ther.* **2000**, *2* (4), 302–317.

7

Molecularly Imprinted Polymers (MIPs)

Gregory T. Rushton and Ken D. Shimizu

CONTENTS

Introduction

Molecularly imprinted polymers (MIPs) are polymers formed in the presence of a template molecule. Removal of the template from the polymer matrix creates complementary binding sites with affinity and selectivity for the template molecule. Molecularly imprinted polymers are attractive materials capable of molecular recognition owing to their versatility, ease of preparation, and robust physical and chemical properties. Over the past 30 years, MIPs have been developed as stationary phases in chromatography, heterogeneous catalysts in organic synthesis, and sensors for a wide array of biologically relevant compounds. This review seeks to highlight the accomplishments of these materials over this period of time, which has led to their rise in popularity and their potential in commercial applications. We begin by describing the primary approaches to preparing MIPs, and then discuss their success in various applications. We conclude with an eye to the future and suggest where the imprinting field may be heading.

Background

Molecular recognition is an important feature of all essential biological processes. For example, enzymes and monoclonal antibodies are able to identify a specific substrate in the presence of many other structurally similar entities and then carry out their catalytic and immune response functions. The study of the recognition properties of these systems has led to insight into the mechanisms of binding and the subsequent rational design of many pharmaceuticals and chemosensors. Both biological and synthetic receptors have been intensely studied as the focus of research groups but each presents a unique set of challenges.[1–4] Biological recognition systems, such as enzymes and antibodies, have a fairly narrow range of chemical, thermal, and physical stability, which limits their "shelf life" and ultimately, their utility in commercial applications. They also typically, require time-consuming preparation from biological sources and the necessity to sacrifice animals. Synthetic molecular receptors, on the other hand, are more chemically and thermally stable and can, in principle, be prepared in large quantities. However, synthetic molecular receptors typically have much lower selectivity and binding affinities when compared to their biological counterparts, and they also require lengthy multistep syntheses.

Since the early 1970s, an alternative route to easily-tailored recognition materials has attracted considerable interest in the form of MIPs.[5,6] Molecularly imprinted polymers are synthetic polymers that contain shape-specific binding cavities that are lined with complementary functional groups, mimicking the recognition properties of biological systems.[7,8] These materials are versatile and easy to prepare. A generalized imprinting scheme is shown below (Scheme 1). First, a template or "guest" molecule is chosen and is linked to functional monomer(s) via either covalent or noncovalent bonds to form a prepolymerization complex. The prepolymerization complex is then preserved by polymerization into a highly cross-linked polymer matrix, when rigid cavities are formed that are complementary in size and shape to the template. Upon removal of the template molecule, the polymer is ready to be used as a recognition material for the host molecule. Imprinted polymers have many attractive characteristics, such as the ability to be prepared and utilized in organic solvents. They retain their recognition properties when stored dry for long periods of time

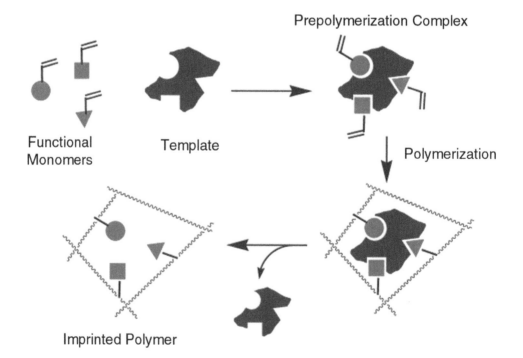

SCHEME 1
The imprinting process begins with the selection of a template molecule and complementary functional monomer(s). (*View this art in color at www.dekker.com.*)

and are thermally stable up to 125°C. Molecularly imprinted polymers are readily accessible in large quantities as they can be formed in a single step from commercially available starting materials. Finally, the recognition properties of imprinted polymers can be easily tailored to new substrates simply by carrying out the polymerization using the appropriate template molecule. This combination of attributes is complementary to the molecular receptors produced by immunological and organic synthesis. For example, MIPs can be made using highly toxic or hydrolytically unstable templates, such as nerve toxins, pesticides, or short peptides for which antibodies cannot be directly elicited. The molecular imprinting process is also more efficient than the preparation of most synthetic or biological receptors.

The first forays into molecular imprinting took place in the 1940s when Frank Dickey, a student of Linus Pauling, prepared silica gels in the presence of organic dyes, such as methyl orange and found that these gels showed improved adsorption of the dyes as compared to a control gel.[9] Although these results were promising, it was not until the early 1970s that Gunter Wulff began conducting research that served as the impetus for the current generation of molecularly imprinted materials. Initial efforts by Wulff et al. focused on the difficult problem of carbohydrate recognition using boronic acid monomers that form covalent boronate linkages with 1,2-diols. Although the boronate ester is a covalent linkage, it is reversible and can be cleaved under hydrolytic conditions (Scheme 2). This covalent imprinting approach yields a homogeneous population of binding sites with a high affinity for the template, which could even resolve enantiomers when the polymer

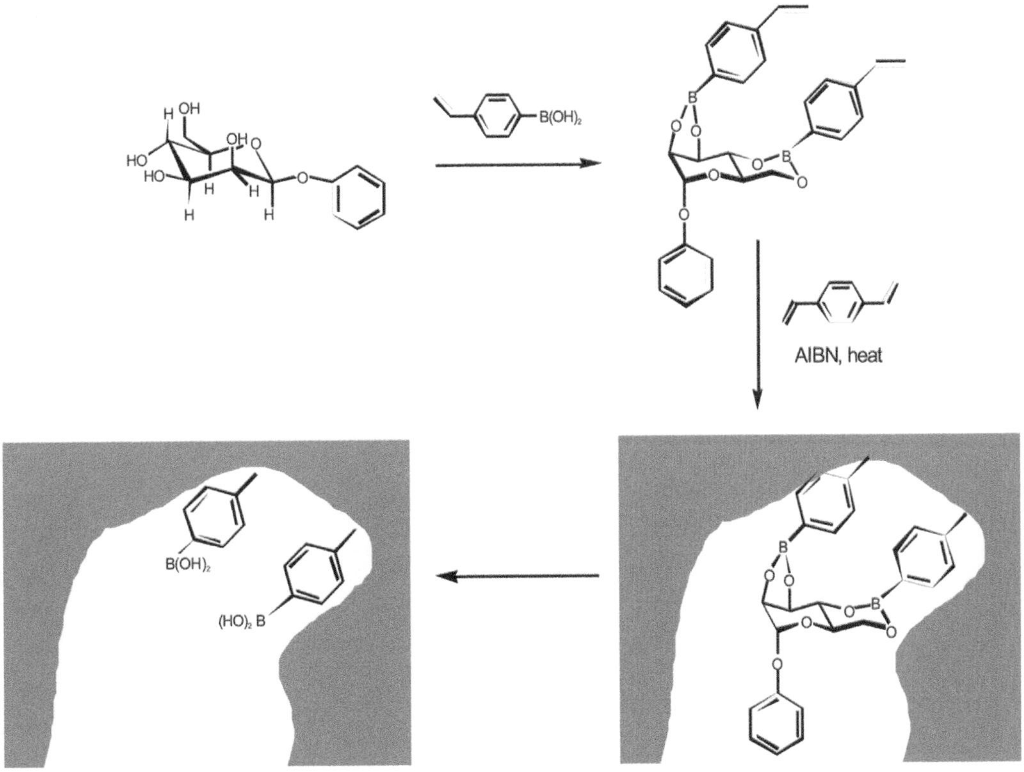

SCHEME 2
Covalent imprinting of α-phenyl-D-mannopyranoside in a DVB/4-vinylphenylboronic acid matrix, via the formation of covalent boronic ester linkages between the 4-vinylphenylboronic acid and the carbohydrate. *(View this art in color at www.dekker.com.)*

was used as the stationary phase in liquid chromatography. For example, when phenyl-α-mannopyranoside was selected as the template and (4-vinyl)-phenylboronic acid was used as the monomer, the resulting MIP could discriminate D- and L-enantiomers with baseline separation and separation factors up to 4.56.[10] The enantioselectivity of the imprinted polymer provided verification that the imprinting process produced template selective binding cavities. The constituent monomers are achiral and, therefore, the enantioselectivity of the imprinted polymer can only arise from the imprinting process. Polymers made without template did not show any enantioselectivity when measured under similar conditions.

While the covalent imprinting approach proved successful, it was limited to templates that could form labile covalent bonds to a functional monomer, such as boronic esters, imines, ketals, and disulfides. In the 1980s, a new noncovalent imprinting approach was introduced by Klaus Mosbach. The noncovalent imprinting approach quickly became the preferred approach owing to the ease of synthesis and the ability to tailor the materials to accommodate a wide range of template molecules. The key feature of the approach was the in situ formation of the prepolymerization complexes via self-assembly of the template and functional monomers using reversible noncovalent interactions, such as hydrogen bonding, electrostatic interactions, solvaphobic, and π-π interactions. This simplifies the imprinting procedure into a more versatile one-pot process (Scheme 3). The self-assembled prepolymerization complex is formed by simply combining the component template and

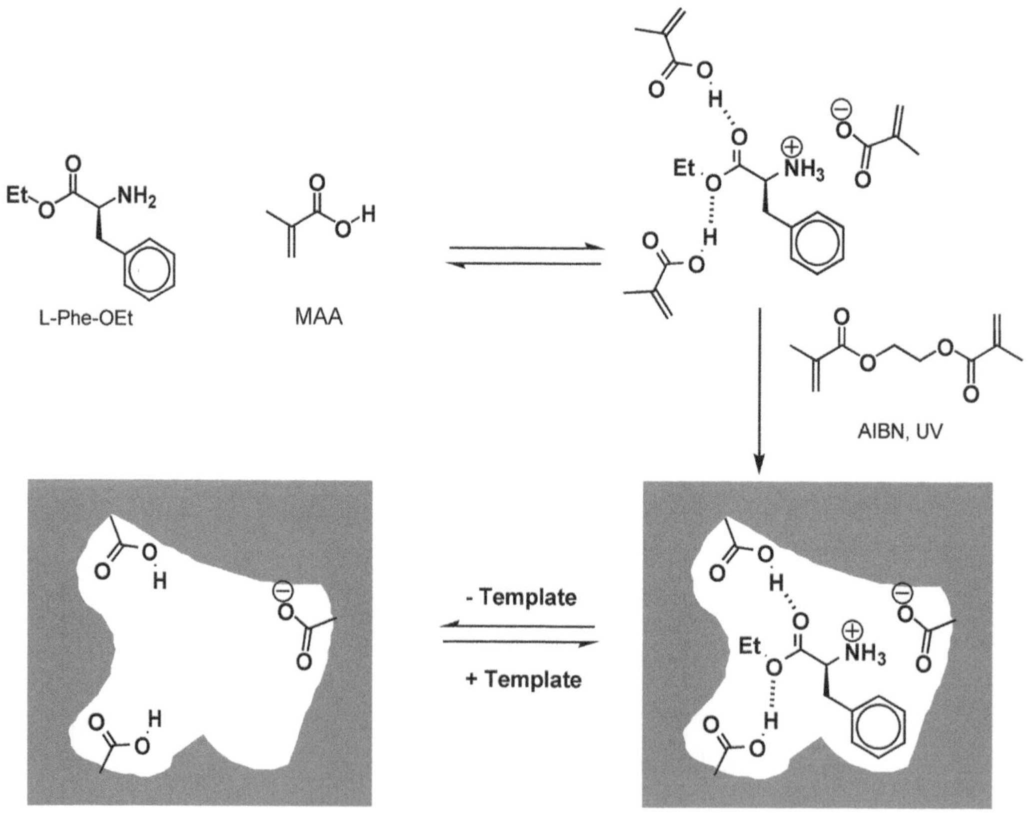

SCHEME 3
Noncovalent imprinting of L-phenylalanine methyl ester in a methacrylic acid (MAA)/ethyleneglycol dimethacrylate (EDMA) polymer matrix. (*View this art in color at www.dekker.com.*)

functional monomer(s) and then is polymerized into a highly cross-linked polymer matrix. The reversibility of the template-functional monomer interactions in the prepolymerization complex allowed a common set of functional monomers and cross-linkers to be used with many different template molecules. The most common pair of functional monomer and cross-linking agent is methacrylic acid (MAA) and ethylene glycol dimethacrylate (EDMA). This copolymer system has been used to imprint amino acids, pharmaceuticals, and herbicides. An additional advantage of the noncovalent imprinting approach is that interactions of the imprinted binding cavity with the template molecule are also noncovalent interactions. This allows for much faster binding kinetics, which is important for chromatographic and sensing applications. One disadvantage of the noncovalent imprinting approach is binding site heterogeneity arising from the dynamic nature of the noncovalent prepolymerization complex. Noncovalently imprinted polymers contain a range of binding sites that vary from low affinity to high affinity.[11,12] Unfortunately, the binding site distribution is heavily weighted toward the low-affinity selectivity binding sites. The MIPs prepared by the covalent approach tend to have a more homogeneous binding distribution that does not extend out as far into the high-affinity region.

More recently, hybrid-imprinting strategies have been developed that seek to combine the advantages of covalent and noncovalent imprinting methodologies. The hybrid imprinting methods attempt to form the prepolymerization complex in a stoichiometric fashion like the covalent imprinting approach, while still producing an imprinted polymer that forms noncovalent interactions with the template molecule. For example, the sacrificial spacer approach developed by Whitcombe et al. uses covalent bonds to connect the monomers and template like the covalent approach. However, hydrolysis of the template unmasks hydrogen-bonding functionality that forms noncovalent interactions to rebind the guest. A specific example of Whitcombe's sacrificial spacer imprinting method is outlined in Scheme 4. A cholesterol containing prepolymerization complex was prepared using (4-vinyl)-phenyl carbonate ester. Polymerization with EDMA followed by hydrolysis

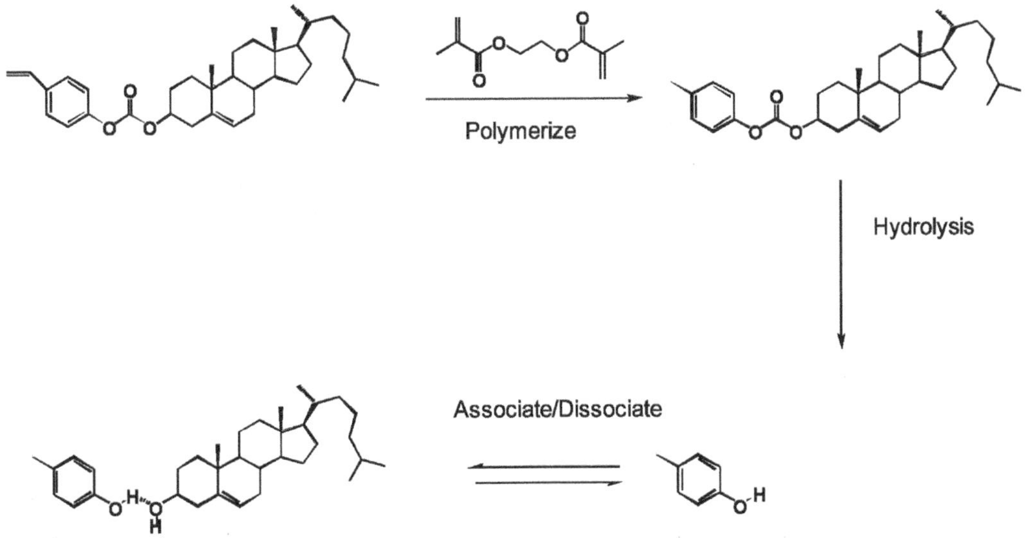

SCHEME 4
A hybrid method of imprinting by Whitcombe et al. uses stoichiometric amounts of monomer and template during polymerization and then relies on noncovalent interactions during the rebinding phase.

of the template from the polymer matrix yielded a phenol moiety positioned in a comple-
mentarily shaped binding cavity that could bind cholesterol through noncovalent hydrogen bonding interactions.[13] This polymer was able to discriminate among cholesterol
analogs, such as epicholesterol and cholesterol acetate.

Another hybrid imprinting method is to use functional monomers that form particularly
strong noncovalent interactions with the template. The "stoichiometric" noncovalent imprinting approach retains the synthetic efficiency "one pot" imprinting procedure of the noncovalent approach but also produces MIPs with a more homogeneous distribution of high-affinity,
high-selectivity binding sites like the covalent approach. One moiety that has been used is the
amidine group, which can form both strong hydrogen bonds and electrostatic interactions
with carboxylic acids, phosphonates, and phosphoric esters. Sellergren used MAA to bind the
pneumonia drug pentamidine and Wulff used the amidine moiety in a monomer to resolve
two enantiomers of N-(4-carboxybenzoyl)-phenylglycine with separation factors up to 2.8.[14]

Composition and Preparation of MIPs

One of the most attractive features of the imprinting process is the ease with which new
MIPs can be prepared for and tailored to specific applications. This is particularly the case
for the noncovalent imprinting approach. In most cases, MIPs can be synthesized and
processed within a few days using readily available starting materials. First, a functional
monomer must be selected that forms reversible interactions with the chosen template
molecule. The most common functional monomers, such as acrylic acids and vinylpyri-
dines, are inexpensive, commercially available, and readily polymerized by free radical
polymerization.[15] Although the imprinting mechanism is not restricted to free radical
polymerization of vinyl monomers, radical polymerizations are the most common method
of preparation of MIPs because of their high yields in a range of solvents and tolerance to
many acidic and basic functional groups that are used in the imprinting process.

The second variable is the cross-linking agent, which serves as a solid support for the
functional monomer, ensuring the proper distance and orientation of the functional
groups around the template molecule. The cross-linking agent also provides shape selectivity by forming a rigid cavity around the template. These cavities must remain intact
during polymerization and subsequent extraction of the template from the matrix for
selective binding properties to be observed. Common cross-linking agents are EDMA,
divinylbenzene (DVB), and the bisacrylamides (Figure 7.1). Wulff systematically examined
the influence of the cross-linker on the selectivity of the resulting imprinted polymer. A
high percentage of cross-linking agent, typically 80%, was required for the matrix to maintain the integrity of the binding cavity after cleavage of the template.

The final component in the polymerization mixture is the solvent. Although the choice
of solvent is often overlooked, it can have dramatic consequences on the surface area,
properties of the materials, and binding affinity and selectivity of the resulting MIP. The
solvent must dissolve the template, monomers, and cross-linker without disrupting the
stability of the prepolymerization complex. Common solvents include acetonitrile, chloroform, and toluene. The solvent also acts as a porogen, creating "macropores" within the
imprinted polymer so that the template can be efficiently removed from the cross-linked
matrix and can also access the binding sites during the rebinding process. Imprinted polymers are typically macroporous monoliths with high surface areas of 100–600 m^2/g. Near

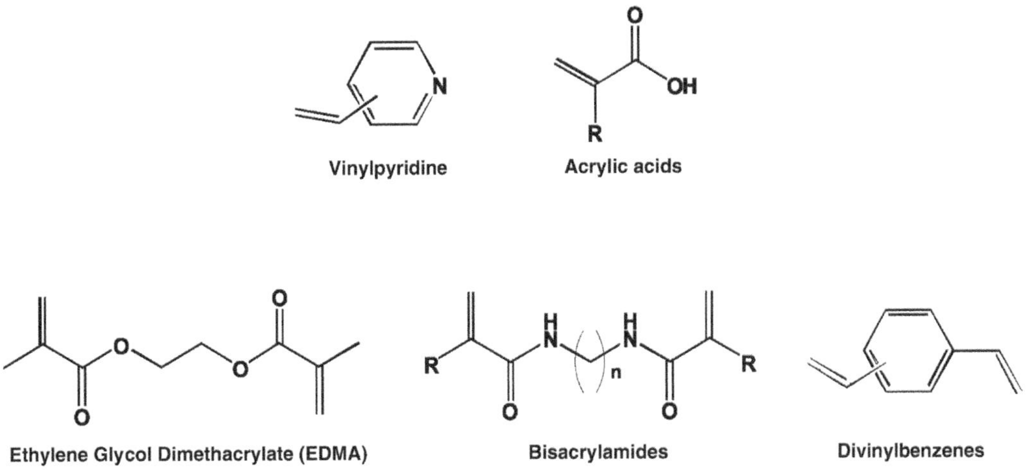

FIGURE 7.1
Common functional monomers (top) and cross-linking agents utilized in noncovalent imprinting.

theta solvents in which short polymer chains are soluble but longer polymer chains are insoluble are particularly effective in yielding highly macroporous morphology because of phase separation during polymerization.

Other variables in the imprinting process are stoichiometry and temperature, which are particularly important in the cases of noncovalent imprinted polymers. Lower temperatures and higher functional monomer to template ratios stabilize noncovalent prepolymerization complexes, resulting in noncovalent MIPs with higher capacities and selectivities. The imprinting process can be carried out at lower temperatures by either UV initiated radical polymerizations or by using azo-based radical initiators.

Once the imprinted polymer has been synthesized, the template must be washed out of the matrix. In most cases, greater than 80% of the template can be effectively removed by Soxhlet extraction. This is sufficient for most applications, and the recovered template can be reused, which is important if the template molecule is particularly valuable. However, a small amount of template usually remains in the highest affinity sites and slowly leaches out over time, which can interfere with sensing applications at very low concentrations (nM). These issues can be circumvented by selecting applications for MIPs that are carried out at micromolar (μM) or higher concentrations or by indirectly monitoring binding of the analyte by radioligand or fluorescently labeled analyte assays.

Polymer Morphologies

As the imprinting field moves toward commercial applications, the need to conveniently and efficiently produce the polymers in different formats is becoming increasingly important. In this section, the primary methods utilized for the preparation of MIPs and their relative merits for various applications will be presented.

By far the most popular approach for preparing MIPs is bulk polymerization. In this strategy, the template, monomer, cross-linker, and initiator are combined in a suitable porogenic solvent and irradiated or heated until a sufficient degree of polymerization has taken place. The macroporous monolith is then ground and sieved to achieve a narrow size distribution of particles. Polymers synthesized in this way are used in chromatography, solid phase extractions, and binding assays.[16] Grinding the polymer does not destroy or alter the binding sites. However, it does enhance the accessibility of the binding sites, which is important for chromatographic applications. The polymer monoliths can also be used directly, which eliminates the grinding and sieving steps that contribute to lower yields. Molecularly imprinted polymer monoliths have been synthesized in chromatography columns and have shown similar levels of affinity and selectivity.

For chromatographic applications, monodisperse spherical MIP particles are preferable to the irregularly shaped MIP particles formed by grinding and sieving the polymer monoliths. Consequently, some researchers have developed methods that produce spherical MIP particles through precipitation polymerization and emulsion polymerization.[17] Monodisperse MIP microspheres were prepared by Mosbach et al. under similar conditions to the MIP monolith preparation but under significantly more dilute precipitation polymerization conditions.[18] The MIP microspheres grew to 1–3 μm diameter with a narrow size distribution. These imprinted microspheres can then be packed more efficiently into chromatography columns or into solid-phase extraction (SPE) cartridges than the particles prepared by bulk polymerization techniques. Larger spherical imprinted polymer particles can be prepared by modification of preformed latex particles either by reswelling with a secondary polymerization mixture or by coating a spherical core particle with an imprinted polymer shell.

Imprinted thin films or membranes have also been prepared for applications in chromatography, sensing, and SPE and can be either freestanding or supported on a solid substrate. The MIP films have been synthesized by bulk thin film polymerization, surface initiated polymerization, and phase inversion precipitation of a preformed linear polymer. For example, Rotello and Penelle et al. fabricated a polychlorinated aromatic quartz crystal microbalance (QCM) sensor by bulk thin polymerization with a hexachlorobenzene imprinted thin film directly on to gold surface. The issues of adherence of the polymer film to the gold substrate and the film integrity on drying were resolved by the choice of a more flexible 1,5-bis(2-acetylaminoacryloyloxy)pentane cross-linker. Sellegren et al. have synthesized MIP thin film stability by preparation of MIPs via surface initiated polymerization. Azo-based radical initiators were anchored to silica surfaces to develop new high surface area MIP stationary phases.[19] Kobayashi et al. have developed a unique preparation of MIP thin films by precipitation of linear polymers in the presence of templating agents. For example, Nylon-6 and L-glutamine dissolved in formic acid were cast onto a solid substrate to a thickness of 100 μm.[20] Rinsing the film with water leads to a phase inversion, yielding a macroporous film with enantioselective recognition properties. The imprinted Nylon-6 membranes formed by this inversion polymerization have been used as the recognition element in a QCM sensor and as a filtration membrane.

Applications of MIPs

The most common application for MIPs has been as stationary phases in chromatographic or SPE formats. The stability of MIPs to a wide range of solvents, temperatures, and pressures makes them particularly well suited to be stationary phases in high-performance liquid chromatography (HPLC) and capillary electrophoresis. Some of the most impressive studies using MIPs are reports of MIP systems that preferentially bind a single enantiomer over

its antipode, such as amino acids and drugs. One of the key advantages of MIP-based stationary phases, in comparison to general enantioselective stationary phases, is that there
is a logical order of elution of enantiomers. The imprinted enantiomer is more strongly
retained and typically elutes as a broader peak after the nonimprinted enantiomer. The
known order of elution is useful in identifying the respective enantiomers in analytical
applications. Mosbach et al., for example, successfully separated a racemic mixture (α =
2.9) of the β-blocker timolol, using an HPLC column packed with an *s*-timolol imprinted
polymer (Figure 7.2).[21] Mosbach et al. also developed stationary phases based on MIP
imprinted with a short peptide and observed separation factors as high as α = 17.8, which
exceeds that of commercial chiral stationary phases.[22]

   Solid-phase extraction is a widely used method for cleanup and preconcentration of a
solution containing a mixture of analytes. The SPE sorbents preferentially bind a specific
analyte while allowing the other analytes to pass through. A solvent capable of disrupting
the polymer-analyte interactions is then introduced to wash off the desired analyte. The
concentrated and purified analyte is then quantified by a secondary method, such as gas
or liquid chromatography. An attractive feature of this imprinting application is that the
polymer need only work in "on" or "off" mode, either strongly adsorbing the template or
releasing the analyte in contrast to the chromatographic mode where the MIP column must
also exhibit good resolution to be effective. Theodoridis, for example, recently demonstrated
the selective extraction of caffeine from beverages and spiked plasma using an SPE cartridge loaded with an MIP.[23] Zhu et al. recovered 96% of the herbicide metsulfuron-methyl
in spiked river water using an MIP-SPE cartridge. The stability of MIP-based SPEs was also
demonstrated as the SPE was shown to be reusable for up to 200 sorption studies.[24] In
another study, an MIP imprinted with 4-nonylphenol showed selectivity for the template
over 11 related phenolic pollutants.[25] These successes of MIPs in SPE technology have led
to the establishment of MIP Technologies, Lund, Sweden (www.miptechnologies.se). MIP
Technologies' MIP4SPE-Triazine10 cartridge was tailored, using the imprinting process,
to selectively adsorb triazine and triazine metabolites from environmental samples. MIP

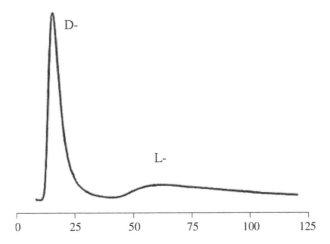

FIGURE 7.2
Elution profile of a racemic mixture of *t*-BOC-D-tyrosine (first peak) and *t*-BOC-L-tyrosine (second peak) on a
molecularly imprinted polymer made using *t*-BOC-L-tyrosine as template and *N,O*-bismethacryloyl ethanol-
amine as the only monomer. HPLC conditions: particle size 45-63 μm; mobile-phase: 99/1, acetonitrile/acetic
acid; flow rate 0.1 mL/min; injected volume 5 μl of a 2.0 mM racemic solution; detection at λ = 270 nm. (Courtesy
of David Spivak, Lousiana State University.) (*View this art in color at www.dekker.com.*)

Technologies have also developed other MIP-based sorbents for clenbuterol, nicotine metabolites, and riboflavin and are working toward others for steroids, peptides, and nicotine.

The MIPs have also been utilized as heterogeneous catalysts.[26] The strategy is analogous to work on catalytic antibodies. The MIPs are imprinted with transition state analogs, and then the imprinted polymers tested for their ability to effect rate and selectivity enhancements. Beach and Shea, for example, prepared an MIP capable of catalyzing the dehydrohalogenation of 4-fluoro-4-(*p*-nitrophenyl)butan-2-one.[27] Sellergren and Shea have developed an MIP that mimics enzyme action by hydrolyzing a protected phenylalanine ester.[28] To date, MIP catalysts have produced only modest rate enhancements. The future of catalytic applications for MIPs will depend on the ability to form binding sites that bind the transition state of the reaction with higher affinity or include catalytic functionality. One interesting development along these lines is transition metal containing MIP catalysts.[29] For example, Severin has prepared MIPs containing ruthenium arene complexes that showed enhanced regioselectivity in the catalytic reduction of 4-acetylbenzophenone.

The MIPs have also been utilized as the recognition elements in pseudoimmunoassays.[30–32] In this approach, MIPs are substituted for antibodies to quantify the amount of analyte in a biological sample, such as blood plasma. Most MIP immunoassays are competitive binding studies in which a radio- or fluorescent-labeled analyte is added to a mixture of the MIP and unlabeled analyte. After equilibrium is reached, some fraction of the labeled species is bound to the polymer surface and thus can be separated from the supernatant. The supernatant is then analyzed via scintillation or fluorescence techniques to determine the concentration of the original unlabeled analyte. Mosbach et al. have demonstrated that MIP-based immunoassays can rival the selectivity of antibody-based assays.[33] Imprinted polymers for the opioid receptor ligands enkephalin and morphine were prepared and showed submicromolar (μM) level selectivity in a radioligand competition assay in aqueous buffers. The analysis by Andersson et al. of *S*-propranolol in human plasma and urine using an MIP assay demonstrated remarkable accuracy and low cross-reactivity with structurally similar analogs.[34] With propranolol, nanomolar selectivity has been demonstrated using an MIP assay. The MIP assays have also been developed for the demanding application of measuring enantiomeric excess (ee).[35] An L-phenylalanine anilide (L-PAA) imprinted polymer was equilibrated with PAA solutions of varying ee. After equilibration, the concentrations of PAA remaining in solution could be correlated to the ee of the original solutions with a standard error of only ±5% ee. Although MIPs have shown comparable selectivity to antibodies at low concentrations, MIPs more typically show high degrees of cross-reactivity and low selectivity. Greene et al. have addressed this issue by grouping a number of different MIPs together into an MIP sensor array.[36] Six aryl amine analytes, which included structurally similar analytes, such as ephedrine and psuedo-ephedrine, were classified with 94% accuracy, using the MIP array assay.

Sensors based on MIPs have also been developed in which the MIP is the recognition element of the sensor, often replacing a polymer thin film or antibody. The MIPs lack signaling functionality and thus the primary difficulty is coupling a singling element to the MIP binding site. Molecularly imprinted polymers containing fluorophores and electrochemically active monomers have been prepared. For example, MIPs synthesized from conductive polymers, such as polypyrrole and polyanaline by electrochemical polymerization have been tested and shown to be able to sense the presence of the template molecule.[37] Alternatively, MIPs can be prepared on signaling surfaces, such as an electrode or a gold QCM or surface plasmon resonance surface. For example, Shoji et al. prepared an MIP-based sensor for atrazine by polymerization on the surface of a gold electrode and measuring the reduction potential of atrazine vs. a Ag/AgCl electrode.[38] The sensor showed good sensitivity for atrazine over other triazines. Kroger et al. developed a sensor for 2,4 dichlorophenoxyacetic acid, and Marx et al. coated glassy carbon electrodes with sol-gel MIP matrices to sense parathion in aqueous solutions.[39]

Future Applications for MIPs

The MIPs are particularly versatile materials, especially as the number and types of templates that have been successfully imprinted increase. For example, imprinted polymers have been prepared using metal ions as templates.[40] Potential applications for these polymeric ionophores include the remediation of toxic metals from the environment, recognition elements for ion sensors for medical diagnostics, catalysis, and chromatography.[29] Other potential applications for MIPs have been as drugs. Researchers from Geltex Pharmaceuticals reported the use of MIPs as cholesterol lowering drugs. An imprinted version of the commercial cholesterol sequestering polymeric drug, colesevelam-HCl, was prepared and shown to have higher capacity and affinity for bile acids than the nonimprinted polymer. The imprinted colesevelam-HCl was synthesized by cross-linking poly(allylammoniumchloride) with epichlorohydrin in the presence of sodium cholate as a templating agent. The nonimprinted polymer had a capacity for bile acids in deionized water of 1.30 mmol/g, and the imprinted polymer had a capacity of 1.97 mmol/g. Other medical applications for MIPs include MIP hydrogels for timed and triggered drug release.

Another limitation of imprinted polymers is the size of the template. Most template molecules are small molecules of less than 20–30 Å in length.[41] Larger templates become physically entrapped in the highly cross-linked matrix and cannot be removed from their binding sites. Thus, important biological macromolecules and structures, such as DNA, proteins, viruses, and cells cannot be imprinted by traditional methods. However, new surface imprinting methods are being developed, which should greatly increase the scope and utility of imprinted materials. For example, Ratner et al. have imprinted proteins using cellulose films.[42] The proteins are first coated with disaccharides. Then, a glow-discharge plasma deposition covalently links the disaccharides into a polymer film around the proteins. A number of different proteins were imprinted by this method including immunoglobulin G, ribonuclease, and streptavidin. The protein-imprinted films could be patterned into a microarray using microcontact printing, opening up the possibility of biomedical applications. Nusslein et al. have imprinted even larger biological structures, specifically bacteria, using a polymer film on the surface of a QCM. The imprinted polymer film displayed recognition abilities for the shape and surface functionality of the imprinted bacteria. The cell-selective QCM was able to differentiate gram-positive and gram-negative cells, cell aggregates, and cell shapes.

Conclusions

The MIPs have already shown great potential in a wide range of applications requiring molecular recognition elements. The MIPs can replace synthetic molecular receptors or antibodies in many applications and can function with similar levels of affinity and selectivity. The MIPs have the advantage that they can be tailored to recognize an ever-increasing pool of templates including chiral amines, carbohydrates, proteins, and bacteria. With improvements in the imprinting process and the development of new imprinted polymer morphologies and formats, MIPs will find new applications in catalysis, separation materials, pharmaceuticals, medical devices, and sensors.

Acknowledgment

The authors are grateful to the National Institutes of Health (GM062593) for support of research on MIPs.

References

1. Breining, S.R. Recent developments in the synthesis of nicotinic acetylcholine receptor ligandsCurr. *Top. Med. Chem.* **2004**, 4 (6), 609629.
2. Hiley, C. R., Ford, W.R. Cannabinoid pharmacology in the cardiovascular system: potential protective mechanisms through lipid signaling. *Biol. Rev. Camb. Philos. Soc.* **2004**, 79 (1), 187–205.
3. Abe, H., Mawatari, Y., Teraoka, H., Fujimoto, K., Inouye, M. Synthesis and molecular recognition of pyrenophanes with polycationic or amphiphilic functionalities: artificial plate-shaped cavitant incorporating arenes and nucleotides in water. *J. Org. Chem.* **2004**, 69 (2), 495–504.
4. Lehn, J.-M. *Supramolecular Chemistry: Concepts and Perspectives*; VCH: New York1995.
5. Sellergren, B., Ed. *Molecularly Imprinted Polymers. Man Made Mimics of Antibodies and Their Applications in Analytical Chemistry*; Elsevier: Amsterdam 2001.
6. Molecularly imprinted materials Sci. Technol. **2005**, 734.
7. Alexander, C., Davidson, L., Hayes, W. Imprinted polymers: artificial molecular recognition materials with applications in synthesis and catalysis. *Tetrahedron* **2003**, 59 (12), 2025–2027.
8. Haupt, K. Molecularly imprinted polymers in analytical chemistry. *Analyst.* **2001**, 126 (5), 747–756.
9. Dickey, F.H. Specific adsorption. *J. Phys. Chem.* **1955**, 59, 695–707.
10. Wulff, G., Vesper, W., Grobe-Einsler, R., Sarhan, A. Enzyme-analog built polymers, 4. The synthesis of polymers containing chiral cavities and their use for the resolution of racemates. *Makromol. Chem.* **1977**, 178 (10), 2799–2816.
11. Umpleby, R.J.,II Bode, M., Shimizu, K.D. Measurement of the continuous distribution of binding sites in molecularly imprinted polymers. *Analyst.* **2000**, 125 (7), 1261–1265.
12. Umpleby, R.J., II, Baxter, S.C., Chen, Y., Shah, R.N., Shimizu, K.D. Characterization of molecularly imprinted polymers with the Langmuir-Freundlich isotherm. *Anal. Chem.* **2001**, 73 (19), 4584–4591.
13. Whitcombe, M.J., Rodriguez, M.E., Villar, P., Vulfson, E.N. A new method for the introduction of recognition site functionality into polymers prepared by molecular imprinting: synthesis and characterization of polymeric receptors for cholesterol. *J. Am. Chem. Soc.* **1995**, 117, 7105–7111.
14. Wulff, G., Schönfeld, R. Polymerizable amidines—adhesion mediators and binding sites for molecular imprinting. *Adv. Mater.* **1998**, 10, 957–959.
15. Sellergren, B., Ed. *Molecularly Imprinted Polymers. Man Made Mimics of Antibodies and Their Applications in Analytical Chemistry*; Elsevier: Amsterdam, 2001.
16. Sellergren, B. Imprinted chiral stationary phases in high-performance liquid chromatography. *J. Chromatogr. A.* **2001**, 906 (1-2), 227–252.

17. Tovar, G.E.M., Krauter, I., Gruber, C. Molecularly imprinted polymer nanospheres as fully synthetic affinity receptors. *Top. Curr. Chem.* **2003**, *227*, 125–144.
18. Ye, L., Weiss, R., Mosbach, K. Synthesis and characterization of molecularly imprinted microspheres. *Macromolecules.* **2000**, *33* (22), 8239–8245.
19. Sulitzky, C., Ruckert, B., Hall, A.J., Lanza, F., Unger, K., Sellergren, B. Grafting of molecularly imprinted polymer films on silica supports containing surface-bound free radical initiators. *Macromolecules.* **2002**, *35* (1), 79–91.
20. Reddy, P.S., Kobayashi, T., Abe, M., Fujii, N. Molecular imprinted Nylon-6 as a recognition material of amino acids. *Eur. Polym. J.* **2002**, *38* (3), 521–529.
21. Fischer, L., Müller, R., Ekberg, B., Mosbach, K. Direct enantioseparation of β-adrenergic blockers using a chiral stationary phase prepared by molecular imprinting. *J. Am. Chem. Soc.* **1991**, *113*, 9358–9360.
22. Ramstrom, O., Nicholls, I.A., Mosbach, K. Synthetic peptide receptor mimics—highly stereoselective recognition in noncovalent molecularly imprinted polymers. *Tetrahedron Asymmetry.* **1994**, *5* (4), 649–656.
23. Theodoridis, G., Manesiotis, P. Selective solid-phase extraction sorbent for caffeine made by molecular imprinting. *J. Chromatogr. A.* **2002**, *948* (1-2), 163–169.
24. Zhu, Q.Z., DeGelmann, P., Niessner, R., Knopp, D. Selective trace analysis of sulfonylurea herbicides in water and soil samples based on solid-phase extraction using a molecularly imprinted polymer. *Environ. Sci. Technol.* **2002**, *365*, 411–5420.
25. Masque, N., Marce, R.M., Borrull, F., Cormack, P.A.G., Sherrington, D.C. Synthesis and evaluation of a molecularly imprinted polymer for selective on-line solid-phase extraction of 4-nitrophenol from environmental water. *Anal. Chem.* **2000**, *72* (17), 4122–4126.
26. Wulff, G. Enzyme-like catalysis by molecularly imprinted polymers. *Chem. Rev.* **2002**, *102* (1), 1–27.
27. Beach, J.V., Shea, K.J. Designed catalysts. A synthetic network polymer that catalyzes the dehydrofluorination of 4-fluoro-(*p*-nitrophenyl)butan-2-one. *J. Am. Chem. Soc.* **1994**, *116*, 379–380.
28. Sellergren, B., Shea, K.J. Enantioselective ester hydrolysis catalysed by imprinted polymers. *Tetrahedron Asymmetry.* **1994**, *5*, 1403–1406.
29. Severin, K. Imprinted polymers with transition metal catalysts. *Curr. Opin. Chem Biol.* **2000**, *4* (6), 710–714.
30. Ansell, R.J. Molecularly imprinted polymers in pseudoimmunoassay. *J. Chromatogr. B.* **2004**, *804*, 151–165.
31. Ansell, R.J., Ramstrom, O., Mosbach, K. Towards artificial antibodies by the technique of molecular imprinting. *Clin. Chem.* **1996**, *42*, 1506–1512.
32. Ansell, R.J. MIP-ligand binding assays (pseudo-immunoassays). *Bioseparation.* **2002**, *10*, 365–377.
33. Andersson, L.I., Muller, R., Vlatakis, G., Mosbach, K. Mimics of the binding-sites of opioid receptors obtained by molecular imprinting of enkephalin and morphine. *Proc. Natl. Acad. Sci. USA.* **1995**, *92* (11), 4788–4792.
34. Bengtsson, H., Roos, U., Andersson, L.I. Molecular imprint based radioassay for direct determination of *S*-propranolol in human plasma. *Anal. Commun.* **1997**, *34*, 233.
35. Chen, Y., Shimizu, K. Measurement of enantiomeric excess using molecularly imprinted polymers. *Org. Lett.* **2002**, *4* (17)2937–2940.
36. Greene, N.T., Morgan, S.L., Shimizu, K.D. Molecularly imprinted polymer sensor arrays. *J. Chem. Soc. Chem. Commun.* **2004**, *10*, 1172–1173.
37. Piletsky, S.A., Turner, A.P.F. Electrochemical sensors based on molecularly imprinted polymers. *Electroanalysis.* **2002**, *14* (5), 317–323.
38. Shoji, R., Takeuchi, T., Kubo, I. Atrazine sensor based on molecularly imprinted polymer-modified gold electrode. *Anal. Chem.* **2003**, *75*, 4882–4886.
39. Marx, S., Zaltsman, A., Turyan, I., Mandler, D. Parathion sensor based on molecularly imprinted sol-gel films. *Anal. Chem.* **2004**, *76*, 120–126.

40. Striegler, S. Designing selective sites in templated polymers utilizing coordinative bonds. *J. Chromatogr. B.* **2004**, *1804*, 183–195.
41. Spivak, D.A., Shea, K.J. Investigation into the scope and limitations of molecular imprinting with DNA molecules. *Anal. Chim. Acta.* **2001**, *435* (1), 65–74.
42. Shi, H.Q., Tsai, W.B., Garrison, M.D., Ferrari, S., Ratner, B.D. Template-imprinted nanostructured surfaces for protein recognition. *Nature.* **1999**, *398* (6728), 593–597.

8

Functionalization of Carbon Nanotubes (CNTs)

David B. Henthorn

CONTENTS

Introduction

Carbon nanotubes (CNTs) were first discovered by Sumio Iijima of NEC in 1991.[1] They consist of sheets of carbon atoms connected in a honeycomb-like pattern and rolled into tiny cylinders. Single-walled carbon nanotubes (SWNTs) consist of one graphene layer, and most have a diameter of approximately 1 nm, while their length can range from several hundred nanometers to microns. Multi-walled carbon nanotubes (MWNTs) are comprised of 10–50 concentric tubes with an average interlayer distance of approximately 3.4 nm (Figure 8.1).

Three types of CNT structures exist and are classified by the way the graphene layers are rolled in. The chiral vector is represented by a pair of indices, n and m, with these two integers corresponding to the number of unit vectors along two directions in the honeycomb crystal lattice of graphene. The nanotube is designated as "zigzag" when $m = 0$, "armchair" when $m = n$, and chiral for all other configurations (Figure 8.2).

CNTs have unusual physical and chemical properties including extraordinary metallic or semimetallic electrical conductivity, extremely high mechanical strength, high hydrogen adsorption capability, and good thermal properties.[2–9] As a result, CNTs drew strong interest from physics, chemistry, materials science, and other high-technology fields

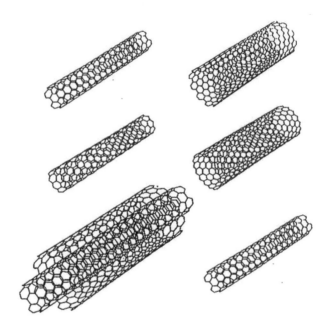

FIGURE 8.1
Single-walled and multi-walled carbon nanotubes (SWNT and MWNT, respectively).

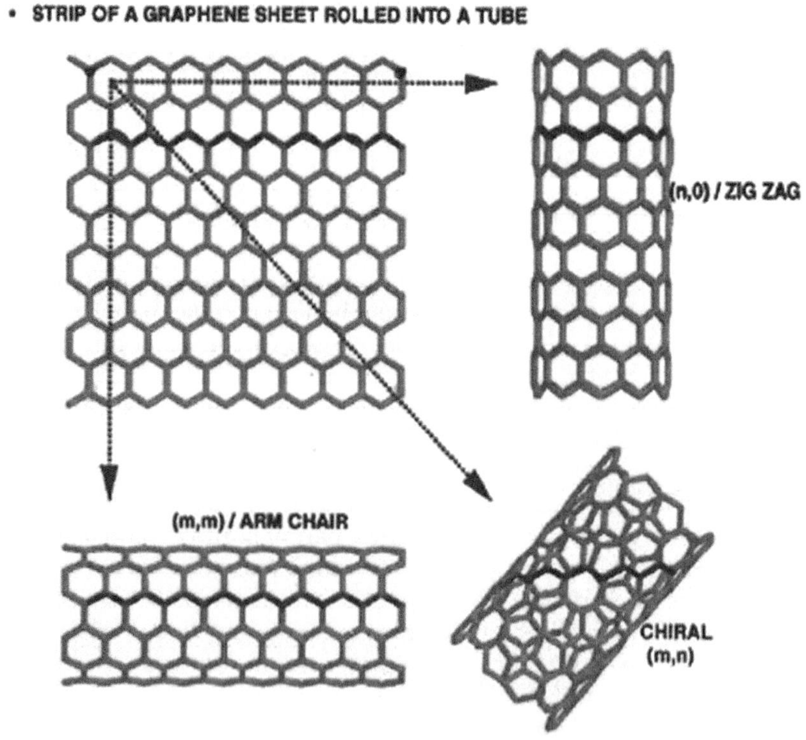

• **STRIP OF A GRAPHENE SHEET ROLLED INTO A TUBE**

(n,0) / ZIG ZAG

(m,m) / ARM CHAIR

CHIRAL
(m,n)

FIGURE 8.2
Description of CNT chirality through use of chiral vector (*n,m*). (Adapted from Dresselhaus and Endo.[2])

almost immediately following their discovery in the 1990s. Significant investment has led to the development of facilities for the production and application of CNTs worldwide.

To date, arc discharge, laser ablation, and chemical vapor deposition (CVD) are the three main methods for CNT production. The popularity of the CVD production method has increased significantly in recent times as CNTs produced via the arc discharge and laser ablation methods contain a significant amount of amorphous carbon byproduct that makes purification difficult. Also, the high yield and low cost for scale-up makes CVD the production method of choice.[10,11]

As-produced CNTs are difficult to utilize and manipulate even after significant purification has been completed. Additionally, some of the most interesting applications proposed for CNTs, such as carriers for biomolecules or as surfaces/sites for chemical reactions, require precise chemical modification of the nanomaterial surface. The lack of aqueous dispersibility and the difficulty associated with manipulation in water and other polar solvents have also imposed significant obstacles to the utilization of CNTs. Therefore, in order to make CNTs useful in various applications, pristine nanotubes are typically functionalized through one of several methods. In this functionalization step, physical and/or chemical methods are used to add active groups or molecules on the surface and/or tips of the nanomaterial. These sites will be used later to introduce species which can passivate the surface, render the surface catalytically active, aid in dispersal of the nanotube in solvent, or provide a chemical link between nanomaterial and biomolecule, for example.

Functionalization techniques of CNTs can be classified as either non-covalent or covalent in nature. Non-covalent functionalization techniques utilize physical adsorption of a chemical species on the CNT surface, are vastly simpler, and require less sample modification. Because the CNT surface is hydrophobic, π–π hydrophobic or van der Waals forces are the most likely interactions used in adsorption and are spontaneously formed under mild conditions. This technique is especially useful for the simple attachment or adsorption of biomolecules on CNT surfaces.[12] Proteins, for instance, have sufficient hydrophobic/hydrophilic balance to directly adsorb on the surface of CNTs, and in doing so add aqueous dispersibility to the protein/nanomaterials complex. An example of this is the addition of enzyme molecules to CNTs for use as bioanalytical platforms.[13]

An alternative to the direct physical adsorption technique is to use a heterobifunctional linker molecule—one functional group interacts with the CNT surface while the other group is available for further modification and linkage with other species. In a recent work, Chen et al.[14] used the linker molecule 1-pyrenebutanoic acid, succinimidyl ester to functionalize CNTs with proteins. This linker molecule presents a pyrene end which strongly adsorbs on the nanotube surface through π-stacking, and a succinimidyl group—a functionality frequently used in bioconjugation—to link with reactive groups on surface-exposed amino acids (Figure 8.3). The affinity of the pyrene group for the hydrophobic CNT surface is significantly higher than the affinity of protein for CNT, allowing for much greater loading. Tailoring of the length of this linker molecule is also possible. This approach has the advantage that reactive groups with greater affinities may be utilized to improve binding when compared to simple physical adsorption.

In covalent CNT functionalization, the linking functionalities are introduced through chemical reactions, allowing pendant molecules to be added to the CNT surface through covalent bonds. This method leads to materials with greater interaction between nanotube and functional structures, an effect that comes at the expense of permanent alteration of the surface structure.[15] Researchers must consider that such alterations may lead to irreversible changes in the desired electrical, thermal, or mechanical properties. The standard

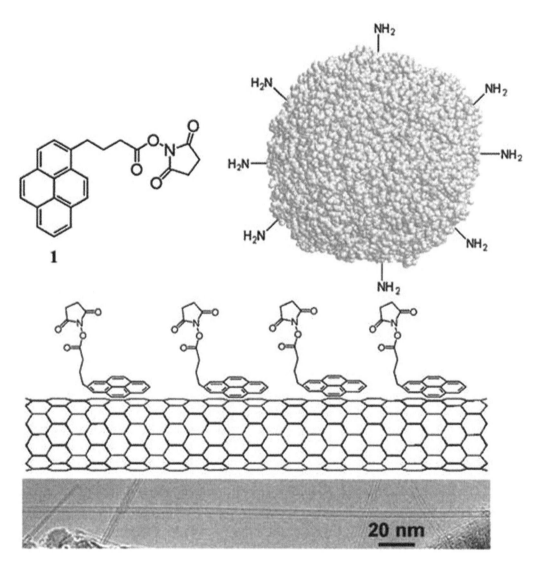

FIGURE 8.3
Functionalization of CNTs with protein via heterbiofunctional linker molecule chemistry. (Adapted from Chen et al.[14])

method of covalent functionalization is through the introduction of oxygen-containing functionalities, such as carboxylic acid functional groups (–COOH).[16] Ultrasonication and reflux of CNTs in the presence of concentrated nitric acid not only eliminates significant amounts of amorphous carbon (reaction by-product) and metallic nanoparticles (residual catalyst from nanotube manufacture), but oxygen-containing functional groups are introduced either on the defects or the tips of the CNTs. These groups may now act as sites for covalent attachment of chains, enzymes, and other biomolecules.

Wang et al.[17] reported a rapid covalent functionalization method to achieve high concentrations of modified CNTs using a microwave oven. CNTs were added to a mixture of nitric and sulfuric acids and were irradiated in a microwave oven. The microwave-assisted functionalization was rapid and was completed in only minutes, whereas traditional sono-chemical treatments require hours of sonication and reflux in acidic solutions.

Functionalization of CNTs with grafted polymer chains allows for high densities of active groups to be added effectively.[18-21] Grafted hydrophilic chains, for instance, may be utilized to improve the aqueous dispersibility of CNTs. Additionally, the steric hindrance associated with high densities of grafted chains leads to lower chance for aggregation once dispersed.[18-21] Furthermore, the large number of repeating units allows for numerous sites for biomolecule attachment, or for interaction with the matrix resin used in the formation of composite materials, e.g. linkage between grafted epoxide groups and epoxy resin.

The interaction between the surface of CNTs and the grafted polymer chains is therefore of utmost importance. Functionalization methods may be broken into two fundamental categories. In "graft to" method, reactive polymer chains or fragments are added to the surface of activated CNTs. Chains compete for available functionalization sites, leading to concerns of steric hindrance from the bulky molecules being attached. Figure 8.4 shows a representative graft to method where reactive polystyrene attacks a SWNT surface in the presence of heat and *ortho*-dichlorobenze (*o*-DCB) solvent.[18]

Alternatively, "graft from" methods rely on the growth of polymer chains from reactive species that originate at the nanomaterial's surface. For instance, a polymerization initiator may be grafted to the surface of a CNT and used to start polymerization of monomer directly from the surface of the monomer. A separation step is then used to remove functionalized materials from a sea of unreacted monomer. Instead of diffusion of macromolecules to the CNT surface, a graft from reaction is therefore dependent on the transport of the more mobile monomer to the reaction site. A photoinitated graft from method was recently developed by Zhang and Henthorn[22,23] with the reaction scheme shown in Figure 8.5. In the first step, photoinitiator is immobilized on the nanotube surface to provide sites for polymer growth. In the first step, irradiation of dissolved photoinitiator leads to their dissociation into radical fragments, which subsequently attack the nanotube surface. In the second step, further irradiation leads to cleavage of this newly formed bond, and the creation of radical species which, in the presence of monomer or macromonomer molecules, leads to the creation of grafted polymer chains with numerous and controllable functional groups.

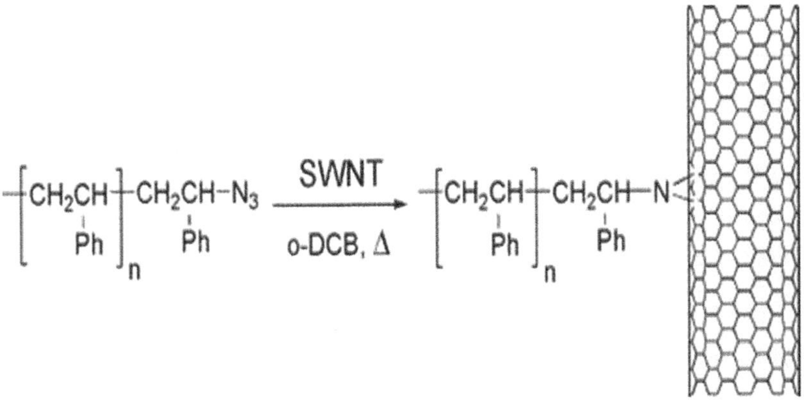

FIGURE 8.4
A "graft to" method for the functionalization of CNTs with polystyrene. A reactive polystyrene chain is added to the surface of a CNT through the addition of heat. (Adapted from Jin et al.[18])

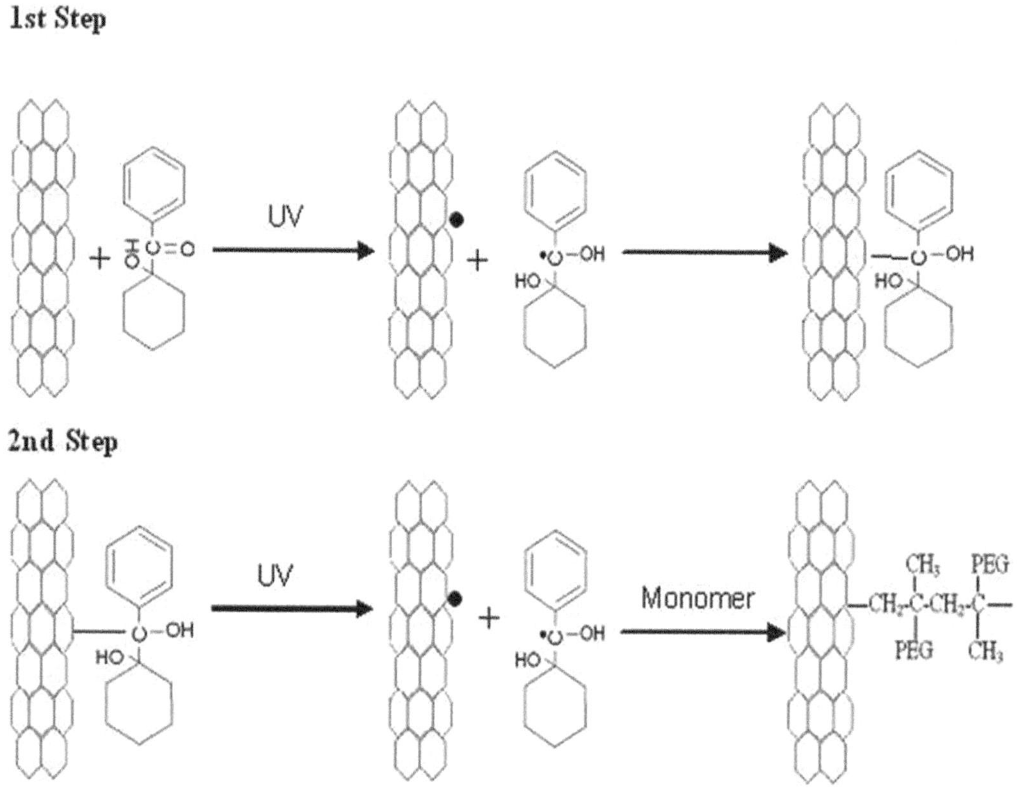

1st Step

2nd Step

FIGURE 8.5
A method of a photoinitiated "graft from" polymerization method to synthesize polymer-functionalized CNTs with varied moieties. In this case, macromonomers of PEG have been grafted to the CNT surface to create dispersible nanotubes.

Applications of Functionalized CNTs

CNTs for Biological and Biomedical Applications

Some of the most important and promising applications of CNTs are in the biological and biomedical fields.[24–26] Because of the extremely high surface area-to-volume ratio of CNTs and other nanomaterials, and because of their stability in aqueous environments, CNTs have been proposed as supports for biochemical reactions,[27] as carriers for pharmaceuticals in drug delivery,[28] and as materials in sensing elements.[29]

CNT Dispersion

The hydrophobic nature of CNTs limits their application as support materials in aqueous media and related environments (buffers, bodily fluids, etc.). A crucial step, therefore, toward utilizing CNTs in a variety of applications is to make CNTs dispersible in water and other desired solvents. Functionalization of CNTs with different molecules, done through either non-covalent or covalent methods, is one of the most common methods for enhancing nanotube dispersibility.[30–35] In one of the simplest examples, Bandyopadhyaya et al.[36] mixed as-grown CNTs with a solution of gum arabic, a polysaccharide derived from the Acacia senegal tree. This method is analogous to the recipe which originated more than 5,000 years ago in ancient Egypt to make carbon black ink. Adsorbed chains radiate outward from the CNT surface and prevent agglomeration through steric hindrance. In a second example of simple strategies for aqueous dispersibility enhancement, Zheng et al.[37] reported that, through π–π interactions, single-stranded deoxyribonucleic acid (DNA) molecules became helically wrapped around the CNT surface.

While there are examples of simple but effective non-covalent methods to enhance aqueous dispersion of CNTs, many methods rely on the formation or reorganization of chemical bonds to impart compatibility with water. Covalent functionalization of nanomaterials is more popular and often gives better results. Sonochemical treatment, leading to the formation of surface-bound carboxyl groups, imparts significantly enhanced dispersibility. This dispersibility can be furthered through addition of hydrophilic species. After the introduction of carboxyl groups through sonochemical treatment, for example, polymer chains may be grafted to the material at these reactive sites. Fernando et al.[38] reported using three different polymers: poly(propionylethylenimine-*co*-ethylenimine) (PPEI-EI), poly(vinyl alcohol) (PVA), and poly(ethylene glycol) (PEG) to increase CNT dispersibility (Figure 8.6). Carboxyl groups were first introduced at the material surface through sonochemical treatment (CNTs placed in strong acids and sonicated while heated). PPEI-EI–CNTs, PVA–CNTs, and PEG–CNTs conjugates were formed through thermal and arylation–amidation reactions. After functionalization, the aqueous dispersibility of these three polymer-nanomaterial composites increased dramatically to more than 13.5, 7.1, and 9.8 mg/ml, compared to almost zero of pristine CNT.

Functionalization of CNTs with Enzymes

Both non-covalent and covalent functionalization methods can be used to modify CNTs with biomolecules such as enzymes, proteins, DNA, etc.[39–42] Noncovalent functionalization techniques, such as simple physical adsorption, are straightforward in practice and do

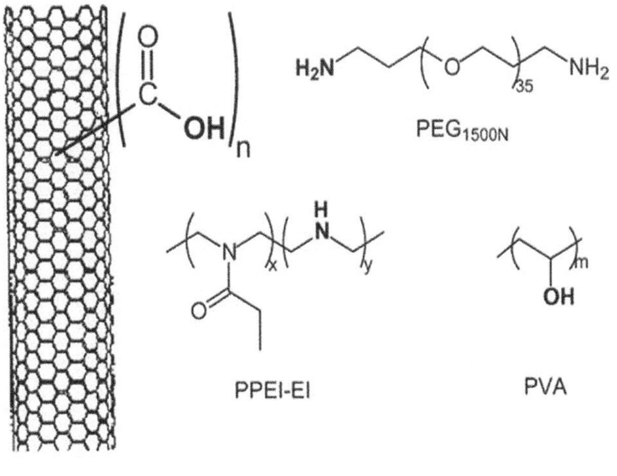

FIGURE 8.6
Illustration of CNTs covalently functionalized by PPEI-EI, PEG, and PVA. (Adapted from Zheng et al.[37])

not require the harsh processing conditions associated with sonochemical treatment. The direct adsorption of biomolecules on CNT materials, while simple, leads to lower amounts of loading than more involved methods, as reported in a study conducted by Zhang and Henthorn.[43] The authors immobilized the model enzyme alkaline phosphatase on CNT carriers using either direct adsorption or through the use of a linker molecule, 1-pyrenebutanoic acid, succinimidyl ester. It was found that the amount of enzyme loading was significantly less for the direct adsorption method (66 µg of enzyme per milligram of dry CNTs) than for the linker molecule method (140 µg/mg). A study of retained enzyme activity was then conducted. Zhang and Henthorn found that only 27% of the alkaline phosphatase activity remained when the conjugate was produced using direct adsorption, while 57% of the enzyme activity was retained when using the linker molecule. Karajanagi et al.[12] reported the formation of two-model enzyme–CNT systems, one utilizing a-chymotrypsin (CT) and the other soybean peroxidase (SBP). Both conjugates were formed through adsorption of protein directly on the CNT surface. The SBP enzyme retained approximately 30% of its bioactivity after immobilization, while adsorbed CT retained only 1%. The authors were then able to show a significant change in the secondary structure of CT enzyme occurred upon physical adsorption, ultimately resulting in the loss of activity. In a similar study, Asuri et al.[44] used circular dichroism spectroscopy and fluorescence to analyze the structural changes that occur upon protein adsorption. Circular dichroism confirmed the distortion of basic enzyme structures, and tryptophan fluorescence further verified the structural change and loss of activity for the CNT–enzyme conjugate.

In covalent CNT functionalization, foreign biomolecules are attached through chemical reactions such as esterification, amidation, or other reactions, utilizing functional groups introduced on the CNT beforehand.[45–47] One frequently used method in the covalent functionalization of CNTs with biomolecules is shown in Figure 8.7 and makes use of

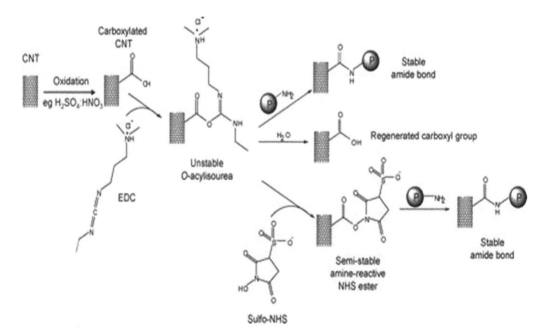

FIGURE 8.7
Adsorption of protein onto CNT surface, promoted through coupling with EDC. (Adapted from Torres and Basiuk.[47])

1-ethyl-3-[3-dimethylaminopropyl] carbodiimide hydrochloride (EDC).[48] EDC reacts with a carboxyl group to form an intermediate capable of reacting with surface-exposed amine groups of the protein, yielding a conjugate of the two molecules joined by a stable amide bond.

By utilizing a covalent functionalization strategy, both biomolecular loading amount and activity can be improved significantly. Zhang and Henthorn[23] reported the use of an ultraviolet light initiated graft from polymerization method to fabricate high-capacity bimolecular carriers from dispersible SWNT-polymer composites. Copolymerized chains, comprised of PEG methyl ether methacrylate (PEGMA) and glycidyl methacrylate (GMA) repeating units, were grafted from the CNT surface. Dispersibility was controlled by altering the ratio of the PEGMA:GMA repeating units, with increasing ratios of PEGMA leading to enhanced dispersibility. Enzyme conjugation was done through reaction with pendant epoxide groups of the GMA units. To study the utility of the method, the authors utilized the same enzyme, alkaline phosphatase, that was studied in their evaluation of non-covalent methods.[43] The graft from method resulted in enzyme loading amounts of more than 409 μg of enzyme per milligram of CNT, a 190% improvement compared to their previous work. In addition, conjugated enzyme retained 50% of their activity.

CNTs as Drug Delivery Vehicles

Kam and Dai[49] showed that CNTs might be used as intercellular transporters, sparking interest in CNTs as drug delivery vehicles. In their study, an anticancer drug was attached on PEG functionalized CNTs for cellular internalization. In another study, Kam et al.[50] covalently attached biotin to the surface of oxidized CNTs, and the CNT–biotin conjugates were incubated with streptavidin. Cellular uptake of the CNT–protein conjugates was studied using both confocal imaging and flow cytometry. It was found that streptavidin was able to penetrate cells when complexed with the nanotubes as a transporter.

Uptake of nanomaterial by cells has been utilized to optically address diseased cells. While biological systems are highly transparent to 700–1100 nm near-infrared irradiation, CNTs readily absorb these wavelengths. As a result, CNTs may absorb intense, focused near-infrared light, resulting in the generation of localized areas of heat, even when in contact with tissues, cells, and proteins. In their investigation, Kam et al.[51] used this property to optically address internalized CNTs, creating an agent for the selective ablation of cancerous cells.

CNTs for Biosensor Applications

CNTs, whether employed as sensing elements or transducers, will likely play a role in the future generation of biological, chemical, and physical sensors. Compared to traditional sensors, the extremely small size of a nanosensor could significantly enhance sensitivity and would couple well with trends in microsampling methods. As a result, the incorporation of CNTs has the potential to address a variety of issues with respect to sensing. To date, significant progress has been made on the fabrication of nanosensors with uses as platforms for the design of biosensors.

Besteman et al.[52] reported using semiconducting CNTs to detect glucose molecules (Figure 8.8). Their group functionalized CNTs with glucose oxidase using the methods proposed by Chen et al.[14] and found that immobilization of glucose oxidase resulted in the decreased conductance of the nanotube material. In addition, the reaction between glucose and glucose oxidase was measured through a change in conductance voltage as

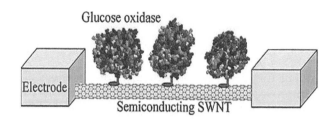

FIGURE 8.8
Scheme of CNT-based glucose biosensor. (Adapted from Besteman et al.[52])

a function of current. At the same time, the authors noticed the glucose oxidase-function-alized nanotube conductance was varied when different pH environments were applied, indicating further potential use for the materials as a nanoscale pH sensor.

Barone et al.[53] also used CNTs to monitor glucose, but they utilized the fact that CNTs luminesce in the near-infrared region (NIR), a region where the body's tissues are transparent. CNTs functionalized with ferricyanide were coated with glucose oxidase. Ferricyanide is reduced by hydrogen peroxide, a product of the oxidase reaction, and a measurable change in the luminescent properties is monitored by a NIR detector. This strategy produces good results since the glucose oxidase adsorbed on the surface of the CNT non-covalently, which does not affect CNT luminescence.

Azamian et al.[54] demonstrated a system with functionalized CNTs distributed on a carbon electrode. To make the system sensitive to the glucose analyte, glucose oxidase was first adsorbed on the surface of the CNTs. Upon the introduction of glucose and fer-rocene, the glucose signal intensity increased by more than an order of magnitude when compared to enzyme immobilized on a carbon electrode alone. This set of experiments demonstrated how nanotubes provided a high surface area to which the glucose oxidase could adsorb, with higher signal intensity resulting from the increased amount of active enzyme on the electrode interface.

Okuno et al.[55] reported a label-free immunosensor for the detection of prostate-specific antigen (PSA), an oncological marker for the presence of prostate cancer, using Pt microelec-trodes modified with a CNT array to monitor antibody–antigen interactions. Measurable signal resulted from the oxidation of tyrosine and tryptophan residues and increased with the interaction between PSA and PSA-monoclonal antibody covalently immobilized on SWNTs. The detection limit for PSA was calculated to be as low as 0.25 ng ml^{-1} (Figure 8.9).

In addition to the antibody–antigen interaction, high specificity biotin–avidin reactions may also be used in the development of CNT-based biosensors. Star et al.[56] fabricated a biosensor by first coating a mixture of two polymers: poly(ethylene imine) (PEI) and PEG. The PEI conjugated with biotin molecules, while PEG was used to reduce nonspe-cific protein adsorption. After the biotin-modified nanotubes were immobilized between two microelectrodes, streptavidin was introduced. An immediate current change was recorded when the biotinylated nanotubes bound streptavidin, and the change intensified when exposed to additional streptavidin.

Construction of Polymer Composites with Carbon Nanotubes

Utilization of CNTs to Improve Mechanical Properties of Polymer Materials

In the field of materials science, CNTs are called "super fibers," because of their extremely high strength.[57] Theoretical predictions on CNTs show that the Young's modulus is close

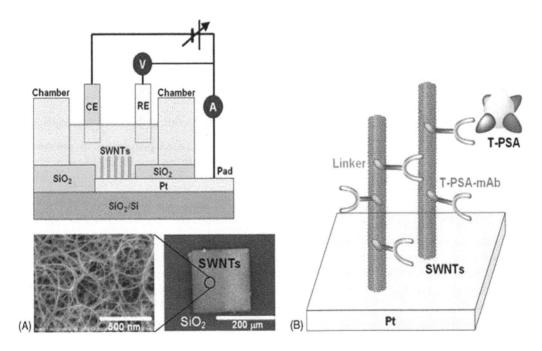

FIGURE 8.9
(A) Illustration for the experimental set-up with SWNT-modified Pt microelectrode as the working electrode, Pt wire as the counter electrode (*CE*), and the miniaturized reference electrode (*RE*, Ag/AgCl) with the electron microscopy images of a SWNT-modified microelectrode; (B) Illustration of the label-free electrochemical immunosensor design. (Adapted from Okuno et al.[55])

to that of diamond.[58] In addition, the material is very light, with a density only 1/6th that of steel. Therefore, CNTs are considered to be an ideal material to reinforce polymer composite materials.[59–64] The effective utilization of CNTs in composite materials strongly depends two things: 1) the dispersion of CNTs homogeneously within the polymer matrix, and 2) the establishment of strong interfacial bonding to achieve efficient load transfer from the polymer matrix to the CNTs. To address these problems, different routes has been developed, including direct high shear physical mixing[65] and other covalent functionalization methods.[66,67]

Schadler et al.[68] employed a direct solution evaporation method assisted by high-powered ultrasonication to prepare CNT-polystyrene composite materials. Through high intensity, long duration ultrasonication processing in toluene, large aggregated CNT bundles were separated into smaller bundles or even individual nanotubes. After the introduction of polymer matrix, the mixture was then cast into a Petri dish and the toluene evaporated with the help of heat. Tensile tests on the composite showed that addition of CNT at 1 wt.% resulted in a 20% increase in elastic modulus and 17% increase in break stress.

Liu et al.[69] reported using CNTs functionalized with hydroxyl groups to reinforce PVA, since the –OH groups would hydrogen bond with the –OH of the PVA. Reasonably good modulus enhancement was observed with an increase from 2.4 to 4.3 GPa on addition of 0.8 wt.% nanotubes, corresponding to a reinforcement value of 305 GPa. A strength increase from 74 to 107 MPa was also observed. These results were explained by the observation of good load transfer by Raman spectroscopy.

Zhu et al.[70] demonstrated using functionalized CNTs with polyamine grafted to the surface. The amino groups in polyamine could react easily with the epoxy groups and act as curing agents for the epoxy matrix. Furthermore, a cross-linked structure was formed through covalent bonds between the tubes and the epoxy polymer. Through this method, not only was effective load transfer seen, but the chains aided in dispersal. Nanomaterial loadings of up to 4 wt.% were created, which helped increase the ultimate strength by 30% and the modulus by 50%, when compared to pure resin.

Utilization of CNTs to Improve Thermal and Electrical Properties of Polymer Materials

CNTs can also be used to improve the thermal and electrical properties of polymeric materials. Guthy et al.[71] investigated the thermal conductivity of SWNT/PMMA nano-composites. It was demonstrated that the thermal conductivity of PMMA improved by approximately 250% at a SWNT loading of 10 wt.%.

Ma et al.[72] successfully fabricated a CNT-poly(urea urethane) (PUU)-poly(dimethylsi-loxane) (PDMS) system which could effectively reduce problems associated with electro-magnetic interference (EMI). Even though polymer materials bear numerous advantages including high strength and high flexibility, EMI is a big problem since most polymer materials are not conductive. As a result, no protection can be provided from external fields. By dispersing a small amount of functionalized CNTs into polymer resins, the EMI shielding effectiveness of the material increased significantly. Ma et al. first functional-ized the nanotubes with nickel and silver, which improved the conductivity of nanotubes after they were oxidized in a mixture of HNO_3 and H_2SO_4. Next, 3–5% of functionalized nanotubes were dispersed in the PUU-PDMS system by high shear stirring at room tem-perature for 1 hr. They concluded that the increase of functionalized CNTs improved the shielding effectiveness on several frequencies and that nickel showed higher EMI shield-ing effectiveness than silver.

Side Effects of Covalent Functionalization and Acid Treatment

The main contaminants associated with CNT manufacture are amorphous carbon and various catalyst nanoparticles. As a result, purification of the as-received nanomaterial is vital before utilization. The most common purification process is chemical oxidation,[73] where the amorphous carbon and most of the catalyst metal nanoparticles are dissolved using a strong oxidative acid, usually sulfuric acid or nitric acid. In addition, CNTs will likely be functionalized to include oxygen containing functional groups (defects) for sub-sequent modification; this acid oxidation step is commonly adopted as part of the covalent functionalization procedure.[74,75]

Through this covalent functionalization process, however, it is easy to destroy the sur-face graphene structures of the CNTs, diminishing the desired electrical and mechani-cal properties of the pristine material. The electronic properties of CNTs depend on the unique symmetry structure[76] for a given chiral vector (n,m). If $n = m$, the nanotube is a very good conductor; if $n–m$ is a multiple of 3, the nanotube is semiconducting with a small band gap; otherwise, the nanotube is a moderate semiconductor. The introduction of defects from acid oxidation treatment will damage the surface structure and cause the loss of these desirable electronic properties.

Mechanical properties are also altered in these harsh processing steps. Sui et al.[77] stud-ied using acid oxidation treatment to aid in the dispersal of CNTs in polymethylmeth-acrylate (PMMA) in order to reinforce its mechanical property. The introduced carboxyl groups, covalently attached to the CNT surface, led to enhanced nanotube dispersion in the polymer. However, it was found that the mechanical properties of the polymer/

nanotube composite fell below those of pure polymer material. The authors concluded that covalent functionalization should be regarded as a double-edged sword. While it is true that functionalized nanotubes disperse in the polymer better, damage of the nanotube surface results in a decrease in overall mechanical properties.

Ultrasonication is commonly accompanied by an acid oxidation treatment to break aggregated CNT bundles into smaller bundles or even individual tubes. During ultrasonication, sound energy generates small bubbles in the CNT solution, and the collapse of these bubbles can cause extremely high localized temperatures and pressures. This processing technique, therefore, has significant detrimental effects on the mechanical properties of the nanomaterials, resulting in the reduction of the effective length and rupture of the CNTs. Lu et al.[78] reported damage on the CNT surfaces after ultrasonication, and thinning of the CNTs, with exfoliation of the graphene layer observed. Lu et al. noticed that the extent of damage also depended on the solvent that was used to disperse the nanotubes. Water and ethanol were found to better protect the nanotube than methylene chloride.

Safety Concerns of CNT Applications

Although CNTs have shown great potential as future tools in biological and biomedical fields, the biocompatibility of these materials is of great importance and concern. The toxicity of CNTs has been controversial for a long time. Magrez et al.[79] reported that incubation of H596 cells with CNT solutions resulted in a decrease in cell proliferation and a higher rate of cell death, especially when high concentrations of CNT solutions were applied. The introduction of carbonyl ($C = O$), carboxyl (COOH), and hydroxyl (OH) groups on CNTs enhanced damage to cells. Results from this work are shown in Figure 8.10. The cytoplasm has been stained pink, while the nuclei, stained by hematoxylineeosine, appeared purple. Initially, as shown in Figure 8.10(a), the observed clusters indicate good contact between cells. However, after the introduction of CNT solution, cells lose their mutual contact, and

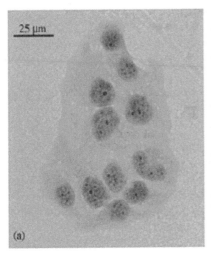
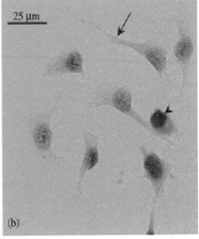

FIGURE 8.10
Cytopathological analysis on H596 cells before (a) and after (b) cultured in CNT solution. (Adapted from Magrez et al.[79])

darker pink and purple indicated cytoplasm retraction and condensed nuclei, as shown in Figure 8.10(b).

In another study, Poland et al.[80] observed that long, thin CNTs behave like asbestos fibers, which have been shown to cause mesothelioma, a deadly cancer of the membrane lining the body's internal organs (in particular, the lungs). Asbestos fibers are especially harmful, because they are small enough to penetrate deep into the lungs but too long for the body's immune system to destroy. They reached their conclusions after studying the exposure of lab mice to CNTs. Results showed significant inflammation of the inside lining of the animal's body cavities, with the formation of lesions.

CNTs have also shown significant environmental concerns. For example, juvenile rainbow trout showed a dose-dependent rise in ventilation rate, gill disease, and mucus secretion.[81] Other ecological indicator species, such as zebrafish, showed delayed hatching, while estuarine crustaceans showed increased mortality.[82] Plants have also recently been shown to react adversely to CNTs and fullerenes. Rice plants exposed to carbon fullerenes transmitted nanomaterials to the next generation. Exposure to both carbon fullerenes and CNTs also delayed the onset of rice flowering by at least one month and reduced the seed set.[83]

It is therefore of great importance to set stricter rules for using CNTs in the biological and biomedical fields and to tighten control over waste CNTs. Zhang and Henthorn are investigating the fabrication of recoverable nanomaterial using "switchable" grafted polymer chains to control dispersibility. A graft from polymerization method is used to fabricate polymer-functionalized SWNTs with pendant pH- or temperature-responsive polymer chains. The attached polymer chains, formed from methacrylic acid and PEGMA monomers, are well established for their pH-responsive swelling/deswelling behavior.[84,85] This special property is utilized here to control the CNT's aqueous dispersibility as a function of solution acidity. Our group has also studied the grafting of poly(N-isopropylacryl-amide) chains from the CNT surface. These chains are temperature-responsive, with a lower critical solubility temperature of around 32°C. As a result, the dispersibility of the CNTs is dependent on the solution temperature. At room temperature, the mixture is stabilized by the extended, solvated chains. At elevated temperatures, the CNTs agglomerate and are easily recovered. Through this method, dispersed CNTs can be easily reclaimed, which is environmentally beneficial.

Conclusions

CNTs have been researched extensively due to extraordinary properties and enormous potential as new materials. Before use in many applications, however, these nanomaterials must be functionalized—either through non-covalent or covalent functionalization methods. The potential applications of CNTs in the biological, biotechnological, and biomedical fields are wide, and a few representative studies have been discussed. In addition, the use of CNTs to improve the mechanical, thermal, and electrical properties of other polymer materials was presented. In the near future, further studies of CNT modification are necessary. Efforts designed to better understand the consequences of enhanced CNT aqueous dispersibility, for instance, will focus on the biocompatibility of CNTs and/or the environmental impact of such materials.

References

1. Iijima, S. *Nature.* **1991**, *354*, 56–58.
2. Dresselhaus, M., Endo, M. *Carbon Nanotubes Synthesis, Structure, Properties, and Application,* Springer, 2000.
3. Dai, L., Mau, A.W.H. *J. Phys. Chem. B* Soft Condens. Matter Biophys. Chem. **2000**, *104*, 1891–1915.
4. Hughes, M.E., Brandin, E., Golovchenko, J.A. *Nano Lett.* **2007**, *7*, 1191–1194.
5. Zhao, B., Hu, H., Yu, A., Perea, D., Haddon, R.C. *J. Am. Chem. Soc.* **2005**, *127*, 8197–8203.
6. Ratchford, N.N., Bangsaruntip, S., Sun, X., Weisher, K., Dai, H. *J. Am. Chem. Soc.* **2007**, *129*, 2448–2449.
7. Bahr, J.L., Tour, J. *J. Mater. Chem.* **2002**, *12*, 1952–1958.
8. Niemeyer, C.M. *Angew. Chem. Int. Ed.* **2001**, *40*, 4128–4154.
9. Wang, J., Liu, G., Jan, M.R. *J. Am. Chem. Soc.* **2004**, *126*, 3010–3011.
10. Zheng, B., Lu, C., Gu, G., Makarovski, A., Finkelstein, G., Liu, J. *Nano Lett.* **2002**, *2*, 895–898.
11. Balogh, Z., Halasi, G., Korbely, B., Hernadi, K. *Appl. Catal. A: Gen.* **2008**, *344*, 191–197.
12. Karajanagi, S.S., Vertegel, A.A., Kane, R.S., Dordick, J.S. *Langmuir.* **2004**, *20*, 11594–11599.
13. Wong, S.S., Joselevich, E., Woolley, A.T., Cheung, C.L., Lieber, C.M. *Nature.* **1998**, *394*, 52–55.
14. Chen, R.J., Zhang, Y., Wang, D., Dai, H. *J. Am. Chem. Soc.* **2001**, *123*, 3838–3839.
15. Gao, J., Zhao, B., Itkis, M., Bekyarova, E., Hu, H., Kranak, V., Yu, A., Haddon, R. *J. Am. Chem. Soc.* **2006**, *128*, 7492–7496.
16. Xing, Y., Li, L., Chusuei, C., Hull, R. Sonochemical oxidation of multiwalled carbon nanotubes. *Langmuir.* **2005**, *21*, 4185–4190.
17. Wang, Y., Iqbal, Z., Mitra, S. *J. Am. Chem. Soc.* **2006**, *128*, 95–99.
18. Jin, H., Choi, H., Yoon, S., Myung, S., Shim, S. *Chem. Mater.* **2005**, *17*, 4034–4037.
19. Chen, W., Wu, J., Kuo, P. *Chem. Mater.* **2008**, *20*, 5756–5657.
20. Alvaro, M., Atienzar, P., Cruz, P., Delgado, J., Garcia, H., Langa, F. *J. Phys. Chem., Part B Soft Condens. Matter Biophys. Chem.* **2004**, *108*, 12691–12697.
21. Czerw, R., Guo, Z., Ajayan, P.M., Sun, Y.P., Carroll, D.L. *Nano Lett.* **2001**, *1*, 423–427.
22. Zhang, P., Henthorn, D.B. *AIChE J.* **2009**, doi:10.1002/aic.12108.
23. Zhang, P., Henthorn, D.B. *Langmuir.* **2009**, *25*, 12308–12314.
24. Erlanger, B.F., Chen, B.X., Zhu, M., Brus, L. *Nano Lett.* **2001**, *1*, 465–467.
25. Collins, P.G., Bradley, K., Ishigami, M., Zettl, A. *Science.* **2000**, *287*, 1801–1804.
26. Joshi, P.P., Merchant, S.A., Wang, Y., Schmidtke, D.W. *Anal. Chem.* **2005**, *77*, 3183–3188.
27. Panhius, M., Salvador-Morales, C., Franklin, E., Chambers, G., Fonseca, A., Nagy, J.N., Blau, W.J., Minett, A.I. *J. Nanosci. Nanotech.* **2003**, *3*, 209–213.
28. Bianco, A., Kostarelos, K., Prato, M. *Curr. Opin. Chem. Biol.* **2005**, *9*, 674–679.
29. Tsai, Y.C., Li, S.C., Chen, J.M. *Langmuir.* **2005**, *21*, 3653–3658.
30. Huang, W., Lin, Y., Taylor, S., Gaillard, J., Rao, A.M., Sun, Y.P. *Nano Lett.* **2002**, *2*, 231–234.
31. Georgakilas, V., Kordatos, K., Prato, M., Guldi, D.M., Holzinger, M. *J. Am. Chem. Soc.* **2002**, *124*, 760–761.
32. Liu, J., Rinzler, A.G., Dai, H., Hafner, J.H., Bradley, R.K. Boul, J., Lu, A., Iverson, T., Shelimov, K., Huffman, C.B., Rodriguez-Macias, F., Shon, Y.S., Lee, T.R., Colbert, D.T., Smalley, R.E. Fullerene pipes. *Science.* **1998**, *280*, 1253–1256.
33. Wang, J., Musameh, M., Lin, Y. *J. Am. Chem. Soc.* **2003**, *125*, 2408–2409.
34. Tasis, D., Tagmatarchis, N., Georgakilas, V., Prato, M. *Chemistry.* **2003**, *9*, 4000–4008.
35. Islam, M.F., Rojas, E., Bergey, D.M., Johnson, A.T., Yodh, A.G. *Nano Lett.* **2003**, *3*, 269–273.
36. Bandyopadhyaya, R., Nativ, R.E., Regev, O., Yerushalmi, R.R. *Nano Lett.* **2002**, *2*, 25–28.
37. Zheng, M., Jagota, A., Semke, E., Diner, B., Mclean, R., Lustig, S., Raymond, E., Richardson, N., Tassi, G. *Nat. Mater.* **2003**, *2*, 338–342.
38. Fernado, K.A.S., Lin, Y., Sun, Y.P. *Langmuir.* **2004**, *20*, 4777–4778.
39. Wang, J., Liu, G., Jan, M.R. *J. Am. Chem. Soc.* **2004**, *126*, 3010–3011.

40. Jiang, K., Schadler, L.S., Siegel, R.W., Zhang, X., Zhang, H., Terrones, M. *J. Mater. Chem.* **2003**, *14*, 37–39.
41. Heller, D.A., Jeng, E.S., Yeung, T.K., Martinez, B.M., Moll, A.E., Gastala, J.B., Strano, M.S. *Science.* **2006**, *311*, 508–511.
42. Jeng, E.S., Moll, A.E., Roy, A.C., Gastala, J.B., Strano, M.S. *Nano Lett.* **2006**, *6*, 371–375.
43. Zhang, P., Henthorn, D.B. *J. Nanosci. Nanotech.* **2009**, *9*, 4747–4752.
44. Asuri, P., Sundhar, S., Ravindra, B., Pangule, C., Shah, D., Kane, R., Dordick, J. *Langmuir.* **2007**, *23*, 12318–12321.
45. Huang, W., Taylor, S., Fu, K., Lin, Y., Zhang, D., Hanks, T.W., Rao, A.M., Sun, Y.P. *Nano Lett.* **2002**, *2*, 311–314.
46. Pantarotto, D., Briand, J.P., Prato, M., Bianco, A. *Chem. Commun.* **2004**, *1*, 16–17.
47. Torres, F.F., Basiuk, V.A. *J. Phys.* **2006**, *61*, 85–89.
48. Gao, Y., Kyratzis, I. *Bioconj. Chem.* **2008**, *19*, 1945–1950.
49. Kam, N.W.S., Dai, H. *J. Am. Chem. Soc.* **2005**, *127*, 6021–6026.
50. Kam, N.W.S., Liu, Z., Dai, H. *Angew. Chem. Int. Ed.* **2005**, *45*, 577–581.
51. Kam, N.W.S., O'Connell, M., Wisdom, J.A., Dai, H. *Proc. Natl. Acad. Sci. U. S. A.* **2005**, *102*, 11600–11605.
52. Besteman, K., Lee, O.J., Wiertz, F.G.M., Heering, H.A., Dekker, C. *Nano Lett.* **2003**, *3*, 727–730.
53. Barone, P., Baik, S., Heller, D., Strano, M. *Nat. Mater.* **2005**, *4*, 86–92.
54. Azamian, B.R., Davis, J.J., Coleman, K.S., Bagshaw, B.C., Green, M.L.H. *J. Am. Chem. Soc.* **2002**, *124*, 12664–12665.
55. Okuno, J., Maehashi, K., Kerman, K., Takamura, Y., Matsumoto, K., Yamiya, E. *Biosens. Bioelectron.* **2007**, *22*, 2377–2381.
56. Allen, B., Kichambare, P., Star, A. *Adv. Mater.* **2007**, *19*, 1439–1451.
57. Bekyarova, E., Thostenson, E., Yu, A., Kim, H., Gao, J., Tang, J., Hahn, H., Chou, T., Itkis, M., Haddon, R. *Langmuir.* **2007**, *23*, 970–3974.
58. Coleman, J.N., Khan, U., Blau, W.J. Small but strong, *Carbon.* **2006**, *44*, 1624–1652.
59. Ding, W., Eitan, A., Fisher, F., Chen, X., Dikin, D., Andrews, R., Brinson, L., Schadler, L.S., Ruoff, R.S. *Nano Lett.* **2003**, *3*, 1593–1597.
60. Ajayan, P.M., Shadler, L.S., Giannaris, C., Rubio, A. *Adv. Mater.* **2000**, *12*, 750–753.
61. Wagner, H.D., Lourie, O., Feldman, Y., Tenne, R. *Appl. Phys. Lett.* **1998**, *72*, 188–190.
62. Thostenson, E.T.; Chou, T.W. *J. Phys. D: Appl. Phys.*, 35, 77–80.
63. Zhang, M., Atkinson, K.R., Baughman, R.H. *Science.* **2004**, *206*, 1358–1361.
64. Baughman, R.H., Zakhidov, A.A., Heer, W.A. *Science.* **2002**, *297*, 787–792.
65. Santos, C.V., Hernandez, A.L., Fisher, F.T., Ruoff, R., Castano, V.M. *Chem. Mater.* **2003**, *15*, 4470–4475.
66. Moniruzzaman, M., Winey, K.I. *Macromolecules.* **2006**, *39*, 5194–5205.
67. Qian, D., Dickey, E.C., Andrews, R., Rantell, T. *Appl. Phys. Lett.* **2000**, *76*, 2868–2870.
68. Schadler, L.S.; Giannaris, S.C.; Ajayan, P.M. *Appl. Phys. Lett.* **1998**, *73*, 3842–3844.
69. Liu, L., Barber, A.H., Nuriel, S., Wagner, H.D. *Adv. Funct. Mater.* **2005**, *15*, 975–980.
70. Zhu, J., Kim, J., Peng, H., Margrave, J., Khabashesku, V., Barrera, E. *Nano Lett.* **2003**, *3*, 1107–1113.
71. Guthy, C., Du, F., Brand, S., Fisher, J.E., Winey, K.I. *Nano Lett.* **2006**, *6*, 1237–1239.
72. Ma, C.C., Huang, Y.L., Kuan, H.C., Chui, Y.S. *J. Polym. Sci. Part B.: Polym. Phys.* **2005**, *43*, 345–358.
73. Haddon, R.C., Sippel, J., Rinzler, A.G., Papadimitrakopoulos, F. *MRS Proc.* **2004**, *29*, 1208–1214.
74. Tchoul, M.N., Ford, W.T., Lolli, G., Resasco, D.E., Arepalli, S. *Chem. Mater.* **2007**, *19*, 5765–5772.
75. Furtado, C.A., Kim, U.J., Gutierrez, H.R., Pan, L., Dickey, E.C., Eklund, P.C. *J. Am. Chem. Soc.* **2004**, *126*, 6095–6105.
76. Ebbesen, T.W., Lezec, H.J., Hiura, H., Bennett, J.W., Ghaemi, H.F., Thio, T. *Nature.* **1996**, *382*, 54–56.
77. Sui, X.M., Giordani, S., Prato, M., Wagner, H.D. *Appl. Phys. Lett.* **2009**, *95*, 233113–233116.
78. Lu, K.L., Lago, R.M., Chen, Y.K., Green, M.L.H. *Carbon.* **1996**, *34*, 814–816.
79. Magrez, A., Kasas, S., Salicio, V., Pasquier, N., Seo, J., Celio, M., Catsicas, S., Schwaller, B., Forró, L. *Nano Lett.* **2006**, *6*, 1121–1125.

80. Poland, C.A., Duffin, R., Kinloch, I., Maynard, A., Wallace, W.A.H., Seaton, A., Stone, V., Brwon, S., MacNee, W., Donaldson, K. *Nat. Nanotech.* **2008**, *3*, 423–428.

81. Smith, C.J., Shaw, B.J., Handy, R.D. Aquat. *Toxicol.* **2007**, *82*, 94–109.

82. Cheng, J.P., Flahaut, E., Cheng, S.H. Environ. *Toxicol. Chem.* **2007**, *26*, 708–716.

83. Lin, S. *Small.* **2009**, *5*, 1128–1132.

84. Zhang, J., Peppas, N.A. *Macromolecules.* **2000**, *33*, 102–107.

85. Zhang, J., Peppas, N.A. *J. Appl. Polym. Sci.* **2001**, *82*, 1077–1082.

9

Animal Cell Culture

Shang-Tian Yang and Shubhayu Basu

CONTENTS

Introduction

Animal cell culture or the ability to continuously grow animal cells in vitro after removing them from animal tissue opens up a plethora of windows in the field of biology and medicine. They provide a platform to investigate the normal physiology and biochemistry of cells, test the effects of drugs and other compounds in vitro, produce artificial tissue for implantation, synthesize valuable products from large-scale cultures, and can even be used as models in studying diseases. Animal cells in culture have been used as end products to provide artificial skin grafts, islet cells, hepatocytes, and bone marrow implants and to produce recombinant and natural proteins like human growth hormone, nerve growth factor, epidermal growth factor (EGF), monoclonal antibodies (MAb), vaccines, interferons, and blood clotting factors. Animal cells are often more advantageous than yeast or bacterial cells for the production of recombinant proteins on a large scale. Proteins that need to be heavily glycosylated for function, or need a proper folding environment because they have a large number of disulfide bonds, are often made in animal cells. Bacterial cells cannot perform glycosylation or phosphorylation, and yeast cells cannot carry out complex glycosylation reactions. Often the glycosylation carried out by yeast cells is not authentic. Animal cells not only secrete the products efficiently but also provide excellent glycosylation and phosphorylation.

Animal cells have very different morphology and characteristics from microbial cells. They secrete an extracellular matrix (ECM), which provides a medium for the cells to interact and migrate. Animal cell lines can be anchorage dependent or independent, and require a suitable medium, temperature, pH, and dissolved oxygen concentration to grow. This chapter discusses the structure of the animal cell, the basic procedures associated with animal cell culture, and the important process parameters governing it. The growth environment is highly important for animal cells as changes in the environmental factors can induce growth arrest or apoptosis or even stimulate proliferation or differentiation. Animal cell culture kinetics follows a similar trend as microbial cultures, but, in general, the cell cycle duration for animal cells is much longer than for prokaryotic cells. General animal cell culture kinetics and the cell cycle are reviewed. The intricacies of the animal cell and its shear sensitivity and anchorage dependence make it difficult to scale up bioprocesses involving animal cells. This chapter discusses these problems and a variety of bioreactors designed to meet these requirements for large-scale animal cell culture processes.

Animal Cell

Cell Structure

A variety of cell types, including epithelial cells, fibroblasts, muscle cells, nerve cells, cardiac cells, mesenchymal cells, endocrine cells, and embryonic stem cells, have been successfully cultured and maintained in vitro. These animal cells can vary widely in shape, size, and function depending on their sources. Figure 9.1 shows a typical animal cell with its internal structure and organelles. In general, animal cells are much larger than microbial cells with a diameter between 10 and 30 μm. Though they lack chloroplasts (and therefore are not photosynthetic), they have many specialized cell organelles, such as Golgi bodies, endoplasmic reticulum, and mitochondria, each surrounded by its own plasma membrane. The nucleus is the most prominent organelle in the animal cell and houses chromosomes and other nuclear proteins. Also present in the nucleus are one or more nucleoli, which are involved in ribosome synthesis. The compartmentalization of different organelles facilitates the separation of specific metabolic functions that are incompatible otherwise. Other membrane bound organelles include lysozomes and peroxisomes, which mainly contain digestive enzymes for specific metabolic processes. It is important to note that, apart from chloroplasts, animal cells also lack vacuoles and a rigid cell wall that are present in plant cells. The lack of a cell wall makes the animal cell highly sensitive to shear forces.

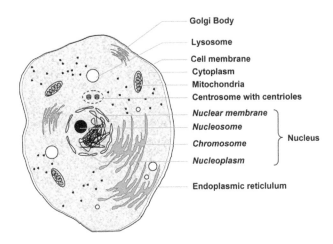

Golgi Body
Lysosome
Cell membrane
Cytoplasm
Mitochondria
Centrosome with centrioles
Nuclear membrane
Nucleosome
Chromosome   } Nucleus
Nucleoplasm
Endoplasmic reticlulum

FIGURE 9.1
The structure of a typical animal cell. (*View this art in color at www.dekker.com.*)

Nonmembranous organelles within the cell include microtubules, microfilaments, and centrioles. The microtubules and microfilaments form a framework called the cytoskeleton that helps the animal cell to maintain its form and keep the organelles in place. The cytoskeleton is also involved in the process of cell division and migration. The larger the cell, the more intricate and elaborate its internal cytoskeletal structure.

Extracellular Matrix

In contrast to prokaryotes, which are unicellular and therefore can live as single entities, animal cells need to socialize—an observation that can be attributed to their origin from higher multicellular organisms. Cells in tissues are usually in contact with a complex network of secreted extracellular proteins referred to as the ECM. This matrix holds the cells together and provides a medium for the cells to interact and migrate.[1] The ECM is secreted mainly by the cells within it and comprises primarily two classes of macromolecules: 1) the glycosaminoglycans, which are linked to proteins in the form of proteoglycans, and 2) the fibrous proteins. Some fibrous proteins are structural proteins such as collagen and elastin, which provide rigidity and elasticity to tissue, respectively, and some are mainly adhesive proteins such as fibronectins and laminins. Fibronectin is a large glycoprotein with binding domains for other ECM proteins and is of primary importance. Mice with their fibronectin gene "knocked out" either die in the embryonic stage or grow up to have multiple morphological defects. Of equal importance are the integrins—transmembrane proteins that interact with the cellular cytoskeleton and thereby anchor the cells to the ECM. They are heterodimers, which mediate bidirectional interaction between the cytoskeleton and the ECM. The fibronectins and the integrins together function to attach the cells to their surroundings. It is because of this interdependence of the integrins and the fibronectins that some animal cells can be cultured only when they have a substratum to grow on. This anchorage dependence of some animal cells helps each cell to attach to the substrate and spread along it to interact with other neighboring cells and thus mimics the functions of a tissue in vitro. Figure 9.2 illustrates how the ECM molecules help cells attach and spread on the surface of the substratum. Also, cells may have different shapes when

(A) Cell attachment and spreading mechanism

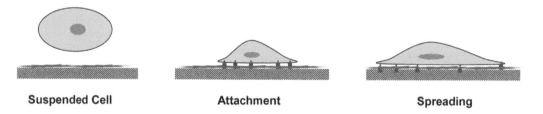

Suspended Cell   **Attachment**   **Spreading**

(B) Cell interactions through ECM proteins

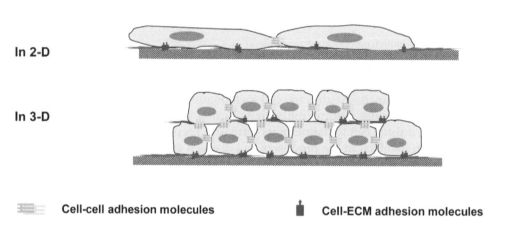

In 2-D

In 3-D

▨▨ **Cell-cell adhesion molecules**   ▮ **Cell-ECM adhesion molecules**

FIGURE 9.2
Cell attachment on surface through cell-ECM interactions. (A) Cell attachment and spreading mechanism.
(B) Cell-cell and cell-substratum interactions in two-dimensional and three-dimensional environments. (*View
this art in color at www.dekker.com.*)

attached on the two-dimensional surface or in a three-dimensional space with cell-cell
interactions.

Transformed and tumorigenic cells are different from normal cell lines in that they are
not usually anchorage dependent. They exhibit a spherical shape, increased life span and
lateral diffusivity of membrane proteins, decreased cell receptors and membrane proteins,
and a different cytoskeletal structure.[2] The decrease in the concentration of the cell adhesion molecules in the cell membrane of these cells causes the anchorage independence.
Transformed cell lines also do not assemble a normal ECM. It is important to note that
some cell lines (e.g., lymphocytes) that are normally anchorage dependent can be induced
and then adapted to become anchorage independent. This is of tremendous importance
to recombinant protein production as discussed later, because the scale-up of suspension
cultures is easier than that of anchorage-dependent cell lines.

Cell Lines

The first stage of tissue culture is the primary cell culture. After isolating a desired piece
of tissue, it is disaggregated either mechanically or enzymatically. The resulting tissue

fragments are then used to inoculate the culture vessel that contains medium. Most normal cells are anchorage dependent. Hence, some of these fragments attach to the vessel wall and migrate out along the surface. Such a culture, before it is first passaged or subcultured, is called a primary culture. As the cells proliferate, they keep spreading out on the culture dish surface until the dish is covered by a single layer of cells. The cells are then said to have reached confluence. Once the cells form this continuous sheet, they stop proliferating because of contact inhibition. Transferring them at low densities (a process termed subculturing or passaging) to new culture vessels that contain fresh medium induces them to resume proliferation. Such a cell population that can continue to grow through many subcultures is called a normal, untransformed cell line.

After about 50 divisions, however, proliferation slows down and the cells show senescence and begin to die. Some of the cells in the culture may undergo some genetic modifications or transformation, which allows them to escape senescence. As long as they are subcultured periodically, they can grow indefinitely. Such cells with an infinite life span are said to have undergone "immortalization" and the cell line is called a transformed cell line. Transformed cells have an enhanced growth rate and may lose anchorage dependence. Tumorigenicity is analogous to transformation but not all transformed cells are tumorigenic or malignant. However, all tumorigenic cells have an infinite life span and enhanced growth rate akin to transformed cells and are mostly anchorage independent.

Table 9.1 shows some animal cell lines that are commonly used in various applications. Creating a stable, permanent cell line is the first critical step in producing recombinant proteins for therapeutic and diagnostic applications. Hybridoma, commonly used in the production of MAb, are generated by fusing antibody-producing spleen cells, which have a limited life span, with cells derived from an immortal tumor of lymphocytes (myeloma). The resulting hybrid is capable of unlimited growth and producing the antibody. Industrial cell lines used for recombinant protein production usually have been genetically engineered to improve the cell in its growth and ability to produce the protein product at a high expression level.

TABLE 9.1

Common Cell Lines and their Application

Cell Line	Origin	Cell Type	Application
BHK	Baby hamster kidney	Fibroblast	Vaccine production
COS	African green monkey	Fibroblast	Transient expression of recombinant
Vero	kidney	Fibroblast	genes
3T3	African green monkey	Fibroblast	Human vaccine production
CHO	kidney	Epithelial	Development of cell culture technique
HeLa	Mouse connective tissue	Epithelial	Recombinant glycoprotein production
MDCK	Chinese hamster ovary	Epithelial	Animal cell model
Namalwa	Human cervical	Lymphoblast	Veterinary vaccine production
NS0	carcinoma	Lymphoblast Lymphoblast	α-Interferon production
MPC-11	(Madin Darby) canine	Hybrid	MAb production
HKB11	kidney	Pluripotent embryonic	Immunoglobulin production
ES-D3	Human lymphoma	stem cell	Recombinant protein production
	Myeloma		Insulin production
	Mouse myeloma		
	Human somatic hybrid		
	Mouse embryo		

(Adapted from Butler.[3])

Culturing Conditions For Animal Cells

Animal cells can be anchorage dependent or independent but all cell lines need nutrients, a suitable temperature, pH, and dissolved oxygen level to grow. The growth environment for animal cells is highly important because environmental stimuli can trigger different responses from the cell. Changes in growth conditions can induce growth arrest or even apoptosis, or may also stimulate proliferation or differentiation. It is therefore crucial to maintain and closely monitor the growth environment so that the cells grow to high cell densities, have stable genotypic and phenotypic expression, and are able to efficiently express recombinant genes if desired. Some important parameters in animal cell culture are discussed here.

Substrate for Cell Attachment

In earlier days, reusable borosilicate glass bottles were used for animal cell cultures. The hydrophilic glass surface was suitable for cell attachment and growth. With the advent of the plastic age, presterilized polystyrene culture apparatus are readily available for cell growth. The polystyrene surface is usually sulfonated to make it hydrophilic and is sterilized by γ-irradiation. They are meant for a "one-time" use and reduce the risk of contamination. Popular forms of culture flasks are T-flasks, Petri dishes, and multiwell plates. Cells grown on the flat or two-dimensional surface such as in a T-flask usually stretch and show a somewhat flattened morphology (Figure 9.3A). Stretching allows cells to migrate on the surface and promotes proliferation.

Microcarriers have also provided a good way to increase the available surface area per unit volume in large-scale bioreactors. Various types of microcarriers have routinely been used for the growth of anchorage-dependent cells.[4] Cylindrical cellulose-based microcarriers (DE-53) were among the first used.[5] Since then, different materials like collagen coated-glass beads, gelatin, DEAE dextran, glass-coated plastic, collagen-coated polystyrene, and polyacrylamide have been used.[6] These beads range from 90 to 330 μm in diameter and can be optimized for different cell lines. Calcium alginate gel beads and new surface modified polystyrene have also been used.[7] Cells attached and grown on microcarriers usually form a monolayer and reach confluence during the culture (Figure 9.3B).

So far, discussions have focused on animal cell growth in monolayers or two dimensions. However, for tissue engineering applications, cell masses need to aggregate into a three-dimensional tissue construct. To aid this process, tissue scaffolds are used as a matrix to guide animal cell growth in three dimensions. In general, a variety of scaffold types have been used. A broad classification would be foam-like scaffolds (e.g., alginate sponge, chitosan, collagen foam, and PLA foam) and fibrous scaffolds like collagen fibers, nonwoven polyethylene terephthalate, and polypropylene mesh. Cells cultivated in the three-dimensional support environment can not only attach to the surface of the substratum but also grow into the three-dimensional space to form aggregates (Figure 9.3C).

Culture Medium

Various types of media have been used to cultivate different cell lines. The choice is mostly empirical, but formulations can be optimized for different cell lines and purposes. Most media, however, have the following essential components: balanced salt solutions (BSS), essential amino acids, glucose, vitamins, buffers, and antibiotics. The BSS provides a

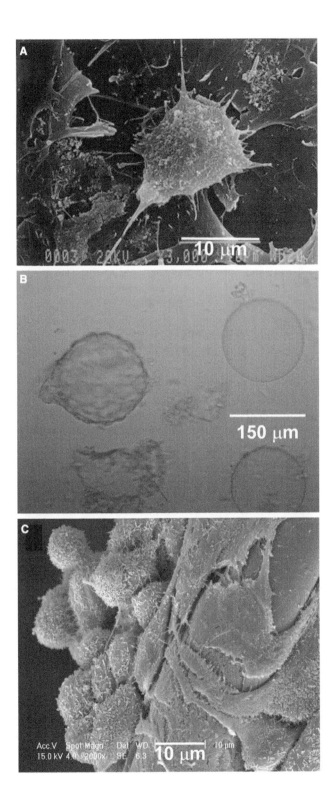

FIGURE 9.3

Different cell morphologies observed for growth on different substrata: (A) cow luteal cells grown on the surface of a T-flask; (B) osteosarcoma cells grown on microcarriers; and (C) osteosarcoma cells grown on a fiber in nonwoven polyester fabrics. (*View this art in color at www.dekker.com.*)

concoction of inorganic salts required by the cells and usually has an osmolality between 260 and 320 mOsm/kg, which is similar in range to that experienced by cells in vivo.[8,9] Balanced salt solution often contains sodium bicarbonate and phosphates, which apart from nutrient value, also act in a buffering capacity.

Glucose is the major carbon and energy source in medium formulation, but its concentration varies in different media. Eagle's minimum essential medium contains 1 g/L of D-glucose, whereas Dulbecco's modification incorporates a higher glucose concentration of 4.5 g/L. Amino acids are important nutrients for cell growth and are additional sources for carbon and energy.[10] Essential amino acids are those that cannot be synthesized by the cell metabolic machinery and, therefore, need to be supplemented through the media. Most cell culture media contain 2–4 mM of L-glutamine.

Buffers are inherent constituents of all media formulations and can maintain the medium pH within an acceptable range.[9] The most commonly used buffer is sodium bicarbonate and CO_2, which is usually provided in air at 5%. Dissolved CO_2 reacts with water to form carbonic acid, which dissociates into the bicarbonate ion [Eq. (1)]:

$$CO_2 + H_2O \leftrightarrow H_2CO_3 \leftrightarrow H^+ + HCO_3^- \tag{1}$$

So, increased amounts of dissolved CO_2 increase the acidity of the medium. This action is countered by the presence of sodium bicarbonate [Eq. (2)]:

$$NaHCO_3 \leftrightarrow Na^+ + HCO_3^- \tag{2}$$

The dissociation of sodium bicarbonate to bicarbonate ions in solution shifts the equilibrium of the reaction in Eq. (1) back, thereby effectively maintaining the pH at 7.4. Good et al. came up with a range of zwitterionic buffers.[11] Such buffers, like N-2-hydroxyethylpiperazine-N'-2-ethanesulfonic acid, have a pK_a of 7.31, which is optimal for cell culture, do not penetrate the cell membrane, and equilibrate with air.[9]

The pH indicator phenol red (phenol sulfonphthalein) is often added to commercially available media. It is pale yellow at pH 6.5, orange at pH 7.0, and red at pH 7.4. It becomes purplish above pH 7.4. Growth media are also often supplemented with antibiotics to promote the growth and propagation of antibiotic resistant strains and also to prevent contamination by micro-organisms. But the presence of antibiotics in the medium does not obviate the use of good aseptic techniques. Nowadays, however, most commercially available media are presterilized as are the polystyrene culture dishes.

Serum

Different types of serum have been used to supplement media with various necessary growth factors and hormones that cells need for their growth. Serum also contains various adhesion factors and antitrypsin activity, which promotes cell attachment. Serum components can act as buffers and as chelators for labile or water insoluble nutrients, bind and neutralize toxins, and provide protease inhibitors. Serum can also reduce oxidative injury to cells caused by ferrous ions.[12] Reduced serum conditions have also been reported to increase the susceptibility of cells to apoptosis.[13]

Various types of serum are fetal bovine serum—the most widely used type, newborn calf serum—which is derived from animals less than 10 days of age, donor calf serum—which is obtained from processed whole blood of calves up to 8 mo old, and horse serum. Human serum and chicken serum have specialized uses for those cells that require a serum derived from similar species, e.g., chicken serum has been repeatedly used for growing various types of avian cells.[9] Serum can be stored safely for over 12 mo at – 20°C and longer storage is possible at – 70°C. Serum can also be heat inactivated (incubation at 56°C for 30 min) to remove toxic compounds or other agents that can interfere with tissue typing assays.

However, as serum contains a wide range of components whose exact concentration is not known, there is a lot of variability from batch to batch. The presence of serum in culture medium is undesirable in cases where a protein or product has to be purified for commercial use. The cost of serum and the possibility of viral contamination also inhibit its use.

Serum-Free Medium

Many attempts have been made to replace serum and add defined amounts of the essential components of serum to form what is known as serum-free medium (SFM). Serum-free medium, generally, consists of a basal medium and additional supplements. The basal medium provides the essential and nonessential amino acids, vitamins, nucleic acids, lipids, inorganic salts, and a carbon source. The additional supplements are: growth promoters such as insulin, insulin-like growth factors, EGF, platelet-derived growth factor, estradiol, and dexamethasone; attachment factors like collagen, fibronectin, and laminin; and transport proteins and detoxifying agents like albumin and transferrin.[14] Ito et al. reported that insulin immobilized on microcarriers promoted growth of anchorage-dependent cells in a protein-free cell culture system.[15] Some serum-free systems have even used the structural heterogeneity of high-density lipoproteins to influence cell proliferation.[16] Serum-free systems are also very useful in reducing downstream processing steps for recombinant protein production using animal cell culture. Unfortunately, the transition to SFM has not been easy. Different cell lines require several growth factors and the specific growth rates of cells are usually slower in SFM. Moreover, SFM can also be expensive.

Temperature and pH

In contrast to micro-organisms, mammalian cells do not show great adaptation to varying temperature or pH ranges. Most cell lines prefer a pH of 7.4. Eagle reported that normal fibroblasts grow within a pH range of 7.4–7.7, while transformed cells prefer a pH between 7.0 and 7.4.[17] Most cell lines grow best at a temperature of 37°C. Cells can tolerate considerably large drops in temperature in that they can be stored cryogenically in liquid nitrogen at –196°C for several months. At temperatures slightly lower than 37°C, the growth rate decreases, but the cellular metabolic activity does not cease totally. Reduction of temperature to 33.5°C resulted in a lowering of the specific growth rate of Chinese hamster ovary (CHO) cells while having no effect on the cell proliferation.[18] This could be attributed in part to the physical state of the lipid bilayer that makes up the plasma membrane. However, cells in general die at temperatures higher than 42–48°C, where the lipid bilayer exhibits a liquid crystal (fluid) behavior.[19] Temperature can also be used as a tool to control recombinant protein production by engineered cell lines. Hendrick et al. (2001) increased the productivity of tissue plasminogen activator under the control of the SV40 promoter in CHO cell, by a shift in temperature from 37°C to 32°C.[20]

Dissolved Oxygen

Oxygen is required for respiration and is thus a key nutrient for animal cell cultures even though requirements vary between cell lines.[21] Because of the low solubility of oxygen in water, oxygen must be provided continuously, usually by aerating the culture medium. Several authors have reported the importance and effects of oxygen in animal cell cultures. It is difficult to conclude a general trend for oxygen dependence vs. cell metabolism. In the case of antibody production using the AB2-143.2 hybridoma cell line, a pO_2 of 50% air saturation was optimum, while the highest immunoglobulin yield from human lymphoblastoid cells (RPMI #7430) was obtained at a low pO_2.[22,23] The trend is opposite for cell biomass production. While more hybridoma cells were obtained at a low pO_2 of 0.5% air saturation, more lymphoblastoid cells were produced at the highest atmospheric pO_2.[22,23] So, each cell line can be studied and grown at its own optimal pO_2 depending on the desired end product. Ma et al. also reported that a low (2%) oxygen tension promoted proliferation while a high (20%) oxygen tension induced differentiation in human trophoblast cells.[24]

Cell Culture Kinetics

The aim of animal cell culture can be either to use the cells as end products or to develop enough biomass (cells) to express a certain target protein of interest in economically viable amounts.

Growth Kinetics

Most animal cell cultures exhibit similar trends in their growth kinetics. While the specific growth rate of each cell line varies, all cell lines do exhibit a characteristic growth curve similar to the one shown in Figure 9.4. Cells go through an initial phase of adjustment, the lag phase, after being either subcultured or dissociated from tissue. After this preliminary lag phase, the cells start growing exponentially and this phase is called the logarithmic phase. Following this phase of active growth, the cell growth rate reduces because of nutrient limitation and product accumulation. The total cell number then ceases to increase, a phenomenon triggered off not only by contact inhibition between cells but also because

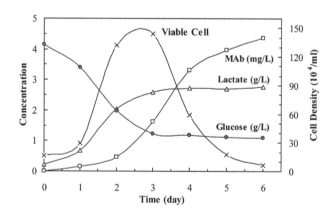

FIGURE 9.4
Typical batch kinetics of a hybridoma culture showing characteristic growth curve and growth-associated production of lactate and nongrowth-associated production of antibodies.

the cell culture reaches a dynamic equilibrium between the rate of cell growth and cell death. The last phase is the death phase, where cell number is reduced because of death caused by either apoptosis or necrosis.

Cell Metabolism

Lactic acid is often a product of glucose metabolism via the glycolytic pathway, while ammonia is produced from the catabolism of glutamine and other amino acids. Cell growth can be inhibited by the accumulation of lactate and ammonia in the culture medium. Production of lactic acid can also lower the medium pH below the physiological range. Reducing or selectively removing toxic metabolites in the culture medium is critical to the efficient production of recombinant proteins by animal cells. The production of primary metabolites such as lactate is usually growth associated, whereas protein expression can be either growth associated or nongrowth associated. Figure 9.4 shows a typical batch hybridoma culture with the production of MAb mainly in the stationary phase.

Cell Cycle

For a cell to grow and reproduce, it must go through the cell cycle shown in Figure 9.5. The two major events in the cell cycle are DNA replication followed by cell division into two new daughter cells. A cell entering the cell cycle first goes through a gap phase (G1). During this phase, it undergoes protein synthesis, which primes the cells for the next phase, the synthesis (S) phase. A second copy of the cellular genome is made during the S phase, thus ensuring that fidelity is maintained when the cell divides. The third phase (G2) is another phase of protein synthesis and it prepares the cell for division, and mitosis finally occurs in the M phase. Essentially, the duration of the cell cycle is an important factor in determining the fraction of dividing cells in a given population. The cell cycle not only controls the rate of cell proliferation and growth, but also may affect protein expression as production of some recombinant proteins by animal cells is cell-cycle dependent.

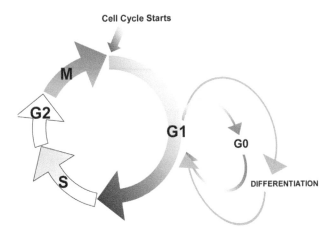

FIGURE 9.5
The cell cycle. A proliferating cell goes through four phases: an initial growth phase (G1), a DNA synthesis phase (S), a second growth phase (G2), and, finally, a mitotic phase (M). During G1 the cell may also choose to differentiate into a new cell type or go into a quiescent state or G0 phase. (*View this art in color at www.dekker. com.*)

Cell lines vary in their cell cycle duration, but all of them follow a basic pattern and are tightly regulated at certain checkpoints. If anything goes wrong, the regulatory machinery promptly causes a cell cycle arrest and eventual cell death. Deprivation of growth factors can cause a cell to exit from a proliferating mode to a quiescent mode (G0). During this phase, the whole cell metabolic machinery is suppressed. Addition of growth factors can again stimulate a cell to re-enter the G1 phase. The cell can also exit the cell cycle to differentiate into a new lineage or head toward programmed cell death or apoptosis.

Apoptosis

Apoptosis or programmed cell death contributes to cell death in in vitro cultures under suboptimal conditions. Nutrient deprivation like glucose, glutamine, serum or oxygen limitation, or mild hydrodynamic stress can induce apoptosis. High levels of apoptosis have been reported following deprivation of glucose and essential amino acids.[25] A number of studies have demonstrated that the suppression of this death pathway, by means of overexpression of survival genes such as *bcl*-2, results in improved cellular robustness and antibody productivity during batch culture.[26]

It is important to know the cell cycle and the points where it can be regulated. This is because cancer basically results when the cell cycle control machinery fails and the cell undergoes rapid proliferation and metastasis. On the other hand, apoptosis is also not desirable in cells growing in bioreactors as it would lower the specific productivity of cells.

There is a plethora of proteins responsible for controlling and regulating the cell cycle and thereby deciding which course the cells should take. They can be broadly divided into the cyclins and cyclin-dependent kinases (Cdks), which interact with each other during regulation. In general, cyclin Ds are associated with the G1 phase, cyclin Es with the transition from G1 to S, cyclin As with the S phase, and both cyclin As and Bs with the transition from G2 to M. The Cdk-cyclin complexes can be inhibited by the Cdk inhibitors, which add an extra level of regulation.

Animal Cell Bioreactors

A variety of bioreactors have been developed and used for animal cell cultures, from simple static T-flasks and roller bottles to more complicated multitray and rotating disk reactors. Because most animal cell lines are anchorage dependent, scale-up for animal cell cultures is usually based on providing the maximum surface area for cell attachment. For this reason, microcarriers have been developed for use in culturing animal cells in conventional stirred-tank and airlift reactors. A recent industrial trend is to adapt animal cells to grow in suspension without needing microcarriers for surface attachment. Immobilized cultures in hollow-fiber, packed-bed, and fluidized-bed reactors also have been used to greatly increase cell density and reactor productivity. Improvements in bioreactor design have focused on increasing oxygen transfer, reducing shear and bubble damages, and increasing cell density by cell recycle or immobilization in perfusion cultures. Figure 9.6 shows different types of animal cell bioreactors.

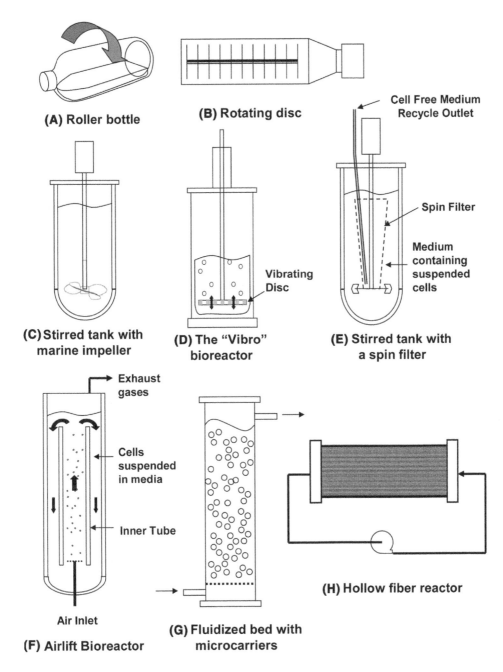

FIGURE 9.6
Various types of animal cell bioreactors: (A) roller bottle; (B) rotating disk; (C) stirred tank with a marine impeller; (D) tank with a pulsating agitator; (E) stirred tank with a spin filter; (F) airlift; (G) fluidized bed; and (H) hollow fiber. (*View this art in color at www.dekker.com.*)

Bioreactors for Suspension Cultures

Bioreactors for suspension cultures are similar to the common microbial fermenters. However, the increased shear sensitivity of animal cells necessitates a few changes. The fermenters for this purpose usually lack baffles and other sharp projections that can cause turbulence.[27] The interior of these fermenters is usually lined with glass or finished to a high grade of smoothness to minimize mechanical damage and to enhance cleanliness. The impeller designs are different too.[28,29] Modified marine and pitched-blade impellers usually cause much less cell damage than conventional disk turbine blades. Nevertheless, cell damage and death caused by mechanical agitation and gas sparging is still a major concern in reactor design and scale-up. For cell lines like hybridoma, which are extremely sensitive to shear stress, novel methods have been developed. The Vibro fermenter (Chemap) uses a plate that vibrates (0.1–3 mm) in the vertical plane to achieve mixing. Airlift bioreactors do away with mechanical agitation and achieve mixing by the process of aeration itself. Airlift bioreactors in general consist of two concentric tubes. The inner tube carries a sintered steel ring or other oxygenation apparatus through which air containing 5% CO_2 is bubbled. Air escapes at the top and the liquid comes down the outer tube. Hybridoma cells have been successfully grown in airlift bioreactors.[30] In general, cell densities in suspension cultures are lower than 10^6–10^7 ml^{-1} because of limited oxygenation at low agitation and aeration rates used to avoid severe cell damages.

Apart from ingenious modifications to the typical bioreactor, rotating wall culture vessels have also been in use (1). But a recent resurgence in their use has been triggered off by the discovery that gravity, or rather the lack of it, plays an important role in the morphology and physiology of tissue constructs.[1,31] The reactor cultures cells in a slowly rotating horizontal cylinder, which produces low shear stress and the continuous rotation keeps the cells always in a state of free fall to simulate microgravity conditions.

Perfusion Cultures

To achieve high cell densities, perfusion cultures have also been used. Perfusion implies the continuous or semicontinuous addition of fresh medium and withdrawal of used medium. This, however, dictates the need for cell separation. This has been achieved by using spin filters, hollow-fiber filters, gravitational settling, or centrifugal separation. Spin filters are usually attached to the stirrer shaft. As they spin, they create a boundary layer effect around them that reduces cell attachment and clogging. Centrifugal methods employ a recycle stream that passes through the centrifuge. Hollow-fiber filters operated under tangential flow allow continuous removal of medium filtrate without significant membrane fouling. Gravitational settlers using inclined tubes allow cells in the outflowing stream to return to the reactor. Bierau et al. used ultrasound to aggregate cells for their fast separation from liquid medium in the settler.[32]

Microcarrier Cultures

As microcarriers provide a good surface area for attachment per unit volume, various types have been routinely used to grow anchorage-dependent cell lines in bioreactors primarily used for suspension cultures. Airlift reactors can also be operated using microcarriers. Wang et al. reported the use of a fluidized-bed bioreactor in which cells were attached to Cytoline-1 macroporous microcarriers.[33] The upward flow of the medium fluidized the beads or carriers and provided a unique perfused system that gave higher erythropoietin (EPO) production than a conventional stirred-tank bioreactor. Fluidized-bed bioreactors have also been used for suspension cultures of hybridoma.[34] Chong et al. used microcarriers to form a packed-bed reactor to grow CHO cells and report cell density as high as 2×10^7 cells/ml.[35]

However, cells grown on solid microcarriers are often subjected to fluid mechanical damages caused by small turbulent eddies as well as by collision between microcarriers and against the impellers and other bioreactor parts.[36] The development of macroporous microcarriers has largely solved this problem of shear damage to cells. As the name suggests, macroporous microcarriers have a network of pores within them, which not only present a larger surface area for cell attachment but also protect the cells from shear damage. Macroporous microcarriers are mostly made of gelatin, collagen, cellulose, polystyrene, and polyethylene, and the cell distribution in them can be studied by confocal laser scanning microscopy.[37,38]

Hollow-Fiber Reactors

Hollow-fiber reactors are widely used in the production of MAb and can reach cell densities higher than 2×10^8 cells/ml of the fiber volume.[39,40] With continuous perfusion and intermittent harvesting, hollow-fiber reactors can give high reactor productivity and produce antibodies at a high concentration comparable to or even higher than that in mouse ascites.[41] The gradients of metabolites and nutrients created along the axis of the hollow-fiber reactor are undesirable. This problem can be overcome with the radial flow hollow-fiber reactor, which consists of a central flow distributor tube surrounded by an annular bed of hollow fibers.[42] The central flow distributor ensures an axially uniform radial convective flow of nutrients across the fiber bed. In this reactor, the cells grow on the outer side of the fibers. However, in conventional hollow-fiber reactors, the cells could be on either side depending on the inoculation method used and the way the process is carried out. Hollow-fiber reactors have also been used for perfusion cultures. A practical problem with these systems is that the cell density cannot be directly monitored. Hollow-fiber reactors are also expensive, and their uses are limited to small-to-medium scale production of antibodies.

Fibrous-Bed Bioreactors

A promising new technology has been the immobilization of animal cells to fibrous beds rather than microcarriers. The CelliGen Plus® bioreactor uses polyester fabric disks packed in a basket inside a stirred tank for immobilizing animal cells. The reactor productivity was reported to be as high as 12-fold of that in static and stirred suspension culture systems for antibody production.[43] Chen et. al. also developed a fibrous-bed bioreactor to successfully grow osteosarcoma cells to cell density as high as 3×10^8 ml^{-1} with 90% cell viability for as long as 4 mo.[44] The three-dimensional structure provided by the fibrous matrix has been shown to have profound effects on cell growth and protein production.[25,45] This three-dimensional culturing method provides a new technique for the scale-up of animal cell culture.

Industrial Applications

Animal cell cultures are increasingly being used for the production of recombinant glycoproteins, viral vaccines, and MAb in the biotechnology industry. They are far superior to yeast and bacterial cells in carrying out the complex post-translational modifications that major recombinant protein products require. Not only can animal cells provide better glycosylation and phosphorylation of complex proteins, but they can also carry out authentic proteolytic cleavage, subunit association, and chemical derivatization. However, large-scale animal cell cultures are more difficult to commercialize than microbial cultures because the growth rate of animal cell cultures is much slower, the nutrient requirements are more complex, and the growth conditions more stringent. In addition, there are several parameters that need to be considered. Not only is the required inoculum size large ($\sim 10^5$ ml^{-1} or 1–5 × 10^4 cm^{-2}), but the cell proliferation rate is also much slower. The productivity of the target protein is also in the milligram per liter range as compared to the higher production (often in g/L) in microbial cells. The medium is usually more expensive and the cells are highly sensitive to toxic metabolites and shear. In the case of suspension cultures, the scale-up is relatively easy as a range of fermentation equipment developed for microbial cultures can be modified to adapt to animal cell cultures. However, anchorage-dependent cells require a large surface area per unit volume and are therefore more difficult to scale up.

In spite of these hurdles, the last two decades have seen an immense leap in animal cell culture technology both at the laboratory scale as well as the industrial scale. A variety of bioreactors and instrumentation have been ingeniously been devised for the scale up and process control of animal cell cultures. Serum-free media development has considerably reduced the downstream processing costs in the recombinant protein production and purification process. The capability to induce some cell lines to lose anchorage dependence has also been an important breakthrough.

Table 9.2 lists various types of biopharmaceutical products from animal cell cultures. Viral vaccines are usually produced by first culturing the host cells (e.g., MRC-5 and WI-38) to form a cell layer on the surface of substratum. Seed virus is then added and incubated for about 3 weeks for replication in the host cells without killing them. After washing to remove the medium components, the cells are lysed to release the virions for harvesting and purification. The inactivated viral vaccine is produced by inactivation with formaldehyde and adsorption onto aluminum hydroxide adjuvant.

Roller bottle reactors have been widely used in the past and can generate cell densities upto 5.4 × 10^6 cells/ml.[46] However, roller bottles are difficult to scale up and cannot meet the growing demand for therapeutic recombinant proteins. Their popularity is on the decline and are largely replaced by microcarriers, and stirred-tank or airlift bioreactors in

TABLE 9.2

Some Important Biopharmaceutical Products from Animal Cell Cultures

Type	Examples
Vaccines	Polio, hepatitis A, measles, mumps, rubella, yellow fever, rabies, and influenza
Glycoproteins	Interferons
	Blood clotting factors (factors VIII, IX)
	Glycoprotein hormones, EPO
	Plasminogen activators, t-PA
MAb	Diagnostics
	Therapeutics—prevention of respiratory syncytial virus (RSV) infection, treatments of inflammation, breast cancer, non-Hodgkin's lymphoma, and treatment or prevention of transplant rejection
Hormones	Human growth hormone, insulin, calcitonin, and parathyroid hormone
Growth factors	Nerve growth factor and EGF
Proteases	Urokinase

process scale-up. Initially, industrial production of EPO by CHO cells is carried out in hundreds of roller bottles in incubation rooms equipped with robots for medium changes and product harvesting. The newer production plant for second-generation EPO employs state-of-the-art bioreactors and has three times the production capacity of the old EPO plant.

Monoclonal antibodies have been widely used in biomedical research and in diagnostics. Because MAb bind to specific cell surface receptors, they can be used for treatments of transplant rejection, cancer, autoimmune and inflammatory diseases, and infectious diseases. Currently, MAb products comprise about 25% of all biotech drugs in clinical development. Commercial MAb production uses two methods: 1) in vivo cultivation in mouse or rabbit ascites and 2) in vitro cell culture in tissue flasks or bioreactors. For mass production of therapeutic antibodies, the latter method is used. For example, recombinant paliviumab is produced by culturing murine myeloma cells (NS0) in a stirred-tank fed-batch bioreactor. The manufacturing process starts with a vial containing about 10 million frozen cells, which are cultured in T-flask and then in spinner flask to expand the number of cells. These cells are subsequently used in the larger-scale bioreactor process with the culture volume increased incrementally to the final volume of 10,000 L. After inoculation of the production bioreactor, the fermentation takes about 20 days to reach the final titer of the MAb product, which is about 1 g of MAb per liter of the culture medium. The high titer of MAb production in this system was accomplished through extensive research and development works on cell line improvement, medium optimization, and process optimization and control.

Conclusions

Animal cells in culture have been widely used to study the physiology, metabolism, and biochemistry of cells; test the effect of compounds on different cell types; in tissue engineering applications; and for the production of recombinant glycoproteins, viral vaccines, and MAb. Though several challenges remain to be overcome, the future of developing more products using animal cell cultures is very bright.

References

1. Alberts, B., Bray, D., Lewis, J., Raff, M., Roberts, K., Watson, J.D. *Molecular Biology of the Cell*, 3rd Ed.; Garland Publishing Inc.: New York, 1994.
2. Prokop, A. Implications of cell biology in animal cell biotechnology. In *Animal Cell Bioreactors*, Ho, C.S., Wang, D.I.C., Ed.; Butterworth-Heinemann: Stoneham, MA, 1991; 21–58.
3. Butler, M.J. *Animal Cell Culture and Technology: The Basics*, Oxford University Press Inc.: New York, 1996.
4. Chu, L., Robinson, D.K. Industrial choices for protein production by large-scale cell culture. *Curr. Opin. Biotechnol.* **2001**, *12*, 180–187.
5. Lazar, A., Silberstein, L., Reuveny, S., Mizrahi, A. Microcarriers as a culturing system of insect cells and insect viruses. *Dev. Biol. Stand.* **1987**, *66*, 315–323.
6. Griffiths, J.B. Scaling up of animal cell cultures. In *Animal Cell Culture: A Practical Approach*, Freshney, R.I., Ed.; IRL Press at Oxford University Press: Oxford, **1992**; 47–93.

7. Zuehlke, A., Roeder, B., Widdecke, H., Klein, J. Synthesis and application of new microcarriers for animal cell culture. Part I: design of polystyrene based microcarriers. *J. Biomater. Sci.* **1993**, *5* (1–2), 65–78.

8. Freshney, R.I. *Culture of Animal Cells—A Manual of Basic Technique*, 4th Ed., Wiley-Liss Inc.: New York, 2000.

9. Harrison, M.A., Rae, I.F. *General Techniques of Cell Culture*; Cambridge University Press: Cambridge, 1997.

10. Butler, M., Christie, A. Adaptation of mammalian cells to non-ammoniagenic media. *Cytotechnology.* **1994**, *15*, 87–94.

11. Good, N.E., Winget, G.D., Winter, W., Connolly, T.N., Izawa, S., Singh, R.M.M. Hydrogen ion buffers and biological research. *Biochemistry.* **1966**, *5*, 467–477.

12. Song, J.H., Harris, M.S., Shin, S.H. Effects of fetal bovine serum on ferrous ion-induced oxidative stress in pheochromocytoma (PC12) cells. *Neurochem. Res.* **2001**, *26* (4), 407–413.

13. Geng, Y., D'Souza, S., Xin, H., Walter, S., Choubey, D. p202 levels are negatively regulated by serum growth factors. *Cell Growth Differ.* **2000**, *11* (9), 475–483.

14. Keenan, J., Meleady, P., Clynes, M. Serum free media. In *Animal Cell Culture Techniques*; Clynes, M., Ed.; Springer: Berlin, **1998**; 54–56.

15. Ito, Y., Uno, T., Liu, S.Q., Imanishi, Y. Cell growth on immobilized cell growth factor. *Biotechnol. Bioeng.* **1992**, *40*, 1271–1276.

16. Chen, J., LaBrake, S., McClure, D. Functional roles of high density lipoproteins in serum-free medium. In *Growth and Differentiation of Cells in Defined Environment, Proceedings of the International Symposium, 1984*, Murakami, H., Ed.; Kodansha: Tokyo, Japan, 1985.

17. Eagle, H. The effect of environmental pH on the growth of normal and malignant cells. *J. Cell Physiol.* **1973**, *82*, 1–8.

18. Ducommun, P., Rueux, P.-A., Kodouri, A., Von Stockar, U., Marison, I.W. Monitoring of temporary effects on animal cell metabolism in a packed bed process. *Biotechnol. Bioeng.* **2002**, *77*, 838–842.

19. Mamdouh, Z., Giocondi, M.-C., Laprade, R., Le Grimelle, C. Temporary dependence of endocytosis in renal epithelial cells in culture. Biochim. *Biophys. Acta Biomembr.* **1996**, *1282*, 171–173.

20. Hendrick, V., Winnepenninckx, P., Abdelkafi, C., Vandeputte, O., Cherlet, M., Marique, T., Renemann, G., Loa, A., Kretzmer, G., Werenne, J. Increased productivity of recombinant tissue plasminogen activator (tPA) by butyrate and shift of temperature: a cell cycle analysis. *Cytotechnology.* **2001**, *36*, 71–83.

21. Spier, R.E., Griffiths, B. An examination of the data and concepts germane to the oxygenation of cultured animal cells. In *Proceedings of the 5th Meeting of ESACT, Copenhagen, Denmark, 1982*, Karger: *Basel*, **1982**; 81–92.

22. Miller, W.M., Wilke, C.R., Blanch, H.W. Effects of dissolved oxygen concentration on hybridoma growth and metabolism in continuous culture. *J. Cell Physiol.* **1987**, *132*, 524–530.

23. Mizrahi, A. Oxygen in human lymphoblastoid cell line cultures and effect of polymers in agitated and aerated cultures. In *Proceedings of the 5th Meeting of the ESACT; Copenhagen, Denmark, 1982.* Karger: *Basel*, **1982**; 93–102.

24. Ma, T., Yang, S.-T., Kniss, D.A. Oxygen tension influences proliferation and differentiation in a tissue engineered model of placental trophoblast-like cells. *Tissue Eng.* **2001**, *7*, 495–506.

25. Luo, J. Three Dimensional Culturing of Animal Cells in Fibrous Bed Bioreactor. Ph.D. Thesis, The Ohio State University, 2002.

26. Fassnacht, D., Rössing, S., Singh, R.P., Al-Rubeai, M., Pörtner, R. Influence of bcl-2 on antibody productivity in high cell density perfusion cultures of hybridoma. *Cytotechnology.* **1999**, *30*, 95–106.

27. Griffiths, B.J. Scale up of suspension and anchorage dependent animal cells. In *Methods in Molecular Biology, Animal Cell Culture*; Pollard, J.W., Walker, J.M., Ed.; Humana Press: Clifton, NJ, **1989**; Vol. 5.

28. Kamen, A.A., Tom, R.L., Caron, A.W., Chavarie, C., Massie, B., Archambault, J. Culture of insect cells in a helical ribbon impeller bioreactor. *Biotechnol. Bioeng.* **1991**, *38*, 619–628.

29. Shi, Y., Ryu, D.D.Y., Park, S.H. Performance of mammalian cell culture bioreactor with a new impeller design. Biotechnol. *Bioeng.* **1992**, *40*, 260–270.
30. Huelscher, M., Onken, U. Influence of bovine serum albumin on the growth of hybridoma cells in airlift loop reactors using serum-free medium. *Biotechnol. Lett.* **1988**, *10*, 689–694.
31. Freed, L., Langer, R., Martin, I., Pellis, N.R., Vunjak-Novakovic, G. Tissue engineering of cartilage in space. *Proc. Natl. Acad. Sci. USA.* **1997**, *94*, 13885–13890.
32. Bierau, H., Perani, A., Al-Rubeai, M., Emery, A.N. A comparison of intensive cell culture bioreactors operating with hybridomas modified for inhibited apoptotic response. *J. Biotechnol.* **1998**, *62*, 195–207.
33. Wang, M.-D., Yang, M., Huzel, N., Butler, M. Erythropoietin production from CHO cells grown by continuous culture in a fluidized bed bioreactor. *Biotechnol. Bioeng.* **2002**, *77*, 194–203.
34. Ray, N.G., Tung, A.S., Ozturk, S.S., Pang, R.H.L. Animal cell culture using fluidized-bed culture technology. *Food Bioprod. Process.* **1993**, *71* (C2), 124–126.
35. Chong, C., Chang, Y., Deng, J., Xiao, C., Su, Z. A novel scale up method for mammalian cell culture in packed-bed bioreactor. *Biotechnol. Lett.* **2001**, *23*, 881–885.
36. Janowski, T., Grajek, W. Mechanical damage of animal cells in bioreactors. Biotechnologia. **1993**, *3*, 166–183.
37. Shiragami, N., Honda, H., Unno, H. Anchorage-dependent animal cell culture by using a porous micro-carrier. *Bioprocess. Eng.* **1993**, *8* (5–6), 295–299.
38. Bancel, S., Hu, W.-S. Confocal laser scanning microscopic examination of cell distribution in macroporous microcarriers. *Biotechnol. Prog.* **1996**, *12*, 398–402.
39. Gloeckner, H., Lemke, H.-D. New miniaturized hollow fiber bioreactor for in vivo like cell culture, cell expansion, and production of cell derived products. *Biotechnol. Prog.* **2001**, *17*, 828–831.
40. Patankar, D., Oolman, T. Wall-growth hollow-fiber reactor for tissue culture: I. Preliminary experiments. *Biotechnol. Bioeng.* **1990**, *36*, 97–103.
41. Gramer, M.J., Britton, T.L. Antibody production by a hybridoma cell line at high cell density is limited by two independent mechanisms. *Biotechnol. Bioeng.* **2002**, *79*, 277–283.
42. Tharakan, J.P., Chau, P.C. A radial flow hollow fiber bioreactor for the large-scale culture of mammalian cells. *Biotechnol. Bioeng.* **1986**, *28*, 329–342.
43. Wang, G., Zhang, W., Jacklin, C., Freedman, D., Eppstein, L., Kadouri, A. Modified Celligen-packed bed bioreactors for hybridoma cell cultures. *Cytotechnology.* **1992**, *9*, 41–49.
44. Chen, C., Huang, Y.L., Yang, S.-T. A fibrous bed bioreactor for continuous production of developmental endothelial Locus-1 by osteosarcoma cells. *J. Biotechnol.* **2002**, *97*, 23–29.
45. Chen, C., Huang, Y.L., Yang, S.-T. Effects of three-dimensional culturing on osteosarcoma cells grown in a fibrous matrix: analyses of cell morphology, cell cycle, and apoptosis. *Biotechnol. Prog.* **2003**, *19*, 1574–1582.
46. Berson, R.E., Pieczynski, W.J., Svihla, C.K., Hanley, T.R. Enhanced mixing and mass transfer in a recirculation loop results in high cell densities in a roller bottle reactor. *Biotechnol. Prog.* **2002**, *18*, 72–77.

10

Tissue Engineering

Shang-Tian Yang and Clayt Robinson

CONTENTS

Introduction

Tissue engineering is a combination of biology and engineering for producing biological substitute structures that can reconstitute cellular or tissue function, which is lost, declining, or insufficient. Theorized for a long time, the field has evolved greatly within recent decades to meet the challenges of disease and a limited organ donor supply. For tissue substitutes to be successful in replacing lost function, they must accurately and reliably perform the desired function that is dependent on a multitude of variables. Parameters that must be carefully investigated include the cell source, the polymeric scaffold support, the tissue culture protocol, and the implantation procedure. Tissue engineering research covers a wide range of applications including many tissue substitutes, cell therapy, and diagnostic modeling. Although faced with as yet unmet challenges, tissue engineering stands to provide momentous advancements to health in the 21st century.

First coined in 1987, tissue engineering, as a field, is defined by two main objectives. The first objective is to apply methods and principles of engineering and life sciences to understand the biological tissue construct.[1] In addition to deciphering the vast catalog of cellular functions, the interconnections of a cellular population, regulated by signaling pathways, and the interactions of a cellular composite with the noncellular surrounding environment must be understood. Since the three-dimensional structure and organization of tissue components are integral in defining proper function of the tissue, this environmental arrangement must be mimicked to produce a tissue substitute that functions appropriately. A fundamental understanding of a tissue environment structure and how it enables or affects the tissue function is essential. The second objective is to apply established knowledge of the tissue construct toward developing biological substitutes to restore, maintain, or improve natural function to tissues or organs that are structurally or physiologically altered.[1]

Included in this chapter is a review of the development of tissue engineering, from theorized concepts and early experiments through advancements made in developing tissue substitutes in the recent past. A review of biological systems and components precedes a thorough discussion of the components utilized in tissue engineering constructs. The chapter concludes with tissue engineering applications and the challenges that remain for full realization of the field's potential.

Motivation

Motivation for further development in the field of tissue engineering is to save the lives of patients suffering from organ failure or loss, who are waiting for donor organs to become available, and those suffering from debilitating diseases, such as Parkinson's disease, in which essential cell or tissue function deteriorates or is lost over time. Further, people are victims to catastrophic events in which tissue repair or replacement is a grave concern with, commonly, no immediate remedy.

As a result, the annual health care costs in the United States alone exceed $400 billion for patients with tissue loss or organ failure. The cost includes 8 million surgical procedures and 40–90 million hospital days annually.[2] In 2002, the waiting list for organ recipients exceeded 81,700 candidates, with over 38,600 people added to the list. However, only 12,800 organs were transplanted in 2002 owing to a severe shortage of donors.[3] With a limited organ supply, accumulating costs, and limited medical procedures available, there is a void to fill.

From Fiction to Reality

There are several examples of tissue engineering well before the field was named in 1987. Stories that include the replacement of a person's body parts with those from another person are found in literature centuries old. Possibly, the first tale of a tissue engineering procedure is found in the Bible, when Eve is given life from the rib of Adam. In 1818, Mary Shelley wrote *Frankenstein* in which the title character is given life through the compilation of body parts garnered from donor corpses. In the legend of St. Cosmas and St. Damien, about 200 A.D., the two physicians performed a procedure in which the gangrenous leg of a man was successfully replaced with the leg of a recently deceased man.[4] More recently, Hollywood has produced films that portray ideas from tissue engineering. In 1991, the aptly titled *Body Parts* was released, in which a man loses his arm in an accident and receives a transplant from a recently executed inmate. Although successfully transplanted, incompatibility becomes an issue as the transplanted arm retains the murderous personalities of its donor.

The collection of stories that involve tissue engineering concepts shows the promise and spectacular possibilities that the future could bring. However, rarely do these stories fully conceptualize or even mention the challenges involved in performing these acts in reality. Until the advancement of tissue engineering in the 1970s, replacements for bodily tissues were prostheses made of wood, ceramics, and plastics. Replaced body parts included arms, legs, eyes, ears, teeth, and noses.

Experimentation with animals became more common for studying the growth of tissue. In the 1930s, the work done by Bisceglie became one of the earliest documented tissue engineering procedures.[2] Two important concepts were displayed including the transplantation of mouse tumor cells into the abdomen of a pig, while the cells were encased within a polymer membrane structure. First, the results showed that cells could survive within a foreign environment without rejection by the host immune system. Second, the procedure introduced the encapsulation approach. The semi-permeable polymer membrane system allowed for nutrient and waste fluxes into and out of the membrane system, respectively. Concurrently, the membrane selectively rejected the passage of immune system molecules and proteins, thus protecting the enclosed cells. In the work performed by Chick and coresearchers, islet cells were encapsulated within a semi-permeable membrane and transplanted into animal models to provide glucose level control as a cure for diabetes.[2]

In the laboratory setting, the first tissue to be reconstituted was skin because of its relatively simple two-dimensional structure. During the late 1970s and 1980s, artificial skin was created using skin cells distributed within natural collagen or collagen–glycosaminoglycan composite support structures. The growth of tissue engineering in the 1990s and early 2000s is due to interdisciplinary advancements in the fields of engineering, genomics, proteomics, cell biology, and material science.[5] As understanding of the important relationship between tissue structure and function became fully realized, three-dimensional synthetic polymers were utilized to mimic the bodily in vivo environment as a support for cells to attach and grow on. The liver was the first tissue to be cultivated using these three-dimensional constructs owing to its relatively simple composition. Utilizing these newly developed tools, a multitude of tissues have been or are being studied and mimicked with tissue-engineered products for possible future usage in a wide array of applications that is discussed later in this chapter.

Cellular Systems Biology

For a tissue substitute to function properly, many biological aspects of the tissue and component cells must be understood. Some aspects include the extracellular matrix, cell-specific gene expression and surface markers, cell growth parameters, population arrangement and behavior, and the immune system.

Tissues and Organs

An organ is a component of the human body system made of one or more tissue types and has specific jobs within the complex network. Tissue is a collection of similar cells and the surrounding supportive environment that together perform specific tasks. There are four basic categories of tissues: epithelium, which constitutes surfaces such as skin, connective tissue, muscle tissue, and nerve tissue.[4]

Extracellular Matrix

Tissue in vivo consists of cells that are engaged in an environment, called the extracellular matrix (ECM), that provides support to allow for proper cell function, cell–scaffold interactions, and tissue morphology. The ECM directly affects or controls cell shape, function, viability, and population structure. The ECM that supports the tissue structure and cells in vivo is a complex network of collagens, glycoproteins, such as fibronectin and laminin, hyaluronic acid, proteoglycans, glycosaminoglycans, and elastins to which cells adhere and interact. The three-dimensional ECM has a two-way interaction with the cells. The ECM surface properties and molecules provide cell-surface receptor-mediated signals to influence cellular spatial organization, migration, growth, differentiation, and death.[6] Important cell receptors that interact with the ECM include integrin and cadherin adhesion receptors.[7] The cells influence the ECM by remodeling the structure and secreting new ECM components.[8]

Genes and Proteins

The genes on chromosomes within each cell are the blueprints of the human body. The pattern in which the genes are expressed determines the cell type and behavior. Gene expression, which leads to protein production, is a highly controlled process dependent on signals originating from other cellular components within the same cell, externally from other cells within the immediate environment, or from distant locations, such as part of the endocrine system. External signals are transduced, or passed, with the aid of surface molecules that are receptors for the signal molecules. Other molecules and proteins are expressed to perform cell-specific functions within the cell, or externally following secretion. The presence of these surface receptors or expressed functional proteins is characteristic of the cell type and can be used as cell markers for identifying proper cell function.

Cell Growth

As cells grow, or proliferate, they proceed through a highly controlled process called the cell cycle. The cell cycle consists of four phases during which the DNA set is accurately replicated and subsequently divided into two complete sets partitioned into two new cells derived from dividing the original cell via mitosis. Regulatory proteins guard progression between phases to ensure cellular readiness and DNA integrity preservation. Errors in this regulatory process can lead to DNA mutations and uncontrolled growth, both characteristics of cancer. With specific signals present, cells exit the cell cycle for a maturation process called differentiation before re-entering the cell cycle. As cells differentiate, their function becomes more defined and limited.

Morphology and Arrangement

Cellular morphology and population arrangement within the tissue culture environment are dependent on the support surface properties. Cells that are highly proliferating appear round and smooth. As cells adhere to a surface, spreading occurs across the surface, and the cells flatten. As population of cells increases, interaction and rearrangement occur among the cells. In a two-dimensional environment, interactions are limited, but in a three-dimensional environment, cellular aggregates can form within the structure of the supporting material. Aggregation and interaction among cells in all directions mimic the body environment and allows for similar cellular function.

Immune System

The immune system defends the body from infection and illness using cellular and molecular components to detect and clear objects from the body that are not identified as normal. When objects such as invading viruses, micro-organisms, implanted materials, and mutated cancerous cells are targeted by the immune system, the resulting immune rejection mechanism kills, disrupts, or encloses the foreign object to prevent further harm to the body. Cells are screened by immune system antibody molecules or T-cells to verify whether the surface molecules are native or foreign. A key determinant of whether a cell is native or foreign is the major histocompatibility complex genes contained within the cell and expressed on the cell surface. Therefore, for cells to be compatible and to avoid immune rejection in a new host, the similarity of the major histocompatibility complex genes must be high. This is important in determining the success of a tissue-engineered product within the host.

Tissue Engineering Construct Components

As summarized in Figure 10.1, the components that must be customized based on the application include cell source, scaffold parameters, and cell culture procedures.

Cell Sources

The cells utilized for producing a tissue substitute come from multiple sources. When a specific cell type is needed to culture a certain tissue type, the availability of the cell type, the means of obtaining the cells, procedures for maintaining and multiplying the cell population in culture, and immune rejection upon transplantation must all be considered. Avoiding rejection is a huge challenge in the development of a tissue substitute, thus autologous cells obtained from the same person to whom the transplant will be given are ideal as the host body does not reject autologous cells. However, supply of autologous cells is frequently the problem. For example, in many situations, a large enough population of healthy cells is unavailable due to the extent of disease.[9] Further, the harvesting of cells from one location may cause long-term harm to the donor site. Once obtained, autologous cells are cultured, or grown, in a laboratory environment, or in vitro in the presence of a proper support structure and nutrient supply until a larger population is present. The autologous cells are then administered to the donor patient at the necessary site.

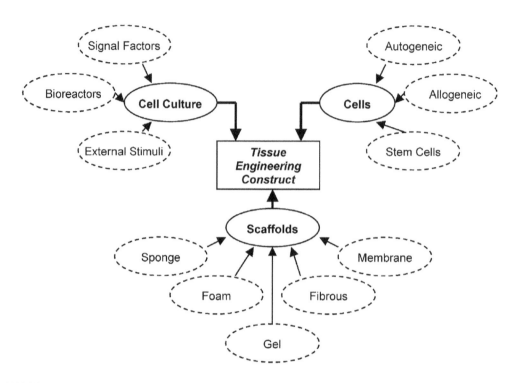

FIGURE 10.1
Overview of the components utilized in tissue engineering constructs.

When autologous cells are not appropriate or available, the cells can come from either another donor of the same species or a different species. The transplantation of cells between similar or dissimilar species is called allotransplantation or xenotransplantation, respectively. Although the supply of these types of grafts is plentiful, immune rejection is very common, so some additional strategy must be utilized to avoid rejection. Cells provided by family members, or allogeneic cells, tend to be the most compatible source in order to avoid rejection due to similar gene sets, including, specifically, the histocompatibility complex genes. Xenografts provide an additional challenge since xenogeneic cells may contain components that are infectious when introduced into a human. This fact, in addition to ethical and moral issues regarding utilizing nonhuman parts in a human, has led to general unpopularity of this procedure.[9]

A fourth cell source is stem cells. Stem cells are characterized by the ability to proliferate indefinitely and develop into different cell types, or pluripotent, depending on the stem cell origin and given the appropriate signals. Embryonic stem cells are present as an embryo first begins to develop and differentiate to form all components of the human body. These stem cells allow for generation of any tissue cell type, however, ethics and regulations limit their usage. An adult retains a limited supply of adult stem cells in the bone marrow and in tissues throughout the body. Most of these progenitor cells are partially differentiated into a lineage of cell types, but remain multipotent to develop into a more limited range of cell types. Examples are neural stem cells, hematopoietic stem cells, and mesenchymal stem cells.

Owing to the limited supply of stem cells in the body, specialized techniques are necessary to find and separate them from the mixture. Stem cells may be isolated from an extracted tissue mass by digesting the ECM structure surrounding the cells, and then detecting the cells based on signature biomolecular expression profiles as a "fingerprint" for stem cells. Once identified, the stem cells can be separated from the total tissue population by passing the cells through a system that selectively removes the stem cells while allowing the remainder of cells to exit the system separately. Selective retention technology systems extract stem cells from the mixed population by customizing the system to have a high affinity for expressed stem cell surface molecules or to secondary molecules previously bound selectively to the cell surface. Example stem cell surface markers are CD34 and CD45 for hematopoietic stem cells and Oct-4 and SSEA-3 for embryonic stem cells.

Once harvested, stem cells can be cultured in specific biological, chemical, and physical stimuli to differentiate into the cell type of interest. Upon expansion in culture to a large enough population, the cells may be transplanted as therapy. However, strategies to avoid rejection are necessary unless the cells are autologous or cultured in such a way to disguise the fact that the cells are from a different source.[9] The difficulty lies in providing the correct composition, amounts, and timing of the stimuli to direct the differentiation to the desired cell type. A current approach is to direct the transformation of the stem cells in vitro until the cells are within one or two steps of the complete differentiation destination. The final transformation steps are accomplished in vivo, after transplantation to the site of the desired cell type where signals are provided to complete the differentiation.[10]

One concern with stem cell usage is that transplanted cells derived from stem cells can be tumorigenic owing to undifferentiated stem cells present in the population that proliferate uncontrollably.[9] The use of stem cells, however, is promising because of the limited cell supply for many tissues. Cells that are not autogeneous must be able to avoid immune rejection. Somatic cell and nuclear transfer procedures could provide cells that function appropriately and yet will not be rejected.[9] The production of a universal donor cell source that can be used in any patient while avoiding immune rejection is a future goal. This probably involves altering the expression of or hiding the histocompatibility complex surface markers. Different types of stem cells are also being investigated as a source for multiple cell types if the differentiation can be precisely directed and controlled.

Scaffolds

Studies performed in vitro with cells growing on a two-dimensional surface have observed isolated cell function performance, such as proliferation, glycolysis, respiration, and gene expression, by optimizing the media nutrient, hormone, and growth factor compositions. However, the proper regulation and control of these functions are dependent on cellular interactions present within a three-dimensional structure.[1] Therefore, scaffolds are essential for creating tissue substitutes that mimic in vivo function.

Scaffolds can be foams, sponges, gels, membranes, or fibrous materials (Figure 10.2). They are categorized as natural, synthetic, or a combination of both. Table 10.1 provides a list of scaffold materials and applications utilized in tissue engineering.[11–14] Natural biomaterials, such as collagen, are inherently equipped for cell interaction, but have the disadvantages of limited adaptability and customizable processing as well as relatively scarce availability compared to synthetic biomaterials.[15] The most commonly utilized synthetic biodegradable materials are poly(glycolic acid) (PGA), poly(lactic acid) (PLA), and poly(lactic co-glycolic acid) (PLGA), a blend of the former two polymers. Synthetic polymers can be used for the culture of many cell types, but it remains difficult to culture some cell types, such as nerve cells, on synthetic polymers. Advancement in three-dimensional

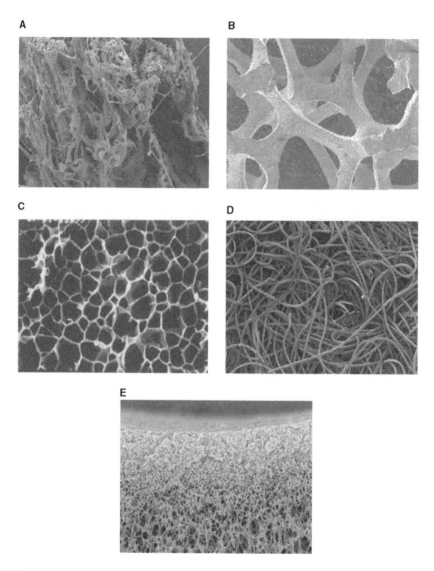

FIGURE 10.2
Five scaffold types: (A) sponge; (B) foam; (C) gel; (D) fibrous; and (E) membrane.

polymer processing makes customization possible for polymer composition, mechanical strength, cell-surface attachment interactions, degradation rates, and high cell density.[15]

Scaffold properties influence a plurality of cell culture aspects including proliferation, differentiation, adhesion, migration, gene expression, and function. Characteristics on three size scales influence these aspects. On the macroscopic scale, the scaffold is conformed to a specific shape and size to direct the formation of a three-dimensional structure. For example, the scaffold utilized for blood vessel regeneration would be tubular in shape in order to direct cell growth and tissue morphology accordingly. A three-dimensional matrix has a high surface area to volume ratio that allows for high-density cell populations and sufficient space for nutrient transfer. The mechanical strength of the scaffold may be

TABLE 10.1

Commonly Utilized Scaffold Materials

Polymer	Application	References
Natural		
Type I collagen	Skin, bone, cartilage, tendon, nerve, kidney, cornea, vessels	[7,11,12]
Alginate	Cartilage, muscle, soft tissue	[12]
Chitosan	Encapsulation, membranes	[11]
Fibrin	Cartilage	[12]
Laminin	Epithelial tissues, islets	[7]
Hyaluronic acid	Medical devices	[11]
Polyhydroxyalkanoates (PHA)	Skin, drug delivery, sutures	[11]
Isolated ECM from bone and small intestine	Bone, blood vessels, ureters	[7]
Synthetic		
Poly(esters) Poly(glycolic acid) (PGA) Poly(lactic acid) (PLA) Poly(caprolactone) (PCL) Poly(lactic-co-glycolic) (PLGA)	Cartilage, bone, muscle, nerve, blood vessel, valves, bladder, liver, cardiac tissue, drug delivery, sutures	[11,12]
Poly(anhydride)	Bone, drug delivery	[12]
Poly(hydroxybutyrate)	Valves	[7]
Poly(vinyl alcohol) (PVA)	Cartilage, nerve	[12]
Poly(ethylene glycol) (PEG)	Cartilage	[12]
Poly(ethylene terephthalate) (PET)	Cornea, blood vessels	[13,14]
Expanded poly(tetrafluoroethylene) (e-PTFE)	Cornea, blood vessels	[13,14]
Poly(propylene fumarate)	Bone, cardiovascular tissue	[12]

an important consideration depending on whether the implanted tissue is subject to a large amount of stress, as for example, with cartilage.

On the microscale, the porosity and pore structure regulate cell penetration, migration, interaction, and growth. Optimal porosities allow for penetration of cell seeding suspensions throughout the scaffold resulting in uniform distributions. The successful mixing and distribution of cells within a scaffold result in chondrocytes functioning properly by producing ECM molecules at high cellularity for enhanced cartilage strength characteristics of the tissue.[16] The morphogenesis of the developing tissue is influenced by the allowable migration of cells. The pore size distribution relates to the migration ability because it determines the amount of space available. The porosity of the scaffold and the size of the pores affect the supply of nutrients and mediation of the waste concentrations via fluid and mass transfer mechanisms. Transfer considerations are increasingly important as high cell density cultures are obtained which limit the available space for fluid and nutrient transport. In fibrous scaffolds, the fiber diameter and affiliated surface curvature affect the spreading ability of attached cells. Spreading allows cells to increase proliferation and this is regulated by fiber dimensions. Additionally, the diameter affects the degree of cell-cell interactions allowable around the fiber which are necessary for proper tissue function. Patterning of the scaffold surface, such as grooves, directs cell adhesion as well as cell growth and function for certain cell types.[17]

On the nanoscale, the surface chemistry of the scaffold must recreate the important cell–ECM properties of adhesion and control. Biocompatibility of the scaffold surface with cells is key for allowing adhesion and migration of cells. The amino acid sequence of arginine–glycine–aspartic acid (RGD) has been identified on fibronectin and other ECM glycoproteins as a key adhesion domain, and the design of synthetic scaffolds incorporating the peptide has been successful in improving adhesion and cytocompatibility.[18] The organization of RGD peptides on the scaffold surface affects adhesion as well, with a clustered arrangement optimal rather than randomly positioned RGD peptides. Scaffolds can be supplemented with binding or signal molecules bound to the polymer surface. For enhanced proliferation and differentiation, scaffolds can be designed to release growth factors efficiently. Neural cells were cultivated in rats on a scaffold equipped with degrading beads that released nerve growth factor in a controlled manner.

If a scaffold is transplanted, the rate of biodegradability is important to ensure that the scaffold remains to support a transplant until a natural ECM replaces it. The biodegradation or resorption rate is a function of the scaffold composition, structure, and the mechanical load present at the site of transplantation.[7] The necessary rate at which the scaffold is degraded varies according to the tissue type. For example, slow degradation is allowable in bone tissue, whereas in other tissues chronic inflammation may occur if the rate is too low.[7] It is important that the degradation by-products are nontoxic to the body.

Cell Culture

To develop a tissue in culture for use as a tissue substitute, the tissue cell density must be high (commonly 10^9 cells/ml) and uniform within the scaffold. In neomorphogenesis, cells are brought into contact with a porous scaffold and they form a structure together. The cell seeding process must be optimized to achieve uniformity of cell distribution within the scaffold, to maximize utilization of the cells, and to minimize seeding process time in order to avoid damage to the cells. To seed a scaffold, cells and scaffolds are incubated together to allow for adhesion to take place. Dynamic seeding protocols incorporate mixing or flow to distribute cells throughout the scaffold efficiently. When introduced, cells attach to the scaffold surfaces if the surface chemistry of the scaffold is compatible. Scaffolds can be pre-treated to alter the surface chemistry thus allowing improved compatibility with the cells. For example, increasing the concentration of hydrophilic compounds of the surface will improve cell adhesion.

The medium utilized to provide complete nutrition for the growth of different cell types is based on a standard minimum essential nutrient composition. Most media consist of a sugar source, minerals, vitamins, and amino acids. Serum, such as fetal bovine serum, commonly supplements the medium to enhance cell growth. Growth factors and cytokines are utilized to accelerate cell growth through interaction with specific cell receptors. Differentiation inducers are added to direct the differentiation pathway of stem cells. Important parameters to control include the pH, pCO_2, and pO_2 of the media.

There are several types of bioreactor designs utilized for cultivating new tissue growth, and four are shown schematically in Figure 10.3. The environment within may be static or mixed using internally designed or externally applied mechanisms. Static environments rely on diffusion as the mass transfer mechanism for nutrient supply. Mixing within a culture vessel provides convective flow of oxygen and nutrients to the cells while removing waste from the surroundings. Mixing-induced shear stress levels are an important consideration since they may cause cell death. The simplest design is the plate or Petri dish. The spinner flask is larger in scale and has an internal agitation mechanism to provide a uniform nutrient concentration within the medium and the enclosed cell–scaffold construct. Perfused bioreactors are culture environments in which media are circulated within a closed system past the immobilized cell–scaffold components. The continuous

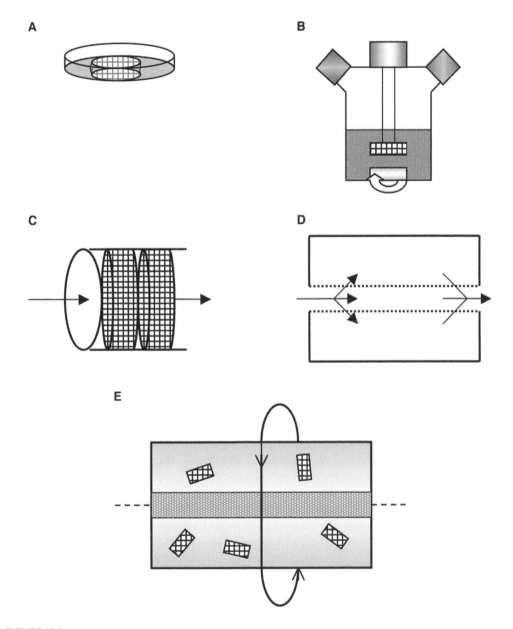

FIGURE 10.3
Tissue engineering culture environment designs: (A) plate; (B) spinner flask; (C) perfusion; (D) hollow fiber; and (E) rotating wall bioreactor.

flow allows for uniform nutrient supply with enhanced mass transfer. Long-term stability of the culture is attained using continuous perfusion reactors.[19] Hollow fiber bioreactors are specialized perfusion bioreactor designs in which a semipermeable membrane in a tubular configuration creates an interior and exterior region. Media can pass through the interior region of the hollow fiber axially, while the membrane allows nutrient and product transfer into and out of the extracapillary region, respectively, where cells are located growing on the exterior surface of the membrane.[20]

As a novel bioreactor design, rotating-wall bioreactors spin on an axis, and the enclosed cell–scaffold constructs tumble within the rotating microgravity environment. Tissue culture in microgravity has been shown to improve cellular aggregation and produce highly differentiated tissue products.[21] Rotating-wall bioreactors avoid the high shear stress found within bioreactors that have agitation devices. As a result, altered gene expression favors improved aggregation resulting in aggregates up to ten times larger in diameter than those attained in conventional bioreactors. Additionally, necrosis of cells within the center of the aggregates due to mass transfer limitations is not seen. The stability of tissue constructs after removal from the microgravity environment remains to be proven in order to realize in vivo utilization.[21]

Other stimuli may be incorporated into the culture environment to cultivate proper tissue function. Mechanical or electrical stimulation, provided at frequencies simulating in vivo conditions, have been shown to improve the resulting properties of the cultivated tissue.[19] For example, pulsatile conditions that simulate a beating heart are utilized in the culture of blood vessels resulting in improved strength and function relative to cultures lacking this stimulus.[2]

In cell culture, the essential nutrients must be present in the medium and be able to flow or diffuse to the cell membranes to allow for a viable culture. The bioreactor design needs to enable the cell population to expand to the cell density of the in vivo tissue; however, nutrient limitations deter this when cell growth decreases the space available within the scaffold for nutrient diffusion. Additionally, stagnant build-up of waste by-products increases acidity and harmful conditions. Therefore, the viability of a culture decreases within heavily populated regions of the scaffold. Supplying nutrients to these interior regions is a challenge, and currently, a tissue thickness of over 1 mm is not maintainable without cell death in the core of the tissue mass. In vivo, nutrient transfer within a tissue is achieved by a process called angiogenesis. As tissue mass increases, vasculature is established by promoting blood vessel growth within a starvation zone. A similar process of angiogenesis established within an in vitro culture would permit large-scale tissue growth. A tissue transplant can promote angiogenesis by providing angiogenic growth factors, such as vascular endothelial growth factor, along with the transplant that are released over time to allow for high cell density tissue regeneration.

Many tissues consist of more than one cell type. The fact that most major organs in the human body consist of more than one tissue and cell type adds to the complexity in recreating a functional organ replacement. The proper physiological function of these tissues depend on the interactions between these multiple cell types, so developing a tissue substitute with the ability to restore function should be a heterogeneous culture consisting of multiple cell types organized appropriately. Coculture of multiple cell types within one environment in order to accomplish this is very difficult. Currently, two different cell types are cultured together by growing each cell type in layers with membranes. The membranes allow for signal passage without direct cell–cell contact. The concept of "organ printing" may permit the generation of heterogeneous, vascularized, and three-dimensional organ constructs using a computer-controlled "printing" device that deposits multiple cell types, biomaterials, and other tissue components layer by layer to form an organized structure.[22]

Transplantation and Cryopreservation

When implanting a tissue substitute or therapy device, immune rejection by the host against the foreign implant is the primary concern unless a biocompatible scaffold with autologous cells is used. Several strategies exist to circumvent this challenge, depending on cell type, source, and desired function. Three specific strategies are immune system therapy before and after the transplantation procedure, gene modification in the cells prior to tissue development to allow immune system acceptance, and immunoisolation.[23] Traditionally, similar to organ transplants, the complete immune system is suppressed for a period of time following the transplant to increase the chances of the transplant eventually being accepted. There is a key 5–10 week period in which the successful integration of the transplant or graft rejection is determined.[23] However, during this period of immunosuppression, the patient is susceptible to other illnesses. A novel strategy involves building a tolerance in the host for a cell type prior to transplantation. By first performing a bone marrow transplant using donor hematopoietic cells, the host will circulate immune system components that match the future donor. Then, when the transplant occurs, the donor tissue is tolerated.[10] Immunoisolation strategies incorporate a semi-permeable membrane enclosing the cell-based device, protecting the cells from immune recognition and yet allowing the transport of therapeutic cell-derived compounds to emanate from the device.

When a tissue-engineered product is not to be used for transplantation upon its creation, cryopreservation allows for long-term storage until future application. Cryopreservation is commonly used for preparing cells for storage, but its utility for storage of tissues remains in development. Cryopreservation is a major focus for researchers, though, as it is a key component for improving the marketability of tissue-engineered products, allowing the development of an on-demand tissue supply, preservation of tissue genetic stability, and establishment of production quality control archives.[24] During the cryopreservation process, both the freezing and thawing procedures are equally important in regulating water displacement and replacement, respectively, while maintaining cellular and tissue integrity. Cryoprotectants and thermal processing protocols are utilized in this process. Commonly utilized cryoprotectants are dimethyl sulfoxide and glycerol, which remove intracellular water to avoid damaging ice crystal formation. Owing to the larger scale of tissues compared to cells, the challenges are greater, including the induction of chemical and thermal gradients within tissue, which must be resolved by utilizing optimized mass and heat transfer operations, respectively. Technologies to improve these processes and to monitor tissue parameters for performance control and modeling will further help to develop applicable cryopreservation protocols.

Applications

Tissue engineering applications can incorporate the aforementioned components in various combinations to achieve specific goals. There are five general applications of tissue engineering:

1. The development of human tissues in vitro for future implantation into the body to replace lost tissue function.

2. In vivo tissue regeneration by transplantation of a seeded or unseeded scaffold to aid in regeneration at a deficient site in the body.

3. Development of an in vivo or extracorporeal device that supplements reduced tissue function. The in vivo device is encapsulated within a semi-permeable membrane to allow for provision of a therapeutic molecule to the site while protecting

the cells from the host immune system. An externally positioned device would provide deficient tissue function compounds through a tube directed to the body site while avoiding cellular contact with the immune system.

4. Establishment of an environment for expanding a cell population that is later extracted from the scaffold for implementation within the body in a cell-based or gene therapy application.

5. Development of a model to promote in vivo-like function of a population of cells in vitro for studying tissue development, pathology, pharmacology, and toxicology projects instead of using animal models.

The future financial outlook for tissue engineering products have varied greatly with some predicting an $80 billion market in 2000.[2] However, owing to increasingly evident challenges involved in developing tissue replacements and a lack of realization of previous tissue engineering product success predictions, more recent market estimates have been around $15 billion annually.[5] Currently, over 20 different tissues have been researched for a tissue engineering application. A selection of these applications is shown in Table 10.2 and Figure 10.4.[26–33] To successfully design a tissue substitute that mimics the normal in vivo counterpart, the cells must perform similar functions with the support of a biomaterial with appropriate biocompatibility, degradation, and strength characteristics and, subsequently, restore functionality of the tissue to the system.

Skin

Skin was the first tissue to be produced as a tissue substitute. The relatively simple two-dimensional, bi-layered structure of skin primarily consists of keratinocytes, fibroblasts, and ECM components. Skin functions include roles as a protective barrier, fluid and heat regulator, and immune system reconnaissance for early warning of dangers. Common features are its flexibility, elasticity, and strength. The major application is to develop skin substitutes for use on burn victims, especially when skin from the same individual is not available for autogenous transplantation. The main role for the regeneration of skin is to re-establish the barrier function with the dermis and epidermis. Other functions can be developed in vivo via migration of other component cells from surrounding areas into the regenerated skin. Wound healing can be achieved by transplanting an unseeded scaffold at the wound site promoting migration of surrounding cells into the scaffold. Artificial skin substitutes were the first tissue engineering products to reach the market. One available dermal substitute on the market is Dermagraft®, produced by Smith & Nephew (Florida, U.S.A.), which consists of fibroblasts, extracellular matrix, and a bioabsorbable scaffold. When applied, usually for healing diabetic foot ulcers, healthy cells surrounding the wound, including keratinocytes, migrate into the scaffold to fully reconstitute healthy skin with natural barrier properties.[5]

Liver

The liver was the first tissue engineered in three-dimensional scaffolding. Proper differentiated function of a tissue-engineered construct containing hepatocytes includes the production of albumin and the completion of urea and bilirubin metabolism. The cells used to seed the scaffold should be highly proliferative in order to develop a high cell density tissue. Owing to the size of the liver, nutrient diffusion limitations are a concern, and, therefore, vascularization of liver construct is necessary for blood to supply nutrients

TABLE 10.2

Selected Tissue Engineering Applications

Tissue	Why	Cells	Scaffold	Application	Goal	References
Blood vessels	Vessel occlusion	Endothelial	PET, e-PTFE	Replacement	Thromboresistance	[13]
Bone	Disease, radiation	Periosteal, osteoblast, mesenchymal stem cells	PGA, PLA, calcium alginate	Repair, replacement, regeneration	Restored mechanical properties	[25,26]
Brain	Parkinson's and Huntington's diseases	Neurons	Membrane	Replacement, encapsulation	Dopamine production, neural network re-established	[27]
Breast	Lumpectomy, mastectomy	Smooth muscle cells (SMC), fibroblast, chondrocytes	PGA, PLA, PCL	Replacement	Flexible transplant with induced angiogenesis	[28]
Heart valves	Valvular disease	Endothelial, fibroblast	PGA	Repair, replacement	Durability over mechanical devices	[29]
Cornea	Disease, trauma	Epithelial	e-PTFE, PMMA, PVA	Repair, replacement	Elasticity, refractive properties, transparency, curvature	[14]
Pancreas	Diabetes	Islets of Langerhans	Alginate/ poly(L-lysine)	Encapsulated, restore	Active glucose concentration control	[30]
Kidney	Renal failure	Endothelial	Hollow fiber	Support device, replacement	Blood filtration, homeostasis maintenance	[31]
Tendons and ligaments	Injury	Fibroblasts	Collagen	Repair, replacement	Mechanical durability	[32]
Red blood cells	Disease, shortage	Artificial with hemoglobin	Membrane	Micro-encapsulation	Avoid removal of artificial cells from circulation	[33]
Liver	Acute liver failure	Porcine hepatocytes	Polysulfone hollow fiber	Extracorporeal support device	Detoxification activity, bridge-to-transplantation	[20]

within the liver mass. When the engineered liver is transplanted to the host site, vascularization promoters can be included to initiate the blood vessel migration into the new tissue. Hepatocytes have also been cultured in hollow-fiber membranes to supply liver function from a device outside the body. The HepatAssist System (Circe Biomedical, Inc., Massachusetts, U.S.A.) is an extracorporeal device supporting liver function in patients waiting for transplantation.[20] The device utilizes porcine hepatocytes immobilized in the outer space of a hollow-fiber design to perform liver functions while plasma is circulated inside the hollow fiber. The polysulfone membrane allows for protein and toxin transport across the membrane while preventing the passage of foreign cells into the patient. Thus, hepatocytes in the device do not have to be compatible with the host immune system since there is no direct contact.

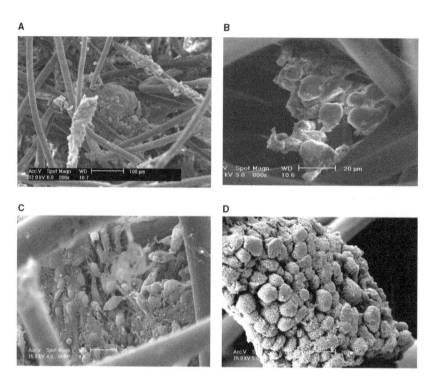

FIGURE 10.4

(A) In vitro modeling of colon cancer; (B) production of cord blood cells; (C) astrocyte culture for cell-based therapy of Parkinson's disease; and (D) placenta model using trophoblast cells for transport.

Bone

Autogeneous bone grafts are difficult because there is a limited supply of cells, and donor sites are subject to post-graft morbidity and bone deformations. Tissue-engineered bone structures consist of osteoblasts and osteocytes, which re-form in three-dimensional scaffolds to produce signature bone structures and features such as load-bearing ability and osteocalcin secretion for osteoblasts. The progenitor cells utilized in the osteogenesis can provide developmental signals such as bone morphogenetic proteins, which induce and promote bone growth. Periosteal cells, from the outer bone membrane periosteum, were seeded on calcium alginate hydrogel to allow for calcium extraction by the cells and to make customized bone-shaped molds for transplantation into bone defect sites.[25]

Gene Therapy

Gene therapy is introducing DNA, which codes for an essential protein, into cells that are lacking productivity of the protein. Gene therapy is an alternative to supplying the deficient region with the protein itself or replacing the defunct tissue with a functional tissue substitute. If stem cells are utilized, the regenerating source of corrected, functional cells will eventually replace expired, deficient cells. DNA is carried and introduced into cells using a delivery system which is commonly viral or plasmid based.[34] Once the gene is expressed within the cell, persistence of the essential protein supply is key. Accompanying the huge potential of this process are important safety concerns with regards to the implementation of the DNA correctly within the chromosomes. Incorrect incorporation of the gene could alter important control genes or other essential genes. Once the successfully transduced cells are selected, the cell population is expanded in vitro.

In Vitro Studies

Tissue engineering can also be used to develop laboratory tools using human cells to perform pathological, developmental, pharmacological, and toxicological studies. Tissue cultured as an in vitro model can be used to study tissue or organ development processes including signaling and control mechanisms. Additionally, toxicology and carcinogen studies on tissue models provide insight without utilizing animal models.[35] Cancer models can be used to study tumor biology, control, and progression as well as use in cancer treatment studies.[36]

Cell-Based Technologies

Additional cell-based technologies are being developed by utilizing the knowledge of cells and their interaction with materials attained from tissue engineering research. Biosensors take advantage of the specificity of cell surface receptors for select target molecules and the high signal amplification for detecting low levels of chemical or biological agents. The brain consists of a neural network that processes a large number of signals. Neurons organized on a support in specific patterns could produce in vitro neural networks that pass signals similarly to the electronic structure on microchips.[37]

Current Challenges

Although there has been some limited success in producing tissue substitutes, especially artificial skin, the overall tissue engineering aim of creating tissue that can replace, maintain, or improve deficient in vivo tissue is not an easy task. Multiple challenges remain in providing tissue products to meet demand. These include creating a greater supply of cells to avoid immune rejection, such as a universal donor cell supply. Advancements in biomaterial production and bioreactor design are necessary to provide customized support and environments to replicate the complex in vivo environment. When developing complex, three-dimensional tissues, challenging issues include promoting angiogenesis to overcome tissue size limitations, the heterogeneous coculture of multiple cell types within one tissue, and the cryopreservation of the resulting tissue before transplantation.

Conclusions

Through the multidisciplinary application of biological and medical knowledge and engineering skills, tissue engineering allows the development of tissues in the human body to restore, repair, or improve deficient biological components. Generally, utilizing cells, supportive scaffolds, and designed culture systems, tissue development processes face many challenges in reconstituting in vivo-like performance. Although many challenges remain, the field is a major research focus, and medical demand is strong, both ensuring that tissue engineering stands to provide momentous advancements to health in the 21st century.

References

1. Ma, T. Fiber-Based Bioreactor Systems In Mammalian Cell Culture and Tissue Engineering Human Trophoblast Cells, Dissertation, The Ohio State University: Columbus, OH, 1999.
2. Langer, R. Tissue engineering. *Mol. Ther.* 2000, 1(1), 12–15.
3. http://www.optn.org/latestData/step2.asp (accessed July 2003).
4. Schulthesis, D., Bloom, D.A., Wefer, J., Jonas, U. Tissue engineering from Adam to the zygote: historical reflections. *World J. Urol.* 2000, *18*, 84–90.
5. Flanagan, N. Engineering methods for tissue and cell repair. *Genet. Eng. News* 2003, *23* (8), 10.
6. Martins-Green, M. Dynamics of cell-ECM interactions. In *Principles of Tissue Engineering*, 2nd Ed., Lanza, R.P., Langer, R., Vacanti, J., Eds., Academic Press: New York, 2000; 33–55.
7. Griffith, L.G. Biomaterials. In *WTEC Panel on Tissue Engineering Research: Final Report*, McIntire, L.V., Ed., Academic Press: New York, 2002; 9–22.
8. Hubbell, J.A. Matrix effects. In *Principles of Tissue Engineering*, 2nd Ed., Lanza, R.P., Langer, R., Vacanti, J., Eds., Academic Press: New York, 2000; 237–250.
9. Heath, C.A. Cells for tissue engineering. *Trends Biotechnol.* 2000, *18*, 17–19.
10. Brower, V. Stem cell research and development advances in face of challenges. *Genet. Eng. News* 2003, *23* (9), 1.
11. Pachence, J.M., Kohn, J. Biodegradable polymers. In *Principles of Tissue Engineering*, 2nd Ed., Lanza, R.P., Langer, R., Vacanti, J., Eds., Academic Press: New York, 2000; 263–277.
12. Elisseeff, J.H., Langer, R., Yamada, Y. Biomaterials for tissue engineering. In *Tissue Engineering and Biodegradable Equivalents: Scientific and Clinical Applications*, Lewandrowski, K.-U., Wise, D., Trantolo, D., Gresser, J.D., Yaszemski, M.J., Altobelli, D.E., Ed., Marcel Dekker: New York, 2002; 1–24.
13. Xue, L., Greisler, H.-P. Blood vessels. In *Principles of Tissue Engineering*, 2nd Ed., Lanza, R.-P., Langer, R., Vacanti, J., Eds., Academic Press: New York, 2000; 427–446.
14. Trinkaus-Randall, V. Cornea. In *Principles of Tissue Engineering*, 2nd Ed., Lanza, R.-P., Langer, R., Vacanti, J., Eds., Academic Press: New York, 2000; 471–491.
15. Langer, R. Selected advances in drug delivery and tissue engineering. *J. Control. Release* 1999, *62*, 7–11.
16. Vunjak-Novakovic, G., Freed, L., Biron, R.J., Langer, R. Effects of mixing on the composition and morphology of tissue-engineered cartilage. *Am. Inst. Chem. Eng. J.* 1996, *42*, 850–860.
17. Saltzman, W.M. Cell interactions with polymers. In *Principles of Tissue Engineering*, 2nd Ed., Lanza, R.P., Langer, R., Vacanti, J., Eds., Academic Press: New York, 2000; 221–235.
18. Griffith, L.G., Naughton, G. Tissue engineering—current challenges and expanding opportunities. *Science* 2002, *295*, 1009–1014.
19. Freed, L.E., Vunjak-Novakovic, G. Tissue engineering bioreactors. In *Principles of Tissue Engineering*, 2nd Ed., Lanza, R.P., Langer, R., Vacanti, J., Ed., Academic Press: New York, 2000; 143–156.
20. Mullon, C., Soloman, B.A. HepatAssist liver support system. In *Principles of Tissue Engineering*, 2nd Ed., Lanza, R.P., Langer, R., Vacanti, J., Eds., Academic Press: New York, 2000; 553–558.
21. Unsworth, B.R., Lelkes, P.I. Tissue assembly in microgravity. In *Principles of Tissue Engineering*, 2nd Ed., Lanza, R.P., Langer, R., Vacanti, J., Eds., Academic Press: New York, 2000; 157–164.
22. Mironov, V., Boland, T., Trusk, T., Forgacs, G., Markwald, R.R. Organ printing: computer-aided jet-based 3D tissue engineering. *Trends Biotechnol.* 2003, *21* (4), 157–161.

23. Hardin-Young, J., Teumer, J., Ross, R.N., Parenteau, N.L. Approaches to transplanting engineered cells and tissues. In *Principles of Tissue Engineering*, 2nd Ed., Lanza, R.P., Langer, R., Vacanti, J., Ed., Academic Press: New York, 2000; 281–291.

24. Karlsson, J.O.M., Toner, M. Cryopreservation. In *Principles of Tissue Engineering*, 2nd Ed., Lanza, R.P., Langer, R., Vacanti, J., Eds., Academic Press: New York, 2000; 293–307.

25. Vacanti, C.A., Bonassar, L.J., Vacanti, J.P. Structural tissue engineering. In *Principles of Tissue Engineering*, 2nd Ed., Lanza, R.P., Langer, R., Vacanti, J., Eds., Academic Press: New York, 2000; 671–682.

26. Bruder, S.P., Caplan, A.I. Bone regeneration through cellular engineering. In *Principles of Tissue Engineering*, 2nd Ed., Lanza, R.P., Langer, R., Vacanti, J., Ed., Academic Press: New York, 2000; 683–696.

27. Wahlberg, L.U. Brain implants. In *Principles of Tissue Engineering*, 2nd Ed., Lanza, R.P., Langer, R., Vacanti, J., Eds., Academic Press: New York, 2000; 773–783.

28. Lee, K.Y., Halberstadt, C.R., Holder, W.D., Mooney, D.J. Breast reconstruction. In *Principles of Tissue Engineering*, 2nd Ed., Lanza, R.P., Langer, R., Vacanti, J., Eds., Academic Press: New York, 2000; 409–423.

29. Love, J.W. Cardiac prostheses. In *Principles of Tissue Engineering*, 2nd Ed., Lanza, R.P., Langer, R., Vacanti, J., Eds., Academic Press: New York, 2000; 455–467.

30. Wang, T.G., Lanza, R.P. Bioartificial pancreas. In *Principles of Tissue Engineering*, 2nd Ed., Lanza, R.P., Langer, R., Vacanti, J., Eds., Academic Press: New York, 2000; 495–507.

31. Humes, H.D. Renal replacement devices. In *Principles of Tissue Engineering*, 2nd Ed., Lanza, R.P., Langer, R., Vacanti, J., Eds., Academic Press: New York, 2000; 645–653.

32. Goulet, F., Rancourt, D., Cloutier, R., Germain, L., Poole, A.R., Auger, F.A. Tendons and ligaments. In *Principles of Tissue Engineering*, 2nd Ed., Lanza, R.P., Langer, R., Vacanti, J., Eds., Academic Press: New York, 2000; 711–722.

33. Chang, T.M.S. Red blood cell substitutes. In *Principles of Tissue Engineering*, 2nd Ed., Lanza, R.P., Langer, R., Vacanti, J., Eds., Academic Press: New York, 2000; 601–610.

34. Fradkin, L.G., Ropp, J.D., Warner, J.F. Gene-based therapeutics. In *Principles of Tissue Engineering*, 2nd Ed., Lanza, R.P., Langer, R., Vacanti, J., Eds., Academic Press: New York, 2000; 385–405.

35. Li, A.P. Screening for human ADME/Tox drug properties in drug discovery. *Drug Discov. Today* **2001**, *6* (7), 357–366.

36. Chung, L.W.K., Zhau, H.E., Wu, T.T. Development of human prostate cancer models for chemoprevention and experimental therapeutics studies. *J. Cell. Biochem. Suppl.* **1997**, *28/29*, 174–181; Supplement.

37. Mrksich, M. Cell-based technologies: non-medical applications. In *WTEC Panel on Tissue Engineering Research: Final Report*, McIntire, L.V., Ed., Academic Press: New York, 2002; 61–69.

11

Bioactive Devices

Sujata K. Bhatia and Surita R. Bhatia

CONTENTS

Introduction

Medical devices have improved millions of lives by providing mechanical and material solutions to biomedical problems. Medical devices are therapeutically applied in a variety of clinical disciplines, including cardiovascular medicine, orthopedics, and neurology, and new materials are in use or under development for nearly every organ system in the body. Traditional medical devices are constructed from polymers, ceramics, glasses, and metals.[1] Although these materials represent a significant advance in medicine, traditional medical devices suffer from several limitations. Implanted medical devices may be prone to cause infection,[2,3] thrombosis, inflammation, and poor healing;[4] these biological responses can limit the lifetime of implanted devices.[5] In addition, implanted devices typically do not respond to changes in the surrounding biological environment and have limited biological activity.

Thus, there is an increasing interest in the development of biologic–device combinations, to render biological activity to implantable materials. Incorporation of biologics can improve the stability, biocompatibility, and efficacy of medical devices.[6] Biologics can also control the biological response and prevent adverse cellular responses to medical implants, promote cellular growth and healing around implants,[7] and facilitate cellular attachment and retention. In addition, biologic therapies may be utilized with implantable devices to impart pharmacologic activity to a medical device for synergistic, targeted treatment of disease states. Such site-directed pharmacologic therapy can avoid adverse effects associated with systemic therapy. Not only will novel medical devices require biologics for optimal efficacy, but novel potent biologics will require medical devices to effect highly localized delivery. A variety of biologic therapies have been included in medical devices, and a few of these combination devices are already in clinical use. With the rising incidence of chronic and degenerative diseases,[8] including cardiovascular disease, arthritis, cancer, diabetes, and neurodegenerative disease, particularly among the aging population, the necessity for innovative biologic-device solutions will continue to grow.

This chapter appraises current advances in biologic–device combinations, including growth factor–device combinations, cytokine–device combinations, enzyme–device combinations, peptide–device combinations, and antibody–device combinations. We will not cover devices where the bioactive components are cells; these are covered in other entries (see "Cross References" at the end of this chapter). This chapter also discusses regulatory issues specific to biologic-device combinations that must be considered during the development and approval of novel implantable bioactive devices.

Growth Factor–Device Combinations

Biological growth factors may be incorporated into medical implants, either through surface display or through controlled release systems. Polypeptide growth factors may regulate a variety of cellular responses, including cell migration, proliferation, survival, and differentiation. Growth factors have been exploited to create implants that deliver angiogenic growth factors to induce vascular repair; neuronal survival and differentiation factors to treat neurodegenerative disease; transforming growth factor-β (TGF-β) to induce bone repair; bone morphogenetic proteins to enhance bone formation; and tissue growth factors to heal chronic ulcers.

Perhaps the most significant advances in growth factor–device combinations have taken place in the field of orthopedics. Implantable bone grafts have been designed to secrete the osteoinductive factor bone morphogenetic protein-2 (BMP-2) to promote bone regeneration. Delivery of BMP-2 from an absorbable collagen sponge induces bone formation and heals bony defects,[9] and BMP-2 protein delivery from a collagen scaffold has demonstrated efficacy in open tibial fractures[10,11] in humans. Medical devices eluting BMP-2 have been Food and Drug Administration (FDA)-approved and are commercially available for orthopedic surgery and oral–maxillofacial surgery. The INFUSE® bone graft system (Medtronic Sofamor Danek, Memphis, Tennessee, U.S.A.) utilizes an absorbable collagen carrier matrix containing recombinant human BMP-2 (rhBMP-2). This system, in combination with metallic cages, is FDA-approved for anterior lumbar interbody fusion surgery to treat degenerative disc disease of the spine. The system is also indicated for treatment of acute, open fractures of the tibia, as well as dental bone-grafting procedures including sinus augmentation and localized alveolar ridge augmentation.

Additional members of the BMP protein family have also been investigated in biologic–device combinations for bone regeneration. Release of osteogenic protein-1 (OP-1™), also known as bone morphogenetic protein 7 (BMP-7), from a collagen matrix induces healing of human tibial non-unions.[12] Controlled delivery of BMP-7 also promotes healing of spinal fusions in pilot human studies, aiding in the treatment of patients with degenerative disc disease.[13,14] The OP-1 system (Stryker, Kalamazoo, Michigan, U.S.A.) combines recombinant human BMP-7 (rhBMP-7) with a granular collagen matrix and has been FDA-approved under a humanitarian device exemption for the treatment of long bone non-unions and revision spinal fusions.

Other osteoinductive factors and scaffold materials have shown promise for induction of bone formation and fracture healing. Administration of basic fibroblast growth factor-1 (FGF-1) in collagen minipellets accelerates bone formation in skull defect models.[15] A bioactive coating of polylactide containing insulin-like growth factor-1 (IGF-1) and TGF-β1 has been applied to biomechanical bone implants, and enhances healing in tibial-fracture models.[16] Mechanical bone implants have also been coated with polylactide containing BMP-2, and such implants stimulate bone healing in tibial-fracture models.[17] The protein BMP-2 has additionally been directly covalently linked to the surface of resorbable fracture plates to facilitate bone growth.[18] These growth factor–device systems for bone regeneration may soon be utilized in human orthopedic surgery.

Surgical devices incorporating growth factors are also being explored in orthopedic surgery for tendon regeneration. Coating of nylon sutures with basic fibroblast growth factor (bFGF) stimulates tendon healing and increases tendon strength in a flexor tendon injury model, and may become a therapeutic tool for use in hand surgery.[19] Coating of polyglactin sutures with recombinant human growth differentiation factor-5 (rhGDF-5) induces tendon healing in an Achilles tendon injury model, and results in thicker tendons.[20] The GDF-5-coated suture accelerates the recovery of biomechanical properties of repaired tendons, including ultimate tensile load and stiffness, and increases the rate of tendon healing.[21]

Outside of orthopedics, growth factor-device combinations are under development for wound healing and vascular growth. A collagen sponge matrix loaded with epidermal growth factor has been shown to boost dermal matrix formation and improve mechanical strength of healing second-degree burn wounds, as compared to untreated wounds.[22] The EGF-impregnated collagen sponge also lowers the rate of wound contraction, and may serve as a novel combination graft for burn victims. For promotion of vascular growth, both vascular endothelial growth factor (VEGF) and FGF have been incorporated into porous collagen scaffolds.[23] Inclusion of both of these growth factors into scaffolds enhances development of an early mature vasculature, and such constructs may have clinical applications for healing of diabetic ulcers and therapy of ischemic heart disease.

Growth factor–device combinations may even show promise for nerve regeneration. Implantation of absorbable collagen sponges containing recombinant human basic fibroblast growth factor (rhbFGF) promotes nerve regeneration in a sciatic-nerve injury model.[24] Nerve growth factor (NGF) has been immobilized onto the surface of the electrically conductive polymer polypyrrole, resulting in a combination device that provides both biological and electrical stimulation to surrounding tissue.[25] The NGF-linked polypyrrole construct improves neurite extension and outgrowth. Growth factors may also have utility for prevention of scar formation surrounding implanted neurological devices. Conjugation of TGF-β1 to the surface of biomaterials decreases astrocyte proliferation[26] and may represent a strategy for reducing glial-scar formation on implanted surfaces.

Cytokine–Device Combinations

A variety of cytokines can be integrated into medical devices to modulate cell and tissue responses or provide a therapeutic effect. The immune-stimulating cytokines interleukin-2 (IL-2) and interferon γ (IFN-γ) have been covalently linked to polyester surgical sutures, to create a combination device that both sutures tissue and stimulates T lymphocytes.[27]

These cytokine-linked sutures promote proliferation of peripheral blood lymphocytes and lymph node lymphocytes in patients with head and neck squamous cell carcinoma, and generate a TH1 immunologic profile of cytokines. The cytokine–suture combination device could be a method for providing cancer immunotherapy to patients and could further enhance wound-healing following surgical resection. Interferons can also be embedded into intracerebral implants composed of ethylene–vinyl acetate polymers for providing cerebral immunotherapy.[28]

Cytokines may additionally be useful for control of cellular adhesion on vascular grafts, as well as cellular proliferation surrounding cardiovascular stents. In the field of vascular grafts, proinflammatory cytokine osteopontin has been immobilized on poly(2-hydroxyethyl methacrylate) surfaces to promote endothelial cell adhesion;[29] this approach is clinically relevant, as it has been shown in clinical studies that enhancement of cell adhesion strength improves the performance of endothelial cell-seeded vascular grafts in high-flow regions.[30] In the field of cardiovascular stents, coating of stainless steel surfaces with a meshwork containing IFN-γ inhibits smooth muscle cell growth without affecting endothelial cell growth.[31] Because smooth-muscle cell hyperproliferation is a main cause of recurrent stenosis following cardiovascular stent implantation, coating of metal stents with IFN-γ may be a promising strategy for preventing restenosis and maintaining stent patency.

Enzyme–Device Combinations

Enzymes and enzyme inhibitors are being engineered into implantable medical devices to provide biosensing capability, prevent device encrustation, and control thrombogenicity. The glucose oxidase enzyme has been immobilized onto gelatin and coated on electrodes, to create glucose biosensors with a detection limit of 0.25 mM glucose;[32] such biosensors may be useful for diabetic patients to prevent hyperglycemia or hypoglycemia. Oxalate-degrading enzymes from the commensal colonic bacterium *Oxalobacter formigenes* have been coated onto urinary implant biomaterials; such enzymes reduce biomaterial encrustation by preventing deposition of calcium oxalate.[33] This enzyme-device combination could lengthen the lifetime of urinary implants, as the long-term placement of devices in the urinary tract is frequently limited by the development of encrustation. Enzyme inhibitors are finding utility in cardiovascular grafts for preventing clotting and restenosis. Tissue factor pathway inhibitor (TFPI) is an endogenous anticoagulant, which is a potent inhibitor of Factor Xa and the tissue factor-Factor VIIa (TF:VIIa) complex. Adsorption of recombinant TFPI (rTFPI) onto synthetic vascular grafts reduces thrombogenicity and intimal hyperplasia, and specifically reduces fibrin deposition on Dacron® grafts.[34]

Peptide–Device Combinations

Peptides can be designed into medical devices to regulate cellular adhesion, cellular migration, and device biodegradation. Adhesion-promoting oligopeptides may be attached to implants and are based on the primary structure of the receptor-binding domains of proteins such as fibronectin and laminin.[35] Often the corresponding peptide sequences display similar receptor specificity and binding affinity, as well as cell signaling activity, as compared to the whole protein.[36,37] The RGD tripeptide from fibronectin may be either immobilized on device surfaces or included directly into the backbone of polymer chains to induce cell adhesion, spreading, focal contact formation, and cytoskeletal organization.[38] Fibronectin itself has been immobilized onto the microporous polypropylene hollow fibers of an artificial lung device, to create a protein-coated artificial lung that allows endothelial cell adhesion;[39] the adhesion-promoting artificial lung device shows promise in a partial cardiopulmonary bypass model. The YIGSR domain and the SIKVAV domain from laminin are migration-promoting peptides and have been incorporated into gels to promote neuronal cell infiltration and nerve regeneration.[40]

Integrin-binding peptides are particularly useful for facilitating adhesion and migration of specific cell types. The tetrapeptide REDV from fibronectin, which specifically binds the integrin receptor $\alpha_4\beta_1$ on endothelial cells, may be employed in vascular grafts to specifically support adhesion and migration of vascular endothelial cells, while also preventing the adhesion of clot-forming platelets.[35] The integrin-binding domain RGDSP, attached in a pendant fashion to implantable hydrogels, may also induce cell migration through the implant via integrin-dependent mechanisms.[41] The GFOGER collagen-mimetic peptide, which selectively promotes $\alpha_2\beta_1$ integrin binding, has been coated onto titanium implants to enhance bone repair.[42] Such GFOGER-modified implants demonstrate improved osseointegration, a critical characteristic for implant success in orthopedic and dental surgeries.

Peptide-treated materials are nearing clinical usage as vascular stent grafts. Stent grafts can be coated with a 15-amino acid cell adhesion peptide (P-15) for promoting endothelialization on the inner graft surface postimplantation.[43] The P-15-coated stent graft has been clinically tested for treatment of patients with saphenous vein graft lesions, as a method for reducing restenosis and distal embolic events in this patient population. The P-15-coated stent graft is not associated with device-related major adverse cardiac events at 3 months postimplantation, and restenosis rates are comparable to those of similar devices. This peptide–device combination may be a safe and effective treatment for saphenous vein graft lesions.

An emerging area of peptide–device combinations is the design of implantable devices containing enzymatic recognition and cleavage sites. Incorporation of enzymatic cleavage sites into a biomaterial allows the degradation rate of the material to be tuned, and also allows the biomaterial to be proteolytically remodeled. Implantable hydrogels containing poly(ethylene glycol) (PEG) chains with central oligopeptide sites that are substrates for collagenase or plasmin are degradable by cell-associated enzymatic activity.[35] Linear oligopeptide substrates for matrix metalloproteinases (MMPs) have been linked into implantable PEG hydrogels and may assist in cell migration through implants via MMP-dependent mechanisms.[41]

Antibody–Device Combinations

An innovative method for facilitating highly specific cell activation and cell adhesion to medical devices is to employ antibodies in implantable materials. Anti-CD3/anti-CD28 monoclonal antibodies have been coated onto nylon surgical sutures, to create a combination device that both sutures tissue and enhances T lymphocyte immune function.[44]

These antibody-linked sutures stimulate proliferation of peripheral blood mononuclear cells and lymph node mononuclear cells in patients with head and neck squamous cell carcinoma, and induce a TH1 immunologic profile of cytokines. The antibody–suture combination device could be another method for providing cancer immunotherapy to patients and could also enhance wound healing following surgical resection.

Several classes of antibody-coated and antibody-eluting coronary artery stents have been developed to improve endothelialization and inhibit thrombogenesis on the inner stent surface, thereby increasing stent patency rates postimplantation. Many of these antibody-containing stents are on the verge of clinical introduction. Coating of stainless steel stents with anti-CD34 antibody allows capture of circulating endothelial progenitor cells (EPCs) onto the stent surface to enhance endothelialization[45]; this may be a critical factor for success in certain patient populations, as the number of circulating EPCs and their migratory activity are reduced in patients with diabetes, coronary artery disease (CAD), or multiple coronary risk factors.[46] The Genous™ BioEngineered R stent™ (OrbusNeich, Hong Kong) utilizes anti-CD34 antibodies coated onto a stainless steel stent; the first human clinical investigation of this technology suggests that the EPC capture stent is safe and feasible for treatment of de novo CAD.[45] A multicenter, worldwide clinical study is currently underway to evaluate the efficacy of the Genous™ device in treating CAD.

Other receptor-targeting antibodies have been investigated for antibody–stent combinations. Abciximab (ReoPro, c7E3-Fab) inhibits the platelet glycoprotein IIb/IIIa receptor as well as the smooth muscle cell $\alpha_v\beta_3$ integrin receptor. Elution of abciximab from polymer-coated stents results in significantly lower platelet deposition.[47] In human coronary arteries, abciximab-eluting stents are associated with significantly decreased neointimal hyperplasia as compared to control stents.[48] In a prospective randomized trial, the abciximab-coated stent demonstrated lower rates of restenosis and target vessel revascularization, indicating that the stent may be effective for prevention of coronary restenosis.[49] In patients with acute myocardial infarction, the abciximab-coated stent is safe and effective, and does not cause stent thrombosis.[50]

Growth factor-targeting antibodies have also been incorporated into stents to modulate the tissue response to stent implantation. The VEGF-targeting antibody bevacizumab inhibits angiogenesis and neovascularization. Because neovascularization is associated with the destabilization of atheromatous plaque, inhibition of neovascularization may be a useful strategy for the treatment of stable and vulnerable plaques. Delivery of bevacizumab from vascular stents results in decreased neointimal hyperplasia and decreased neovascularization in an iliac artery model, without compromising endothelialization.[51] Such devices show promise for preventing stent restenosis and stabilizing arterial plaque.

Regulatory Considerations

Biologic–device combinations may present unique regulatory challenges, resulting from their dual function as biologically active therapeutics and implantable medical devices. Combination products involve components that would normally be regulated under different types of regulatory authorities, and frequently by different centers of the U.S. FDA. For example, biologic therapeutics are usually regulated by the FDA's Center for Biologics

Evaluation and Research (CBER), while medical devices are typically regulated by the FDA's Center for Devices and Radiological Health (CDRH).

Formal regulatory recognition and development of a regulatory strategy for dealing with combination devices is relatively new. Up until 2002, combination products were not well defined in the Code of Federal Regulations. The FDA's Office of Combination Products (OCP) was established on December 24, 2002, as required by the Medical Device User Fee and Modernization Act of 2002 (MDUFMA). The law gives the Office broad responsibilities covering the regulatory life cycle of drug–device, drug–biologic, and device–biologic combination products. In 2002, combination products were defined in 21 CFR §3.2(e) as follows:

The term combination product includes:

1. A product comprised of two or more regulated components, i.e., drug/device, biologic/device, drug/biologic, or drug/device/biologic, that are physically, chemically, or otherwise combined or mixed and produced as a single entity;

2. Two or more separate products packaged together in a single package or a unit and comprised of drug and device products, device and biological products, or biological and drug products;

3. A drug, device, or biological product packaged separately that according to its investigational plan or proposed labeling is intended for use only with an approved individ-ually specified drug, device, or biological product where both are required to achieve the intended use, indication, or effect and whereupon approval of the proposed product the labeling of the approved product would need to be changed, e.g., to reflect a change in intended use, dosage form, strength, route of administration, or significant change in dose; or

4. Any investigational drug, device, or biological product packaged separately that according to its proposed labeling is for use only with another individually speci-fied investigational drug, device, or biological product where both are required to achieve the intended use, indication, or effect.

While the OCP is responsible for overview of the regulatory life cycle of biologic–device combinations, the primary regulatory responsibilities for, and oversight of, specific combination products still remains in one of the three product centers—the Center for Drug Evaluation and Research, the Center for Biologics Evaluation and Research, or the Center for Devices and Radiological Health—to which they are assigned. The OCP is thus responsible for assigning the FDA center responsible for reviewing a combination product based on its "primary mode of action." For example, if the primary mode of action of a device–biologics combination product is attributable to the biological product, the agency component responsible for premarket review of that biological product would have primary jurisdiction for the combination product. A final rule defining the primary mode of action of a combination product was published in the August 25, 2005, Federal Register, and is available at http://www.fda.gov/oc/combination/. The final rule defines primary mode of action as "the single mode of action of a combination product that provides the most important therapeutic action of the combination product."

In some cases, neither FDA nor the product sponsor can determine the most important therapeutic action at the time a request is submitted. A combination product may also have two independent modes of action, neither of which is subordinate to the other. To resolve these types of questions, the final rule describes an algorithm FDA will follow to determine the center assignment. The algorithm directs a center assignment based on consistency with other combination products raising similar types of safety and effectiveness questions, or to the center with the most expertise to evaluate the most significant safety and effectiveness questions raised by the combination product. The final rule is effective November 23, 2005. In addition, the FDA has developed intercenter agreements (http:// www.fda.gov/oc/combination/intercenter.html); these are working agreements developed between the FDA centers, which outline certain categories of products and how the FDA has regulated these products. These intercenter agreements are useful in determining jurisdiction when the characteristics of a combination product are clearly evident or specifically listed in the intercenter agreements.

Given the regulatory complexities associated with biologic–device combination therapies, it is essential for those developing such novel products to initiate a dialogue with the FDA early in the development process, to clarify the designation and approval process for each novel biologic–device combination product.

Conclusion

Biologic–device combinations are powerful clinical tools that unite the bioactivity of biologic agents with the mechanical and structural activity of medical devices. Significant advances have occurred in biologic–device development, and biologics have been utilized in medical devices ranging from surgical sutures to cardiovascular stents to orthopedic implants. Biologic growth factors, cytokines, peptides, enzymes, and antibodies can retain their functionality when incorporated into medical devices and allow implanted devices to interact with the biological environment by influencing cellular proliferation, tissue healing, cellular adhesion, cellular migration, and device biodegradation. Carefully chosen biologics impart improved properties of biocompatibility, tunability, and biological functionality on medical devices. In turn, medical devices enable highly targeted and controlled delivery of potent biologics to the desired site of action. Further development of biologic–device combinations will require an increased understanding of cell–material interactions, as well as reliable model systems for the biological environment. In addition, the development of biologic–device combinations will be boosted by continuous progress in processing and manufacturing of biologics, so that biologic devices can be produced in a cost-effective manner. Multidisciplinary collaborations will be essential for the successful design of biologic–device combinations, which will rely on the expertise in chemical engineering, medicine, molecular biology, chemistry, materials science, biotechnology, bioengineering, and biomechanics. Biologic–device combinations represent the newest generation of medical devices, and innovative biologic–device solutions will have a significant impact on clinical practice and patient outcomes.

Acknowledgment

S.R.B. acknowledges support from the National Science Foundation under Grant No. CMMI-0531171 in preparation of this material.

Cross References to Other Entries

Biodegradable Polymers, Biomaterials, Biopolymers, Functional Biomaterials, Hydrogels, Hydrophilic Polymers for Biomedical Applications, Tissue Engineering

References

1. Williams, D.F. On the mechanisms of biocompatibility. *Biomaterials* **2008**, *29* (20), 2941–2953.
2. Gristina, A.G., Shibata, Y., Giridhar, G., Kreger, A., Myrvik, Q.N. The glycocalyx, biofilm, microbes, and resistant infection. *Semin. Arthroplasty* **1994**, *5* (4), 160–170.
3. Donlan, R.A. Biofilms and device-associated infections. *Emerg. Infect. Dis.* **2001**, *7* (2), 277–281.
4. Anderson, J.M., Rodriguez, A., Chang, D.T. Foreign body reaction to biomaterials. *Semin. Immunol* **2008**, *20* (2), 86–100.
5. Kaplan, S.S. Biomaterial-host interactions: Consequences, determined by implant retrieval analysis. *Med. Prog. Technol* **1994**, *20* (3-4), 209–230.
6. Helmus, M.N., Gibbons, D.F., Cebon, D. Biocompatibility: Meeting a key functional requirement of next-generation medical devices. *Toxicol. Pathol* **2008**, *36* (1), 70–80.
7. Babensee, J.E., McIntire, L.V., Mikos, A.G. Growth factor delivery for tissue engineering. *Pharm. Res.* **2000**, *17* (5), 497–504.
8. Michaud, S.M., Murray, C.J.L., Bloom, B.R. Burden of disease: Implications for future research. *JAMA.* **2001**, *285* (5), 535–539.
9. Li, G., Bouxsein, M.L., Luppen, C. Bone consolidation is enhanced by rhBMP-2 in a rabbit model of distraction osteogenesis. *J. Orthop. Res.* **2002**, *20* (4), 779–788.
10. Govender, S., Csimma, C., Genant, H.K. Recombinant human bone morphogenetic protein-2 for treatment of open tibial fractures: A prospective, controlled, randomized study of four hundred and fifty patients. *J. Bone Joint Surg. Am.* **2002**, *84-A* (12), 2123–2134.
11. Swiontkowski, M.F., Aro, H.T., Donell, S. Recombinant human bone morphogenetic protein-2 in open tibial fractures. A subgroup analysis of data combined from two prospective randomized studies. *J. Bone Joint Surg. Am.* **2006**, *88* (6), 1258–1265.
12. Friedlaender, G.E., Perry, C.R., Cole, J.D. Osteogenic protein-1 (bone morphogenetic protein-7) in the treatment of tibial non-unions. *J. Bone Joint Surg. Am.* **2001**, *83-A* (Suppl 1 (Pt 2)), S151–S158.
13. Vaccaro, A.R., Patel, T., Fischgrund, J. A pilot study evaluating the safety and efficacy of OP-1 Putty (rhBMP-7) as a replacement for iliac crest autograft in posterolateral lumbar arthrodesis for degenerative spondylolisthesis. *Spine* **2004**, *29* (17), 1885–1892.
14. Vaccaro, A.R., Anderson, D.G., Patel, T. Comparison of OP-1 Putty (rhBMP-7) to iliac crest autograft for posterolateral lumbar arthrodesis: A minimum 2-year follow-up pilot study. *Spine* **2005**, *30* (24), 2709–2716.
15. Kimoto, T., Hosokawa, R., Kubo, T. Continuous administration of basic fibroblast growth factor (FGF-2) accelerates bone induction on rat calvaria: An application of a new drug delivery system. *J. Dent Res.* **1998**, *77* (12), 1965–1969.
16. Schmidmaier, G., Wildemann, B., Ostapowicz, D. Long-term effects of local growth factor (IGF-I and TGF-beta 1) treatment on fracture healing: A safety study for using growth factors. *J. Orthop. Res.* **2004**, *22* (3), 514–519.
17. Schmidmaier, G., Lucke, M., Schwabe, P. Collective review: Bioactive implants coated with poly(d,l-lactide) and growth factors IGF-1, TGF-beta1, or BMP-2 for stimulation of fracture healing. *J. Long Term Eff. Med. Implants* **2006**, *16* (1), 61–69.

18. Shibuya, T.Y., Wadhwa, A., Nguyen, K.H. Linking of bone morphogenetic protein-2 to resorbable fracture plates for enhancing bone healing. *Laryngoscope* **2005**, *115* (12), 2232–2237.
19. Hamada, Y., Katoh, S., Hibino, N. Effects of monofilament nylon coated with basic fibroblast growth factor on endogenous intrasynovial flexor tendon healing. *J. Hand. Surg. [Am.]* **2006**, *31* (4), 530–540.
20. Rickert, M., Jung, M., Adiyaman, M. A growth and differentiation factor-5 (GDF-5)-coated suture stimulates tendon healing in an Achilles tendon model in rats. *Growth Factors* **2001**, *19* (2), 115–126.
21. Dines, J.S., Weber, L., Razzano, P. The effect of growth differentiation factor-5-coated sutures on tendon repair in a rat model. *J. Shoulder Elbow Surg* **2007**, *16* (5 Suppl), S215–S221.
22. Lee, A.R. Enhancing dermal matrix regeneration and biomechanical properties of 2nd degree-burn wounds by EGF-impregnated collagen sponge dressing. *Arch. Pharm. Res.* **2005**, *28* (11), 1131–1136.
23. Nillesen, S.T., Geutjes, P.J., Wismans, R. Increased angiogenesis and blood vessel maturation in acellular collagen-heparin scaffolds containing both FGF2 and VEGF. *Biomaterials* **2007**, *28* (6), 1123–1131.
24. Yao, C.C., Yao, P., Wu, H. Absorbable collagen sponge combined with recombinant human basic fibroblast growth factor promotes nerve regeneration in rat sciatic nerve. *J. Mater. Sci. Mater. Med.* **2007**, *18* (10), 1969–1972.
25. Gomez, N., Schmidt, C.E. Nerve growth factor-immobilized polypyrrole: Bioactive electrically conducting polymer for enhanced neurite extension. *J. Biomed. Mater. Res. A.* **2007**, *81* (1), 135–149.
26. Klaver, C.L., Caplan, M.R. Bioactive surface for neural electrodes: Decreasing astrocyte proliferation via transforming growth factor-beta1. *J. Biomed. Mater. Res. A.* **2007**, *81* (4), 1011–1016.
27. Shibuya, T.Y., Kim, S., Nguyen, K. Covalent linking of proteins and cytokines to suture: Enhancing the immune response of head and neck cancer patients. *Laryngoscope.* **2003**, *113* (11), 1870–1884.
28. Wiranowska, M., Ransohoff, J., Weingart, J.D. Interferon-containing controlled-release polymers for localized cerebral immunotherapy. *J. Interferon. Cytokine Res.* **1998**, *18* (6), 377–385.
29. Ratner, B.D., Bryant, S.J. Biomaterials: Where we have been and where we are going. *Ann. Rev. Biomed. Eng.* **2004**, *6*, 41–75.
30. Meinhart, J., Deutsch, M., Zilla, P. Eight years of clinical endothelial cell transplantation. *ASAIO J.* **1997**, *43* (5), M515–M521.
31. Kipshidze, N., Moussa, I., Nikolaychik, V. Influence of Class I interferons on performance of vascular cells on stent material in vitro. *Cardiovasc. Radiat. Med.* **2002**, *3* (2), 82–90.
32. Sungur, S., Emregül, E., Günendi, G. New glucose biosensor based on glucose oxidase-immobilized gelatin film coated electrodes. *J. Biomater. Appl.* **2004**, *18* (4), 265–277.
33. Watterson, J.D., Cadieux, P.A., Beiko, D.T. Oxalate-degrading enzymes from *Oxalobacter formigenes*: A novel device coating to reduce urinary tract biomaterial-related encrustation. *J. Endourol.* **2003**, *17* (5), 269–274.
34. Chandiwal, A., Zaman, F.S., Mast, A.E. Factor Xa inhibition by immobilized recombinant tissue factor pathway inhibitor. *J. Biomater. Sci. Polym. Ed.* **2006**, *17* (9), 1025–1037.
35. Hubbell, J.A. Bioactive biomaterials. *Curr. Opin. Biotechnol.* **1999**, *10* (2), 123–129.
36. Yamada, K.M. Adhesive recognition sequences. *J. Biol. Chem.* **1999**, *266* (2), 12809–12812.
37. Hubbell, J.A. Biomaterials in tissue engineering. *Biotechnology (N.Y.).* **1995**, *13* (6), 565–576.
38. Urry, D.W., Pattanaik, A., Xu, J. Elastic protein-based polymers in soft tissue augmentation and generation. *J. Biomater. Sci. Polym. Ed.* **1998**, *9* (10), 1015–1048.
39. Takagi, M., Shiwaku, K., Inoue, T. Hydrodynamically stable adhesion of endothelial cells onto a polypropylene hollow fiber membrane by modification with adhesive protein. *J. Artif. Organs.* **2003**, *6* (3), 222–226.
40. Borkenhagen, M., Clemence, J.F., Sigrist, H. Three-dimensional extracellular matrix engineering in the nervous system. *J. Biomed. Mater. Res.* **1998**, *40* (3), 392–400.

41. Lutolf, M.P., Hubbell, J.A. Synthesis and physicochemical characterization of end-linked poly(ethylene glycol)-co-peptide hydrogels formed by Michael-type addition. *Biomacromolecules.* **2003**, *4* (3), 713–722.

42. Reyes, C.D., Petrie, T.A., Burns, K.L. Biomolecular surface coating to enhance orthopaedic tissue healing and integration. *Biomaterials.* **2007**, *28* (21), 3228–3235.

43. Hamm, C.W., Schäachinger, V., Münzel, T. et al. Peptide-treated stent graft for the treatment of saphenous vein graft lesions: First clinical results. *J. Invasive Cardiol.* **2003**, *15* (10), 557–560.

44. Shibuya, T.Y., Wei, W.Z., Zormeier, M. Anti-CD3/anti-CD28 monoclonal antibody-coated suture enhances the immune response of patients with head and neck squamous cell carcinoma. Arch. Otolaryngol. *Head Neck Surg.* **1999**, *125* (11), 1229–1234.

45. Aoki, J., Serruys, P.W., van Beusekom, H. Endothelial progenitor cell capture by stents coated with antibody against CD34: The HEALING-FIM (Healthy Endothelial Accelerated Lining Inhibits Neointimal Growth-First In Man) Registry. *J. Am. Coll. Cardiol.* **2005**, *45* (10), 1574–1579.

46. Kawamoto, A., Asahara, T. Role of progenitor endothelial cells in cardiovascular disease and upcoming therapies. *Catheter Cardiovasc. Interv.* **2007**, *70* (4), 477–484.

47. Baron, J.H., Gershlick, A.H., Hogrefe, K. In vitro evaluation of c7E3-Fab (ReoPro) eluting polymer-coated coronary stents. *Cardiovasc. Res.* **2000**, *46* (3), 585–594.

48. Hong, Y.J., Jeong, M.H., Kim, W. et al. Effect of abciximab-coated stent on in-stent intimal hyperplasia in human coronary arteries. *Am. J. Cardiol.* **2004**, *94* (8), 1050–1054.

49. Kim, W., Jeong, M.H., Hong, Y.J. et al. The long-term clinical results of a platelet glycoprotein IIb/IIIa receptor blocker (Abciximab: Reopro) coated stent in patients with coronary artery disease. *Korean J. Int. Med.* **2004**, *19* (4), 220–229.

50. Kim, W., Jeong, M.H., Kim, K.H. et al. The clinical results of a platelet glycoprotein IIb/IIIa receptor blocker (abciximab: ReoPro)-coated stent in acute myocardial infarction. *J. Am. Coll. Cardiol.* **2006**, *47* (5), 933–938.

51. Stefanadis, C., Toutouzas, K., Stefanadi, E. et al. Inhibition of plaque neovascularization and intimal hyperplasia by specific targeting vascular endothelial growth factor with bevacizumab-eluting stent: An experimental study. *Atherosclerosis.* **2007**, *195* (2), 268–276.

12

Cell Encapsulation

James Blanchette

CONTENTS

Introduction

Cell encapsulation is a research area which has emerged as an exciting approach for numerous biomedical applications including delivery of therapeutics and regenerative medicine. The seminal study of Lim and Sun[1] published in 1980, using encapsulated islets of Langerhans to temporarily restore euglycemia in diabetic rats, offered a glimpse at the potential of combining living tissue with materials for transplantation. One of the primary motivations for cell encapsulation is isolation of cells from components of the immune system following transplantation. This could facilitate transplantation of allogeneic or xenogeneic tissue or reduce the need for immunosuppressive drugs. Materials can also be used to support cell function, and this is a primary requirement in tissue engineering. Materials can not only provide the desired shape for the tissue being formed but can also be designed to guide differentiation of encapsulated cells toward the desired phenotypes.

Key Material Properties

There are a number of important properties to consider when selecting or designing a material for cell encapsulation. Their optimal values will vary with the intended application, but these critical properties include toxicity/immunogenicity, diffusivity of nutrients and waste products, degradation rate, presentation of cell-binding sequences, and mechanical properties. Figure 12.1 shows a schematic of encapsulated cells and these important material properties.

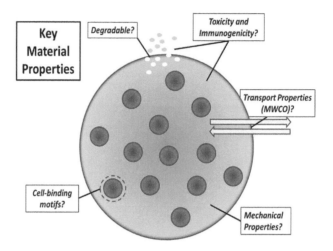

FIGURE 12.1
Schematic of encapsulated cells with five key material properties that must be considered to design a successful system.

Obviously, the material selected should not cause loss of viability for either the encapsulated cells or surrounding tissue at the implantation site. This trait can be expanded to include the desire that the material not elicit a vigorous immune response as well. Many capsules are designed to prevent contact between immune cells and surface antigens on the transplanted cells. This only requires that the material have pores small enough to prevent cell motility or that another strategy (such as a conformal coating on cells) is used to physically interfere with the potential interaction.[2] If immune cells are activated by the surface of the material, secreted cytokines could pass through the material and damage encapsulated cells. Cytokine transport can also go in the other direction, with cytokines secreted by the encapsulated cells triggering an immune response if they are not autologous cells. What can and cannot freely diffuse through the material is represented by the material's molecular weight cutoff (MWCO). This represents a molecular mass, commonly expressed in kilodaltons, above which diffusion is impaired by the material. Materials with a low MWCO may be able to prevent transport of antibodies, but it is not possible to block all components of the immune system while still allowing nutrient transport. Hydrophilic materials tend have a higher biocompatibility as they resist protein adsorption. Protein adsorption can initiate a cascade of events culminating in fibrotic encapsulation of the capsule as discussed in a recent review.[3] The use of natural materials for the capsule increases the risk of triggering an immune response due to antigens in the material or insufficient removal of immunogenic compounds from the material such as endotoxin.

Another potential consequence of activating the innate immune system is fibrous encapsulation of the material. This would limit the diffusion of nutrients and waste products to and from the encapsulated cells. The material itself must not present a barrier to this diffusion or insufficient access to nutrients, such as oxygen and glucose, combined with accumulation of metabolic waste would lead to loss of cell viability. Because many encapsulation materials do not allow cell migration, blood vessels and the lymphatic system cannot penetrate the capsule to deliver nutrients and eliminate waste. Highly porous materials or degradable materials may allow access of the circulatory and lymphatic systems to the cells eventually, but there will be a period when the encapsulated cells will be reliant on diffusion. Selecting a material which will allow maintenance of viability during this period is critical. In addition to allowing nutrient transport, the MWCO for a material must be selected to ensure passage of any molecules essential to function. For example, materials for encapsulation of islets must have a MWCO large enough to allow insulin (5.8 kDa) diffusion. The high water content of hydrogels aids in the diffusion of these compounds and is one reason for their popularity as a cell encapsulation material. MWCO can be modified by adjusting the pore size in the material. For hydrogels, this can be accomplished by modifying the polymer volume fraction in the gel or the cross-linking density, for example.

As mentioned above, some capsules should degrade to allow access of the encapsulated cells to the surrounding tissue. This is often the case in tissue engineering applications. It is ideal for the degradation rate of the capsule to match the rate at which the desired tissue replaces it. For applications like encapsulation of islets, the integrity of the capsule must not be compromised. Nondegradable materials are therefore desired to maintain the immunoprotective effect of the material. Materials which can be modified to control their degradation rate can be optimized for a wide range of cell encapsulation strategies. Degradation products also need to be evaluated for their potential toxicity and immunogenicity to avoid issues mentioned above.

The use of a synthetic material for the capsule presents a foreign microenvironment to encapsulated cells. Anchorage-dependent cells require engagement of membrane proteins like integrins to prevent apoptosis. This is the motivation behind functionalizing these materials to include ECM proteins or peptide sequences. Many natural materials like collagen are selected because such sequences are already present. By presenting the appropriate sequences in the appropriate ratios to the cells, the physiological environment can be recreated more accurately inside the capsule. This can impact not only cell viability but also differentiation of multipotent cells. These sequences can also be used to allow migration of cells into a material which is a component of some tissue engineering strategies. Apoptosis resulting from insufficient or improper cell–cell and cell–extracellular matrix (ECM) interactions is termed anoikis. Failure to provide these interactions to anchorage-dependent cells within the capsule will lead to cell death and failure of the implant. Encapsulating cells as multi-cell aggregates is another strategy to prevent anoikis, but as aggregates increase in dimensions, nutrient transport to central cells may be insufficient. The relationship between these potential causes of cell death is shown in Figure 12.2A and images of adipose-derived stem cells encapsulated in poly(ethylene glycol) as dispersed or aggregated cells are shown in Figure 12.2B.

The relationship between aggregate dimensions and cell stress can be seen when encapsulating islets. Islets are roughly 200 μm in diameter when isolated from the pancreas. The inability to revascularize these cell masses following encapsulation can lead to insufficient nutrient supply for central cells. If oxygen can only penetrate through a tissue layer with a small number of cells, any tissue mass with a radius greater than this will see hypoxic or anoxic conditions in its core. Studies have shown that when native islets are transplanted they experience central necrosis, and that islet size plays an important role in performance.[4,5] A number of research groups have investigated the ability of islets to be dissociated and reaggregated into smaller tissue masses. By breaking down the native islet structure into single cells or small aggregates of cells, the hope is that no necrotic core will be formed due to insufficient oxygen reaching the center of large islets. Small aggregates or single cells are more likely to experience anoikis, which makes interaction with the encapsulation material perhaps more important. This is an example where both the material's transport properties and ability to interact with encapsulated cells are critical to success.

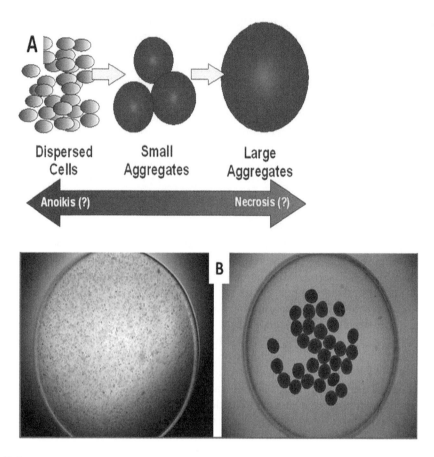

FIGUER 12.2

The relationship between how cells are encapsulated and the potential role of anoikis and necrosis due to insufficient nutrient supply is shown in part A. Images of 300,000 cells encapsulated in a PEG matrix as dispersed celss or thirty 10,000 cell aggregates are shown in part B.

The mechanical properties of the capsule must be carefully considered. Not only does the capsule need sufficient strength to resist mechanical failure and rupture but the elasticity of cell substrates have been shown to influence cell behavior. The mechanism by which substrate elasticity influences differentiation of multipotent cells is not clear. A number of recent studies have analyzed this in both two- and three-dimensional (3D) culture systems.[6–9] Integrin engagement is thought to be required to transmit the elasticity of the capsule material to the cells inside. Matrix elasticity has been shown to influence the lineage commitment of naïve, mesenchymal stem cells.[10] Another important consideration is how the mechanical properties of a capsule impact the transmission of external mechanical forces to the cells. Many cells respond to mechanical loads and this can be necessary for maintenance of phenotype or to guide differentiation.[11]

Common Encapsulation Materials

Alginate showed promising early results and is the most widely used material for cell encapsulation. A number of synthetic polymers are attractive choices due the wide range of functionalities which can be introduced into the material to customize the material's properties. Specific ECM components or decellularized ECM isolated from human or non-human sources are gaining in popularity due to their close match to the physiological environment from which the cells were obtained. For tissue engineering applications, cells are sometimes seeded onto the material rather than encapsulated within it. These studies are still instructive to guide the design of encapsulation systems with that material as the desired material properties do not change significantly. Some of the most common encapsulation materials are discussed in detail below with a focus on ways to control key properties.

Alginates

Alginates are natural polymers obtained from brown seaweed and bacteria.[12,13] Alginate is a collective term for a family of polymers with a wide range in chemical composition, sequential structure and molecular size. This variation leads to a range of different properties; the origin and composition of the most common alginates are summarized by Thu et al.[14] Alginate is a linear polysaccharide copolymer of 1–4-linked β-D-mannuronic acid (M) and α-L-guluronic acid (G) monomers. The monomers are arranged in a block-wise pattern along the chain with homopolymeric regions of M and G termed M- and G-blocks, respectively. These M- and G-blocks are interspaced with regions of alternating structure (MG-blocks). Since alginates are derived naturally, there is a potential lot-to-lot variability of the material as it is isolated from different sources. Another issue with the use of alginates is activation of the immune response from samples that have not been sufficiently purified to remove compounds like endotoxins, polyphenols, and various proteins.[15] The purification and sterilization processes are time-consuming and typically accomplished by filtration techniques.[16,17] Removal of these compounds also leads to a modest increase in the hydrophilicity of the alginate.

Alginates are able to form gels by the binding of divalent cations to the G-units, and this interaction can influence the material's degradation rate. The affinity for various divalent cations for the G-unit is: $Pb^{2+} > Cu^{2+} > Cd^{2+} > Ba^{2+} > Sr^{2+} > Ca^{2+} > Co^{2+} = Ni^{2+} = Zn^{2+} > Mn^{2+}$. Alginate gelling conditions are mild, and the gelation process can be reversed by extracting calcium ions. This is done by adding citrate or by rinsing the gel with a sodium chloride solution.[18] Alginate hydrogels can be enzymatically degraded in a controlled and tunable fashion. Humans do not produce alginases, however, so these enzymes need to be introduced with a protein delivery system. A study[19] developed alginate scaffolds with tunable degradation rates by incorporating alginate lyase-loaded microspheres that released the enzymes over time to degrade the scaffold. These scaffolds were successfully used to culture neural progenitor cells and increased their proliferation rate compared to when such cells were cultured in alginate scaffolds without microspheres.

In 1980, Lim and Sun[1] originally described the treatment of diabetes *in vivo* with alginate-based capsules. The study exhibited a complete correction of the condition for up to 3 weeks, while the morphology and functionality of the cells was maintained for over 15 weeks. Furthermore, Sun et al.[20] showed induced normoglycemia without immunosuppression up to 800 days with the transplantation of encapsulated porcine islets into spontaneously diabetic monkeys. Both adipose-derived adult stem cells as well bone marrow-derived mesenchymal stem cells have been shown to survive and differentiate in these scaffolds.[15,21] Alginate scaffolds have also been used in combination with embryonic stem cells to generate hepatocytes and endothelial cells.[22,23] Alginate has been used extensively in the culturing of chondrocytes, hepatocytes, and Schwann cells for nerve regeneration. [24–28] These studies show that the transport properties of alginate are sufficient to maintain the encapsulated cells for days or perhaps weeks, but improved oxygen transport can lead to extension of cell survival and function. A common modification to alginate capsules to achieve this is incorporation of perfluorocarbon (PFC) into the capsule. A recent study by Johnson et al.[29] tracked viability of islets cultured in low oxygen conditions encapsulated in alginate or alginate containing a PFC emulsion and demonstrated less necrosis in central regions of the islets when PFC was present.

A potential limitation in using alginate gels in tissue engineering is the lack of cellular interaction. Alginates discourage protein adsorption due to their hydrophilic character and are unable to specifically interact with mammalian cells.[30] The cell adhesion ligand RGD can be covalently coupled into alginate gels to enhance cell adhesion and address this stress. These modified alginate gels have been demonstrated to provide for the adhesion and proliferation of encapsulated skeletal muscle cells.[31] Covalent binding of avidin to alginate capsules allows subsequent functionalization of the capsule with biotinylated molecules. This strategy was employed to modify an alginate capsule with lectin and is a flexible technique to introduce a range of cell-binding molecules.[32] Hydroxyapatite has also been blended with alginates to create an environment more closely matching native bone in bone regeneration strategies. Ca^{2+} ions present in the apatite cross-link with the alginate form a composite which supports viability and maintained the osteoblast phenotype for encapsulated cells.[33]

The cross-linking density of the ionic cross-linked gels can be readily manipulated by varying the M-to-G ratio and molecular weight of the polymer chain. This will impact both the pore size and the mechanical properties of the material. In general, alginates rich in G residues form strong, brittle gels, while M-rich alginates form softer, more elastic gels. The elastic modulus of an alginate gel depends on the number and the strength of the cross-links and the length and stiffness of chains between cross-links.[14,34–36] The modulus also depends on the cross-linking ions as the gel strength scales with the affinity between polymer and cross-linking ions. Alginate hydrogels lose more than 60% of their initial mechanical strength within 15 hr of exposure to physiological buffers due to the loss of divalent ions during ion exchange.[35] This problem can be avoided by adding calcium ions to the surrounding fluid. Typically, calcium or barium is used to form rigid, biocompatible beads from alginate droplets. An ionic cross-link is formed between the carboxylic acid group (i.e., guluronic acids), found on the polymer backbone, and the cation.[37] Barium, which forms strong cross-links with alginate, produces stronger gels when compared to calcium. The capsule strength can also be tuned by adjusting other factors including: polymer purity, polymer molecular weight, and gelation time.[38,39]

The stability of alginate capsules can be improved by adding a polycation layer to the outer surface. While successful transplants have been shown with barium alginate capsules in the absence of a polycation layer, most alginate gels, in the absence of additional polycation coatings, are not stable for long-term use due to the slow leakage of the cations out of the capsule.[40–42] Poly-L-lysine (PLL) is the most common cation used to form these surface coatings. While PLL was originally applied to provide an immunoprotective outer membrane as it has been shown to improve capsule stability, studies have shown that it can be toxic for certain cell types, can induce fibrotic encapsulation, can activate macrophages, and can also trigger the compliment system.[1,43–47]

Poly(ethylene glycol)

Poly(ethylene glycol) (PEG) is a polyether which is also referred to as poly(ethylene oxide) or polyoxyethylene depending on the chain's molecular weight. The terminal groups are initially hydroxyls which are replaced when the chains are cross-linked. To form microcapsules, the terminal groups are replaced with specific active groups that can cross-link chains under specific conditions such as the incidence of UV or visible light.[48] PEG is approved by the Food and Drug Administration (FDA) for use in a number of medical materials and is typically selected due to its biocompatibility and high degree of swelling in aqueous environments. This hydrophilic nature decreases the adsorption of proteins, which consequently reduces activation of an immune response following implantation and formation of a fibrous capsule around the material. Synthetic polymers, such as PEG, avoid the issue of source variability that accompanies natural polymers. Due to its properties, PEG has also been used to coat a number of materials and has also been attached to molecules to increase their circulation time in the bloodstream. PEG has been used to coat alginate/PLL capsules to shield the positive PLL charge from the in vivo environment to reduce toxicity. PEG has also been used to form conformal coatings, where a thin PEG barrier is directly attached to the outer membrane of cells.[49] Surface, shape, and pore size may vary between coated cell masses, thereby reducing reproducibility of results.

PEG-based hydrogels have been used for encapsulation of a number of cell types in addition to islets. The list includes marrow stromal cells,[50] chondrocytes,[51–53] osteoblasts,[54] and valvular interstitial cells.[55] The PEG hydrogel network structure is a flexible platform allowing physical properties, such as the swelling ratio and mesh size, to be easily manipulated through modification of the fabrication parameters. Two of these parameters are the macromer molecular weight and percentage of macromer in solution.[53,56–58] The network cross-linking density of a hydrogel controls many of its properties, such as diffusion coefficients, mechanical behavior, and rate of degradation, with significant effects on the behavior of entrapped cells.[59,60] Hydrogels with low cross-linking density have a larger mesh size, or the distance between cross-links, which allows faster diffusion of nutrients and waste to and from encapsulated cells.[60] The swelling ratio of a hydrogel, the ratio of its swollen weight to its dry weight, is related to the cross-linking density and is a measure of how much water is retained by the hydrogel.[52] Covalently cross-linked PEG networks are appealing candidates for an encapsulation material due to the "stealth" properties of PEG. In addition to the immunological concerns mentioned above, interaction with proteins could encourage cell binding to the surface of the capsule. This layer of cells on the surface would create additional diffusion limitations for the encapsulated tissue. For encapsulated islets, the response to glucose changes would be slowed and more of the encapsulated tissue may fail due to insufficient nutrient supply. Cells and biological molecules are readily encapsulated via solution photopolymerization at room temperature and physiological pH in PEG gels with multiple geometric configurations. Using previously developed models for diffusion in PEG gels,[61,62] the average diffusivity in the gel can be estimated relative to that in water as a function of the solute size and gel mesh size. The hydrophilic properties of PEG will facilitate transport of critical solutes to maintain cell viability.

In many applications, biodegradable or bioerodible photopolymerized hydrogels are required. Sawhney et al.[63] developed photopolymerizable, bioerodible hydrogels based on PEG-*co*-poly(α-hydroxy acid) diacrylate macromers. As PEG does not degrade in the body naturally, proteolytically degradable, photopolymerized hydrogels have also been developed for tissue engineering applications. Proteolytically degradable peptides that are cleaved by enzymes involved in cell migration, such as collagenase and plasmin, can be copolymerized with PEG as a BAB block copolymer (with the degradable site as the A block), then terminated with acrylate groups. The resultant polymers were found to rapidly degrade in the presence of the targeted protease in a dose-dependent manner, but remained stable in the presence of other proteases.[64] Furthermore, cells were able to degrade these materials during migration.[65] This scheme may allow the rate of material degradation to match that of tissue formation. If the degradable peptide sequence in the hydrogel is a target sequence of matrix metalloproteinases (MMPs), the degradation of the capsule will occur by a mechanism similar to turnover of the ECM.

The major limitation of PEG hydrogels as scaffolds for tissue engineering is lack of cell-specific adhesion. The resistance to protein adsorption creates a blank environment surrounding the encapsulated cells which can trigger anoikis for anchorage-dependent cells. To overcome this limitation, PEG macromers have been modified with integrin-binding moieties found in the ECM surrounding the encapsulated cell to make the hydrogels more biomimetic. For example, hydrogels been modified with a variety of cell adhesion ligands such as RGD, a fibronectin-derived peptide KQAGDV, or a laminin-derived peptide YIGSR to enhance cell adhesion.[66–70] The pathways initiated by integrin binding may also be triggered by genetic modification to extend viability of encapsulated cells.[71]

In combination with stem cells, PEG scaffolds have been evaluated for their suitability as potential replacements for bone, cartilage, nerve, liver, and vascular tissue. A great deal of research has been published using PEG scaffolds seeded with stem cells to generate bone and cartilage.[50,72–78] For bone applications, studies explored using mesenchymal stem cells combined with a variety of cues, including RGD peptides, bone morphogenetic protein (BMP), and heparin, to promote osteogenic differentiation.[50,72–75] Similar approaches were used to produce cartilage by the addition of bioactive molecules such as chondroitin sulfate, transforming growth factor-β, and BMPs.[76–78] For the treatment of Parkinson's disease or spinal cord injury, nerve tissue was engineered using PEG scaffolds.[79–82] PEG scaffolds were used for coculture of neural progenitor cells and endothelial cells to engineer the nerve tissue.[79] The addition of endothelial cells allowed formation of microvasculature.[80] The addition of bFGF (basic fibroblast growth factor) and collagen was used to promote neuronal differentiation of precursor cells when cultured in PEG scaffolds.[81,82]

The permeability and mechanical properties of the hydrogels depend on the length of the PEG chains, the polymer volume fraction in the gel, and the cross-linking density. The cross-linking density of PEG hydrogels can be increased by increasing the concentration of macromer in PEG solutions, decreasing the molecular weight of the PEG macromers, or by using branched PEG structures instead of linear structures, with corresponding increases in compressive modulus.[52,60,83] The PEG macromer has also been modified with fumaric acid to form hydrogels made of oligo(poly(ethylene glycol) fumarate), which are photo-cross-linkable, injectable, and can be prepared with compressive moduli as high as cartilage.[60,84] One challenge is that the mechanical properties of PEG hydrogels formed through photopolymerization (a cytocompatible method commonly used for cell encapsulation) are inhomogeneous when tested at different locations in the material. This increases when cells are present and interact with the scaffold. These issues are discussed in detail in a recent article discussing the use of PEG hydrogels for 3D culture of cells.[85]

Chitosan

Chitosan is a natural biopolymer that has many desirable characteristics as a scaffold or encapsulation material. It is a biodegradable, semicrystalline polysaccharide obtained by N-deacetylation of chitin, which is harvested from the exoskeleton of marine crustaceans. Chitosan is composed of glucosamine and N-acetyl glucosamine which are linked by glycosidic bonds. Being structurally similar to ECM components, chitosan provides cell–ECM interactions which guide cell behavior. The degree of deacetylation, indicating the free amine groups along the chitosan backbone, is a key parameter, which changes its physico-chemical properties such as solubility, chain conformation, and electrostatic properties.[86] Chitosan possesses a number of characteristics which make it suitable for implantation, such as inducing a minimal foreign body reaction, an intrinsic antibacterial nature, and the ability to be molded into highly porous structures of varied geometry suitable for cell growth and osteoconduction.[87,88]

VandeVord et al.[89] examined the biocompatibility of chitosan in mice. Their data imply that chitosan has a chemotactic effect on immune cells, but that effect does not lead to a humoral immune response. Also, results suggest that the specific responses reported may have been caused by contaminating proteins/polysccharides from the source organism. This type of contamination has been reported with other polysaccharides studied for implant use. As discussed earlier with alginates, the need to use highly purified grades of these biomaterials is critical to avoid toxicity and fibrotic encapsulation.[90,91]

Chitosan exhibits pH-sensitive behavior due to the large quantities of amino groups on its chains. It is a pH-dependent, cationic polymer, which is insoluble in aqueous solutions above pH 7. However, in dilute or weak acids (pH < 6), the protonated free amino groups of glucosamine increase the solubility of the molecule.[92] Above pH 6.2, chitosan solutions form a hydrated gel-like precipitate.[93,94] Due to its cationic nature and predictable degradation rate, chitosan-based materials bind growth factors and release them in a controlled manner.[95] Chitosan is degraded, depending on degree of deacetylation, by enzymes such as lysozyme, N-acetyl-D-glucosaminidase, and lipases.[96] *In vivo*, chitosan is degraded by enzymatic hydrolysis, primarily by lysozyme which appears to target acetylated residues.[97,98] Degradation kinetics seems to be inversely related to the degree of deacetylation.[97] Lysozyme breaks down the chitosan polymer chain, diminishing its molecular weight until it becomes short enough to be processed by cells. Glucosamines, the final degradation products of chitosan, are nontoxic, nonimmunogenic, and noncarcinogenic.[98] *In vivo*, the final degradation products undergo normal metabolism pathways and may be incorporated into glycoproteins or excreted as carbon dioxide gas during respiration.[99,100]

Due to its cationic amine groups, chitosan provides a suitable environment for cell adhesion through interaction with negatively charged ECM components like glycosaminoglycans and proteoglycans.[101] Chondrocytes encapsulated in injectable, chitosan hydrogels repaired non-weight-bearing defects in sheep, with good integration with the surrounding tissue.[102] This can be attributed to chitosan's structural similarity to major components of the ECM of bone and cartilage. To customize the adherent ability for seeding cells, chitosan allows for a wide range of molecules to be introduced within the material. Conjugation of chitosan with biologically active, RGD-containing peptides or laminin peptides allows further customization to develop desirable scaffold materials for tissue regeneration.[103] Kuo and Lin[104] hybridized chitin and chitosan through genipin cross-linking and subsequently freeze-dried the constructs to enhance chondrocytic attachment and growth. The resultant scaffold was then coated with hydroxyapatite to modify the surface chemistry that generated positive effects on the cell number, the content of glycosaminoglycans, and the collagen level for 28-day cultivation of bovine knee chondrocytes.

The cell-binding properties of chitosan have led to numerous studies where chitosan is used as a surface coating. Chitosan was employed as a surface modification of poly (ε-caprolactone) scaffolds and *in vitro* studies with fibroblasts showed significantly improved cell attachment and proliferation when chitosan was present.[105] In another study, polyurethane scaffolds were prepared and surface-modified with chitosan for the same reason.[106] This study showed that on the modified scaffold a monolayer of endothelial intima was formed. The incorporation of collagen with chitosan as a chitosan–collagen scaffold enhanced the resultant scaffold's ability to support cell attachment and is a strategy similar to the incorporation of peptide sequences.[107]

Chitosan can be incorporated into composite materials to create scaffolds with cell-binding and mechanical properties that vary from pure chitosan. Chitosan cross-linked with collagen has been investigated as a candidate for use as a matrix to support a bioartificial liver.[108] The study showed good compatibility of the scaffold with the hepatocytes as the presence of chitosan provided amino groups for cell adhesion. The composition of the cross-linked chitosan and collagen scaffold is closer to the native tissue and has been shown to exhibit higher mechanical strength than chitosan alone. The addition of collagen creates a scaffold with a tensile strength of 1.91 MPa and a Young's modulus of 7.11 MPa. Hydrated, porous chitosan membranes have a tensile modulus typically below 1 MPa. Zhang et al. reinforced chitosan scaffolds by addition β-tricalcium phosphate (β-TCP) to increase their mechanical strength. Pure chitosan scaffolds were soft, spongy, and very flexible with a maximum strain of ~20% before the loss of elasticity. With the addition of β-TCP, the compressive modulus was increased from 0.967 to 2.292 MPa and yield strength from 0.11 to 0.208 MPa.[109] Chitosan–alginate composites can also be formed due to the negative charge on alginate.[110] The strong ionic bonding between the amine groups of the chitosan and the carboxyl group of the alginate stabilize the scaffold despite a high porosity (~92%). This is achieved while reinforcing the scaffold; it displays a compressive modulus of 8.16 MPa and a yield strength of 0.46 MPa, which is about three times the value for pure chitosan scaffold. Also, cell–material interaction studies indicated that osteoblasts seeded on the chitosan–alginate scaffold attached well and promoted the deposition of minerals for bone formation relative to pure chitosan scaffolds.

Poly(lactic-co-glycolic acid)

Poly(lactic-co-glycolic acid) (PLGA) is a copolymer that consists of monomers of glycolic acid and lactic acid connected by ester bonds. It is an FDA-approved polymer that is attractive for cell encapsulation and as a tissue engineering scaffold. Two key properties are its biocompatibility (as it degrades into natural compounds) and the ability to modulate its degradation rate. This degradation rate controls many of the key properties for use of PLGA with cells. The transport and mechanical properties will vary significantly once PLGA is placed in a physiological environment. This material would only be used for cell encapsulation when immunoisolation and/or cell support is only desired for a period of weeks.

PLGA is hydrolytically unstable, and although insoluble in water, they degrade by hydrolytic attack of their ester bonds resulting in the formation of lactic and glycolic acids, which can be removed from the body by normal metabolic pathways.[111–114] Other factors that affect degradation include hydrophobicity and molecular weight.[115–117] The degradation of PLGA is therefore affected by the ratio of hydrophilic poly(glycolic acid) (PGA) to hydrophobic poly(lactic acid) (PLA).[116] The biocompatibility of these polymers also has been demonstrated in biological applications.[118–121] These reports implied that the rate of degradation might affect cellular interaction including cell proliferation, tissue synthesis, and host response.[122,123] However, details of the potential effects of the acidic by-products on the 3D cell culture or upon *in vivo* host response have not been studied sufficiently. Conventionally, the rate of hydrolytic degradation for these biopolymers is controlled by altering their physical properties such as their molecular weights, degree of crystallinity, and glass transition temperature.[114,124]

Numerous studies have been conducted using PLGA scaffolds seeded with mesenchymal stem cells to differentiate into osteogenic, cartilage, neural, liver, or adipose cells.[125–133] Studies suggest that PLGA can possibly be used as a bioscaffold which can help in maintaining the viability of the cells as well as help in differentiation of stem cells when incorporated with ligands, peptides, or growth factors. PLGA scaffolds have been modified with bioactive ligands, such as galactose (a specific ligand for the asialoglycoprotein receptor on hepatocytes), to help binding and attachment of cells. Galactosylated PLGA has been processed to examine hepatocyte-specific cellular binding to the surface.[134] The results demonstrated that conjugation of galactose on PLGA supported cell adhesion as well as cell viability as compared to control PLGA. In another study, the RGD peptide was immobilized onto the surface of PLGA for enhancing cell adhesion and function for bone regeneration. The extent of cell adhesion was substantially enhanced when RGD was present and the level of alkaline phosphatase activity, a marker of osteoblast function, was also elevated.[135]

References

1. Lim, F., Sun, A. M. Microencapsulated islets as bioartificial endocrine pancreas. *Science*. **1980**, *210*, 908–910.
2. Cruise, G. M., Hegre, O. D., Lamberti, F. V., Hager, S. R., Hill, R., Scharp, D. S., Hubbell, J. A. In vitro and in vivo performance of porcine islets encapsulated in interfacially photopolymerized poly(ethylene glycol) diacrylate membranes. *Cell Transplant*. **1999**, *8*, 293–306.
3. Hernández, R. M., Orive, G., Murua, A., Pedraz, J. L. Microcapsules and microcarriers for in situ cell delivery. *Advanced Drug Delivery Reviews*. **2010**, *62*, 711–730.
4. Giuliani, M., Moritz, W., Bodmer, E., Dindo, D., Kugelmeier, P., Lehmann, R., Gassmann, M, Groscurth, P., Weber, M. Central necrosis in isolated hypoxic human pancreatic islets: Evidence for postisolation ischemia. *Cell Transplant*. **2005**, *14*, 67–76.
5. Lehmann, R., Zuellig, R. A., Kugelmeier, P., Baenninger, P. B., Moritz, W., Perren, A., Clavien, P. A., Weber, M., Spinas, G. A. Superiority of small islets in human islet transplantation. *Diabetes*. **2007**, *56*, 594–603.
6. Du, J., Chen, X., Liang, X., Zhang, G., Xu, J., He, L., Zhan, Q., Feng, X. Q., Chien, S., Yang, C. Integrin activation and internalization on soft ECM as a mechanism of induction of stem cell differentiation by ECM elasticity. *Proceedings of the National Academy of Sciences of the United States of America*. **2011**, *108*, 9466–9471.
7. Parekh, S. H., Chatterjee, K., Lin-Gibson, S., Moore, N. M., Cicerone, M. T., Young, M. F., Simon, C. G. Jr. Modulus-driven differentiation of marrow stromal cells in 3D scaffolds that is independent of myosin-based cytoskeletal tension. *Biomaterials*. **2011**, *32*, 2256–2264.
8. Kumachev, A., Greener, J., Tumarkin, E., Eiser, E., Zandstra, P. W., Kumacheva, E. High-throughput generation of hydrogel microbeads with varying elasticity for cell encapsulation. *Biomaterials*. **2011**, *32*, 1477–1483.
9. Zemel, A., Rehfeldt, F., Brown, A. E., Discher, D. E., Safran, S. A. Cell shape, spreading symmetry and the polarization of stress-fibers in cells. *Journal of Physics: Condensed Matter*. **2010**, *22*, 194110.
10. Engler, A. J., Sen, S., Sweeney, H. L., Discher, D. E. Matrix elasticity directs stem cell lineage specification. *Cell*. **2006**, *126*, 677–689.
11. Shav, D., Einav, S. The effect of mechanical loads in the differentiation of precursor cells into mature cells. *Annals of the New York Academy of Sciences*. **2010**, *1188*, 25–31.
12. Smidsrod, O. & Skjak-Baek, G. Alginate as immobilization matrix for cells. *Trends in Biotechnology*. **1990**, *8*, 71–78.

13. Johnson, F. A., Craig, D. Q. M., Mercer, A. D. Characterization of the block structure and molecular weight of sodium alginates. *Journal of Pharmacy and Pharmacology*. **1997**, *49*, 639–643.
14. Thu B., Smidsrod, O., Skjak-Baek, G. Alginate gels—some structure–function correlations relevant to their use as immobilization matrix for cells. *Progress in Biotechnology*. **1996**, *11*, 19–30.
15. de Vos, P., Andersson, A., Tam, S. K., Faas, M. M., Halle, J. P. Advances and barriers in mammalian cell encapsulation for treatment of diabetes. *Immunology, Endocrine & Metabolic Agents — Medical Chemistry*. **2006**, *6*, 139–153.
16. Vandenbossche, G. M. R., Remon, J.-P. Influence of the sterilization process on alginate dispersion. *Journal of Pharmacy and Pharmacology*. **1993**, *45*, 484–486.
17. Zimmermann U., Klock, G., Federlin, K., Hannig, K., Kowalski, M., Bretzel, R. G., Horcher, A., Entenmann, H., Sieber, U., Zekorn, T. Production of mitogen-contamination free alginates with variable ratios of mannuronic acid to guluronic acid by free-flow electrophoresis. *Electrophoresis*. **1992**, *13*, 269–274.
18. LeRoux, M. A., Guilak F., Setton, L. A. Compressive and shear properties of alginate gel: Effects of sodium ions and alginate concentration. *Journal of Biomedical Materials Research*. **1999**, *47*, 46–53.
19. Ashton, R. S., Banerjee, A., Punyani, S., Schaffer, D. V., Kane, R. S. Scaffolds based on degradable alginate hydrogels and poly(lactide-*co*-glycolide) microspheres for stem cell culture. *Biomaterials*. **2007**, *28*, 5518–5525.
20. Sun, Y., Ma X., Zhou, D., Vacek, I., Sun, A. M. Normalization of diabetes in spontaneously diabetic cynomologus monkeys by xenografts of microencapsulated porcine islets without immunosuppression. *Journal of Clinical Investigation*. **1996**, *98*, 1417–1422.
21. Awad, H. A., Wickham, M. Q., Leddy, H. A., Gimble, J. M., Guilak, F. Chondrogenic differentiation of adipose-derived adult stem cells in agarose, alginate, and gelatin scaffolds. *Biomaterials*. **2004**, *25*, 3211–3222.
22. Maguire, T., Novik, E., Schloss, R., Yarmush, M. Alginate-PLL microencapsulation: Effect on the differentiation of embryonic stem cells into hepatocytes. *Biotechnology and Bioengineering*. **2006**, *93*, 581–591.
23. Gerecht-Nir, S., Cohen, S., Ziskind, A., Itskovitz-Eldor, J. Three-dimensional porous alginate scaffolds provide a conducive environment for generation of well-vascularized embryoid bodies from human embryonic stem cells. *Biotechnology and Bioengineering*. **2004**, *88*, 313–320.
24. Hannouche, D., Terai, H., Fuchs, J. R., Terada, S., Zand, S., Nasseri, B. A., Petite, H., Sedel, L., Vacanti, J. P. Engineering of implantable cartilaginous structures from bone marrow-derived mesenchymal stem cells. *Tissue Engineering*. **2007**, *13*, 87–99.
25. Dvir-Ginzberg, M., Gamlieli-Bonshtein, I., Agbaria, R., Cohen, S. Liver tissue engineering within alginate scaffolds: Effects of cell-seeding density on hepatocyte viability, morphology and function. *Tissue Engineering*. **2003**, *9*, 757–766.
26. Mosahebi A., Wiberg, M., Terenghi, G. Addition of fibronectin to alginate matrix improves peripheral nerve regeneration in tissue-engineered conduits. *Tissue Engineering*. **2003**, *9*, 209–218.
27. Mosahebi A., Fuller, P., Wiberg, M., Terenghi, G. Effect of allogeneic Schwann cell transplantation on peripheral nerve regeneration. *Experimental Neurology*. **2002**, *173*, 213–223.
28. Prang, P., Muller, R., Eljaouhari, A., Heckmann, K., Kunz, W., Weber, T., Faber, C., Vroemen, M., Bogdahn, U., Weidner, N. The promotion of oriented axonal regrowth in the injured spinal cord by alginate-based anisotropic capillary hydrogels. *Biomaterials*. **2006**, *27*, 3560–3569.
29. Johnson, A.S., O'Sullivan, E., D'Aoust, L.N., Omer, A., Bonner-Weir, S., Fisher, R. J., Weir, G. C., Colton, C. K. Quantitative assessment of islets of langerhans encapsulated in alginate. *Tissue Engineering Part C Methods*. **2011**, *17*, 435–449.
30. Smentana, K. Cell biology of hydrogels. *Biomaterials*. **1993**, *14*, 1046–1050.
31. Rowley, J. A., Madlambayan, G., Mooney, D. J. Alginate hydrogels as synthetic extracellular matrix materials. *Biomaterials*. **1999**, *20*, 45–53.
32. Sultzbaugh, K. J., Speaker, T. J. J. A method to attach lectins to the surface of spermine alginate microcapsules based on the avidin biotin interaction. *Microencapsulation*. **1996**, *13*, 363–376.

33. Tampieri, M. S. A., Landi, E., Celotti, G., Roveri, N., Mattioli-Belmonte, M., Virgili, L., Gabbanelli, F., Biagini, G. HA/alginate hybrid composites prepared through bio-inspired nucleation. *Acta Biomaterialia.* **2005**, *1*, 343–351.

34. Draget K. I., Skjak-Baek, G., Smidsrod, O. Alginate based new materials. *International Journal of Biological Macromolecules.* **1997**, *21*, 47–55.

35. Lee K. Y., Rowley, J. A., Eiselt, P., Moy, E. M., Bouhadir, K. H., Mooney, D. J. Controlling mechanical and swelling properties of alginate hydrogels independently by cross-linker type and cross-linking density. *Macromolecules.* **2000**, *33*, 4291–4294.

36. de Groot, M., Schuurs, T., van Schilfgaarde, R. Causes of limited survival of microencapsulated pancreatic islet grafts. *Journal of Surgical Research.* **2004**, *121*, 141–150.

37. Martinsen, A., Skjak-Braek, G., Smidsrod, O. Alginate as immobilization material: I. Correlation between chemical and physical properties of alginate gel beads. *Biotechnology and Bioengineering.* **1989**, *33*, 79–89.

38. Grant, G. T., Morris, E. R., Rees, D. A., Smith, P. J. C., Thom, D. Biological interactions between polysaccharides and divalent cations: The egg-box model. *FEBS Letters.* **1973**, *32*, 195–198.

39. Smidsrod, O. Molecular basis for some physical properties of alginates in the gel state. *Journal of Chemical Society, Faraday Transactions.* **1974**, *57*, 263–274.

40. Stabler, C., Wilks, K., Sambanis, A., Constantinidis, I. The effects of alginate composition on encapsulated betaTC3 cells. *Biomaterials.* **2001**, *22*, 1301–1310.

41. Simpson, N. E., Stabler, C. L., Simpson, C. P., Sambanis, A., Constantinidis, I. The role of the $CaCl_2$-guluronic acid interaction on alginate encapsulated betaTC3 cells. *Biomaterials.* **2004**, *25*, 2603–2610.

42. Morch, Y. A., Donati, I., Strand, B. L., Skjak-Baek, G. Effect of Ca^{2+}, Ba^{2+}, and Sr^{2+} on Alginate Microbeads. *Biomacromolecules.* **2006**, *7*, 1471–1480.

43. Benson, J. P., Papas, K. K., Constantinidis, I., Sambanis, A. Towards the development of a bioartificial pancreas: Effects of poly-L-lysine on alginate beads with BTC3 cells. *Cell Transplant.* **1997**, *6*, 395–402.

44. de Vos P., De Haan, B., van Schilfgaarde, R. Effect of the alginate composition on the biocompatibility of alginate–polylysine microcapsules. *Biomaterials.* **1997**, *18*, 273–278.

45. Darquy, S., Pueyo, M. E., Capron, F., Reach, G. Complement activation by alginate–polylysine microcapsules used for islet transplantation. *Artificial Organs.* **1994**, *18*, 898–903.

46. Pueyo, M. E., Darquy, S., Capron, F., Reach, G. In vitro activation of human macrophages by alginatepolylysine microcapsules. *Journal of Biomaterials Science, Polymer Edition.* **1993**, *5*, 197–203.

47. Strand, B. L., Ryan, T. L., In't Veld, P., Kulseng, B., Rokstad, A. M., Skjak-Brek, G., Espevik, T. Poly-L-lysine induces fibrosis on alginate microcapsules via the induction of cytokines. *Cell Transplant.* **2001**, *10*, 263–275.

48. Lee, K., Mooney D. Hydrogels for tissue engineering. *Chemical Reviews.* **2001**, *101*, 1869–1877.

49. Hill, R. S., Cruise, G. M., Hager, S. R., Lamberti, F. V., Yu, X., Garufis, C. L., Yu, Y., Mundwiler, K. E., Cole, J. F., Hubbell, J. A., Hegre, O. D., Scharp, D. W. Immunoisolation of adult porcine islets for the treatment of diabetes mellitus. The use of photopolymerizable polyethylene glycol in the conformal coating of mass-isolated porcine islets. *Annals of the New York Academy of Sciences.* **1997**, *831*, 332–343.

50. Nuttelman, C. R., Tripodi, M. C., Anseth, K. S. In vitro osteogenic differentiation of human mesenchymal stem cells photoencapsulated in PEG hydrogels. *Journal of Biomedical Materials Research.* **2004**, *68*, 773–782.

51. Bryant, S. J., Anseth, K. S. The effects of scaffold thickness on tissue engineered cartilage in photocrosslinked poly(ethylene oxide) hydrogels. *Biomaterials.* **2001**, *22*, 619–626.

52. Bryant, S., Anseth, K. Hydrogel properties influence ECM production by chondrocytes photoencapsulated in poly(ethylene glycol) hydrogels. *Journal of Biomedical Materials Research.* **2002**, *59*, 63–72.

53. Rice, M. A., Anseth, K. S. Encapsulating chondrocytes in copolymer gels: Bimodal degradation kinetics influence cell phenotype and extracellular matrix development. *Journal of Biomedical Materials Research A.* **2004**, *70*, 560–568.

54. Burdick, J. A., Anseth, K. S. Photoencapsulation of osteoblasts in injectable RGD-modified PEG hydrogels for bone tissue engineering. *Biomaterials.* **2002**, *23*, 4315–4323.
55. Masters, K. S., Shah, D. N., Leinwand, L. A., Anseth, K. S. Crosslinked hyaluronan scaffolds as a biologically active carrier for valvular interstitial cells. *Biomaterials.* **2005**, *26*, 2517–2525.
56. Metters, A. T., Anseth, K. S., Bowman, C. N. A statistical kinetic model for the bulk-degradation of PEG-b-PLA hydrogel networks. *Journal of Physical Chemistry B.* **2000**, 104, 7043–7049.
57. Martens, P. J., Bryant, S. J., Anseth, K. S. Tailoring the degradation of hydrogels formed from multivinyl poly(ethylene glycol) and poly(vinyl alcohol) macromers for cartilage tissue engineering. *Biomacromolecules.* **2003**, *4*, 283–292.
58. Temenoff, J. S., Park, H., Jabbari, E., Sheffield, T. L., LeBaron, R. G., Ambrose, C. G., Mikos, A.G. In-vitro osteogenic differentiation of marrow stromal cells encapsulated in biodegradable hydrogels. *Journal of Biomedical Materials Research A.* **2004**, 70, 235–244.
59. Lowman, A. M. Peppas, N. A., Hydrogels, in *Encyclopedia of Controlled Drug Delivery*, Mathiowitz, E., Ed., Wiley, New York, **1999**, 1, 397–418.
60. Bryant, S.J., Anseth, K. S., Lee, D. A., Bader, D. L. Crosslinking density influences the morphology of chondrocytes photoencapsulated in PEG hydrogels during the application of compressive strain. *Journal of Orthopedic Research.* **2004**, *22*, 1143–1149.
61. Mason, M. N., Metters, A. T., Bowman, C. N., Anseth, K. S. Predicting controlled-release behavior of degradable PLA-b-PEG-b-PLA hydrogels. *Macromolecules.* **2001**, *34*, 4630–4635.
62. Watkins, A. W. Anseth, K. S. Investigation of molecular transport and distributions in poly(ethylene glycol) hydrogels with confocal laser scanning microscopy. *Macromolecules.* **2005**, *38*, 1326–1334.
63. Sawhney, A. S., Pathak, C. P., Hubble, J. A. Bioerodible hydrogels based on photopolymerized poly(ethylene glycol)-*co*-poly(*a*-hydroxy acid) diacrylate macromers. *Macromolecules.* **1993**, *26*, 581–587.
64. West, J. L., Hubbell, J. A. Polymeric biomaterials with degradation sites for proteases involved in cell migration. *Macromolecules.* **1999**, *32*, 241–244.
65. Mann, B. K., Gobin, A. S., Tsai, A. T., Schmedlen, R. H., West, J. L. Smooth muscle cell growth in photopolymerized hydrogels with cell adhesive and proteolytically degradable domains: Synthetic ECM analogs for tissue engineering. *Biomaterials.* **2001**, *22*, 3045–3051.
66. Peyton, S. R., Raub, C. B., Keschrumrus, V. P., Putnam, A. J. The use of poly(ethylene glycol) hydrogels to investigate the impact of ECM chemistry and mechanics on smooth muscle cells. *Biomaterials.* **2009**, *27*, 4881–4893.
67. Hubbell, J. A., Massia, S. P., Desai, N. P., Drumheller, P. D. Endothelial cell-selective materials for tissue engineering in the vascular graft via a new receptor. *Biotechnology.* **1991**, *9*, 568–572.
68. Massia, S. P., Hubbell, J. A. Human endothelial cell interactions with surface-coupled adhesion peptides on a nonadhesive glass substrate and two polymeric biomaterials. *Journal of Biomedical Materials Research.* **1991**, *25*, 223–242.
69. Hern, D. L., Hubbell, J. A. Incorporation of adhesion peptides into nonadhesive hydrogels useful for tissue resurfacing. *Journal of Biomedical Materials Research.* **1998**, *39*, 266–276.
70. Mann, B. K., Tsai, A. T., Scott-Burden, T., West, J. L. Modification of surfaces with cell adhesion peptides alters extracellular matrix deposition. *Biomaterials.* **1999**, *20*, 2281–2286.
71. Blanchette, J. O., Langer, S. J., Sahai, S., Topiwala, P. S., Leinwand, L. L., Anseth, K. S. Use of integrin-linked kinase to extend function of encapsulated pancreatic tissue. *Biomedical Materials.* **2010**, *5*, 061001.
72. Benoit, D. S. W., Anseth, K. S. Heparin functionalized PEG gels that modulate protein adsorption for hMSC adhesion and differentiation. *Acta Biomaterialia.* **2005**, *1*, 461–470.
73. Benoit, D. S. W., Collins, S.D., Anseth, K. S. Multifunctional hydrogels that promote osteogenic human mesenchymal stem cell differentiation through stimulation and sequestering of bone morphogenic protein 2. *Advanced Functional Materials.* **2007**, *17*, 2085–2093.

74. Shin, H., Zygourakis, K., Farach-Carson, M. C., Yaszemski, M. J., Mikos, A. G. Modulation of differentiation and mineralization of marrow stromal cells cultured on biomimetic hydrogels modified with Arg–Gly–Asp containing peptides. *Journal of Biomedical Materials Research*. **2004**, *69*, 535–543.

75. Buxton, A. N., Zhu, J., Marchant, R., West, J. L., Yoo, J. U., Johnstone, B. Design and characterization of poly(ethylene glycol) photopolymerizable semi-interpenetrating networks for chondrogenesis of human mesenchymal stem cells. *Tissue Engineering*. **2007**, *13*, 2549–2560.

76. Varghese, S., Hwang, N. S., Canver, A. C., Theprungsirikul, P., Lin, D. W., Elisseeff, J. Chondroitin sulfate based niches for chondrogenic differentiation of mesenchymal stem cells. *Matrix Biology*. **2008**, *27*, 12–21.

77. Salinas, C. N., Cole, B. B., Kasko, A. M., Anseth, K. S. Chondrogenic differentiation potential of human mesenchymal stem cells photoencapsulated within poly(ethylene glycol)–arginine–glycine–aspartic acid–serine thiol–methacrylate mixed-mode networks. *Tissue Engineering*. **2007**, *13*, 1025–1034.

78. Hwang, N. S., Kim, M. S., Sampattavanich, S., Baek, J. H., Zhang, Z., Elisseeff, J. Effects of three dimensional culture and growth factors on the chondrogenic differentiation of murine embryonic stem cells. *Stem Cells*. **2006**, *24*, 284–291.

79. Royce Hynes, S., McGregor, L. M., Ford Rauch, M., Lavik, E. B. Photopolymerized poly(ethylene glycol)/poly(L-lysine) hydrogels for the delivery of neural progenitor cells. *Journal of Biomaterials Science*. **2007**, *18*, 1017–1030.

80. Ford, M. C., Bertram, J. P., Hynes, S. R., Michaud, M., Li, Q., Young, M., Segal, S. S., Madri, J. A., Lavik, E. B. A macroporous hydrogel for the coculture of neural progenitor and endothelial cells to form functional vascular networks in vivo. *Proceedings of the National Academy of Sciences of the United States of America*. **2006**, *103*, 2512–2517.

81. Mahoney, M. J., Anseth, K. S. Three-dimensional growth and function of neural tissue in degradable polyethylene glycol hydrogels. *Biomaterials*. **2006**, *27*, 2265–2274.

82. Mahoney, M. J., Anseth, K. S. Contrasting effects of collagen and bFGF-2 on neural cell function in degradable synthetic PEG hydrogels. *Journal of Biomedical Materials Research A*. **2007**, *81*, 269–278.

83. Sontjens, S. H., Nettles, D. L., Carnahan, M. A., Setton, L. A., Grinstaff, M. W. Biodendrimer-based hydrogel scaffolds for cartilage tissue repair. *Biomacromolecules*. **2006**, *7*, 310–316.

84. Suggs, L. J., Kao, E. Y., Palombo, L. L., Krishnan, R. S., Widmer, M. S., Mikos, A. G. Preparation and characterization of poly(propylene fumarate-*co*-ethylene glycol) hydrogels. *Journal of Biomaterials Science*. **1998**, *9*, 653–666.

85. Kloxin, A. M., Tibbitt, M. W., Anseth, K. S. Synthesis of photodegradable hydrogels as dynamically tunable cell culture platforms. *Nature Protocols*. **2010**, *5*, 1867–1887.

86. Kumar M. N. A review of chitin and chitosan applications. *Reactive and Functional Polymers*. **2000**, *46*, 1–27.

87. Kim, I. Y., Seo, S. J., Moon, H. S., Yoo, M. K., Park, I. Y., Kim, B. C., Cho, C. S. Chitosan and its derivatives for tissue engineering applications. *Biotechnological Advances*. **2008**, *26*, 1–21.

88. Aimin, C., Chunlin, H., Juliang, B., Tinyin, Z., Zhichao, D. Antibiotic loaded chitosan bar: An in vitro, in vivo study of a possible treatment for osteomyelitis. *Clinical Orthopaedics and Related Research*. **1999**, *366*, 239–247.

89. VandeVord, P. J., Matthew, H. W., DeSilva, S. P., Mayton, L., Wu, B., Wooley, P. H. Evaluation of the biocompatibility of a chitosan scaffold in mice. *Journal of Biomedical Materials Research*. **2002**, *59*, 585–590.

90. de Vos P., Wolters, G. H., Fritschy, W. M., van Schilfgaarde R. Obstacles in the application of microencapsulation in islet transplantation. *International Journal Artificial Organs*. **1993**, *16*, 205–212.

91. de Vos P., De Haan, B., van Schilfgaarde, R. Effect of the alginate composition on the biocompatibility of alginate–polylysine microcapsules. *Biomaterials*. **1997**, *18*, 273–278.

92. Madihally, S. V., Matthew, H. W. T. Porous chitosan scaffolds for tissue engineering. *Biomaterials.* **1999**, *20*, 1133–1142.
93. Ruel-Gariépy, E., Leroux, J. C. In situ-forming hydrogels—review of temperature-sensitive systems. *European Journal of Pharmaceutics and Biopharmaceutics.* **2004**, *58*, 409–426.
94. Di Martino A., Sittinger, M., Risbud, M. V. Chitosan: A versatile biopolymer for orthopaedic tissue engineering. *Biomaterials.* **2005**, *26*, 5983–5990.
95. Muzzarelli R. A., Mattioli-Belmonte, M., Tietz, C., Biagini, R., Ferioli, G., Brunelli, M. A., Fini, M., Giardino, R., Ilari. P., Biagini, G. *Biomaterials.* **1994**, *15*, 1075–1081.
96. Muzzarelli, R. A. A. Chitins and chitosans for the repair of wounded skin, nerve, cartilage and bone. *Carbohydrate Polymers.* **2009**, *76*, 167–182.
97. Tomihata, K., Ikada, Y. In vitro and in vivo degradation of films of chitin and its deacetylated derivatives. *Biomaterials.* **1997**, *18*, 567–575.
98. Muzzarelli, R. A. Human enzymatic activities related to the therapeutic administration of chitin derivatives. *Cellular and Molecular Life Sciences.* **1997**, *53*, 131–140.
99. Abarrategi, A., Civantos, A., Ramos, V., Sanz Casado, J. V., López-Lacomba, J. L. Chitosan film as rhBMP-2 carrier: Delivery properties for bone tissue application. *Biomacromolecules.* **2008**, *9*, 711–718.
100. Ma, J., Wang, H., He, B., Chen, J. A preliminary in vitro study on the fabrication and tissue engineering applications of a novel chitosan bilayer material as a scaffold of human neofetal dermal fibroblasts. *Biomaterials.* **2001**, *22*, 331–336.
101. Zhang, Z., Wang, S., Tian, X., Zhao, Z., Zhang, J., Lv, D. A new effective scaffold to facilitate peripheral nerve regeneration: Chitosan tube coated with maggot homogenate product. *Medical Hypotheses.* **2010**, *74*, 12–14.
102. Hao, T., Wen, N., Cao, J. K., Wang, H. B., Lu, S.H., Liu, T., Lin, Q. X., Duan, C. M., Wang, C.Y. The support of matrix accumulation and the promotion of sheep articular cartilage defects repair in vivo by chitosan hydrogels. Osteoarthritis Cartilage/OARS, *Osteoarthritis Research Society.* **2010**, *18*, 257–265.
103. Mochizuki, M., Kadoya, Y., Wakabayashi, Y., Kato, K., Okazaki, I., Yamada, M., Sato, T., Sakairi, N., Nishi, N., Nomizu, M. Laminin-1 peptide-conjugated chitosan membranes as a novel approach for cell engineering. *FASEB Journal.* **2003**, *17*, 875–877.
104. Kuo, Y. C. Lin, C. Y. Effect of genipin-crosslinked chitin–chitosan scaffolds with hydroxyapatite modifications on the cultivation of bovine knee chondrocytes. *Biotechnology and Bioengineering.* **2006**, *95*, 132–44.
105. Mei, N., Chen, G., Zhou, P., Chen, X., Shao, Z. Z., Pan, L. F., Wu, C. G. Biocompatibility of poly(ε-caprolactone) scaffold modified by chitosan—the fibroblasts proliferation in vitro. *Journal of Biomaterials Applications.* **2005**, *19*, 323–339.
106. Zhu, Y., Gao, C., He, T., Shen, J. Endothelium regeneration on luminal surface of polyurethane vascular scaffold modified with diamine and covalently grafted with gelatin. *Biomaterials.* **2004**, *25*, 423–430.
107. Cuy, J. L., Beckstead, B. L., Brown, C. D., Hoffman, A. S., Giachelli, C. M. Adhesive protein interactions with chitosan: Consequences for valve endothelial cell growth on tissue-engineering materials. *Journal of Biomedical Materials Research A.* **2003**, *67*, 538–547.
108. Wang, X. H., Li, D. P., Wang, W. J., Feng, Q. L., Cui, F. Z., Xu, Y. X., Song, X. H., van der Werf, M. Crosslinked collagen/chitosan matrix for artificial livers. *Biomaterials.* **2003**, *24*, 3213–3220.
109. Zhang,Y., Zhang, M. Microstructural and mechanical characterization of chitosan scaffolds reinforced by calcium phosphates. *Journal of Non-Crystalline Solids.* **2001**, *282*, 159–164.
110. Li, Z., Ramay, H. R., Hauch, K. D., Xiao, D., Zhang, M. Chitosan–alginate hybrid scaffolds for bone tissue engineering. *Biomaterials.* **2005**, *26*, 3919–3928
111. Griffith, L. G. Polymeric biomaterials. *Acta Materialia.* **2000**, *48*, 263–277.
112. Tice, T. R., Tabibi, E. S., Parenteral drug delivery: Injectables. *Treatise on Controlled Drug Delivery: Fundamentals Optimization, Applications*, Marcel Dekker, New York, **1991**, 315–339.
113. Wu, X. S., *Encyclopedic Hand Book of Biomaterials, Bioengineering*, Marcel Dekker, New York, **1995**, 1015–1054.

114. Lewis, D. H., *Biodegradable Polymers as Drug Delivery Systems*, Marcel Dekker, New York, **1990**, 1–41.
115. Lu, L., Garcia, C. A., Mikos, A. G. In vitro degradation of thin poly(DL-lactic-co-glycolic acid) films. *Journal of Biomedical Materials Research.* **1999**, *46*, 236–244.
116. Lu, L., Peter, S. J., Lyman, M. D., Lai, H.-L., Leite, S. M., Tamada, J. A., Uyama, S., Vacanti, J. P., Langer, R., Mikos, A. G. In vitro and in vivo degradation of porous poly (DL-lactic-co-glycolic acid) foams. *Biomaterials.* **2000**, *21*, 1837–1845.
117. Miller, R. A., Brady, J. M., Cutright, D. E. Degradation rates of oral resorbable implants (poly-lactates and polyglycolates): Rate modification with changes in PLA/PGA copolymer ratio. *Journal of Biomedical Materials Research.* **1977**, *11*, 711–719.
118. Zentner, G. M., Rathi, R., Shih, C., McRea, J. C., Seo, M. H., Oh, H., Rhee, B. G., Mestecky, J., Moldoveanu, Z., Morgan, M., Weitman, S. Biodegradable block copolymers for delivery of proteins and water-insoluble drugs. *Journal of Controlled Release.* **2001**, *72*, 203–215.
119. Hasirci, V., Lewandrowski, K., Gresser, J. D., Wise, D.L., Trantolo, D. J. Versatility of biodegradable biopolymers: Degradability and an in vivo application. *Journal of Biotechnology.* **2001**, *86*, 135–150.
120. Kweon, H., Yoo, M.K., Park, I.K., Kim, T. H., Lee, H. C., Lee, H. S., Oh, J. S., Akaike, T., Cho, C. S. A novel degradable polycaprolactone networks for tissue engineering. *Biomaterials.* **2003**, *24*, 801–808.
121. Rizzi, S. C., Heath, D. J., Coombes, A. G., Bock, N., Textor, M., Downes, S. Biodegradable polymer/hydroxyapatite composites: Surface analysis and initial attachment of human osteoblasts. *Journal of Biomedical Materials Research.* **2001**, *55*, 475–486.
122. Babensee, J. E., Anderson, J. M., McIntire, L. V., Mikos, A. G. Host response to tissue engineered devices. *Advanced Drug Delivery Reviews.* **1998**, *33*, 111–39.
123. Lewandrowski, K. U., Grosser, J. D., Wise, D. L., Trantolo, D. J., Hasirci, V. Tissue responses to molecularly reinforced polylactide–coglycolide implants. *Journal of Biomaterials Science, Polymer Edition.* **2000**, *11*, 401–414.
124. Cohen, S., Alonso, M. J., Langer, R. Novel approaches to controlled release antigen delivery. *International Journal Technology Assessment in Health Care.* **1994**, *10*, 121–130.
125. Chastain, S. R., Kundu, A. K., Dhar, S., Calvert, J. W., Putnam, A. J. Adhesion of mesenchymal stem cells to polymer scaffolds occurs via distinct ECM ligands and controls their osteogenic differentiation. *Journal of Biomedical Materials Research.* **2006**, *78*, 73–85.
126. Graziano, A., d'Aquino Cusella-De, R., Angelis, M. G., Laino, G., Piattelli, A., Pacifici, M., De Rosa, A., Papaccio, G. Concave pit-containing scaffold surfaces improve stem cell-derived osteoblast performance and lead to significant bone tissue formation. *PLoS ONE.* **2007**, *2*, e496.
127. Kim, H., Kim, H. W., Suh, H. Sustained release of ascorbate-2-phosphate and dexamethasone from porous PLGA scaffolds for bone tissue engineering using mesenchymal stem cells. *Biomaterials.* **2003**, *24*, 4671–4679.
128. Sun, H., Qu, Z., Guo, Y., Zang, G., Yang, B. In vitro and in vivo effects of rat kidney vascular endothelial cells on osteogenesis of rat bone marrow mesenchymal stem cells growing on polylactide–glycoli acid (PLGA) scaffolds. *Biomedical Engineering Online.* **2007**, *6*, 41.
129. Yoon, E., Dhar, S., Chun, D. E., Gharibjanian, N. A., Evans, G. R. **2007**In vivo osteogenic potential of human adipose-derived stem cells/poly lactide-*co*-glycolic acid constructs for bone regeneration in a rat critical-sized calvarial defect model. *Tissue Engineering.* **2007**, *13*, 619–627.
130. Levenberg, S., Huang, N. F., Lavik, E., Rogers, A. B., Itskovitz-Eldor, J., Langer, R. Differentiation of human embryonic stem cells on three-dimensional polymer scaffolds. *Proceedings of the National Academy of Sciences of the United States of America.* **2003**, *100*, 12741–12746.
131. Teng, Y. D., Lavik, E. B., Qu, X., Park, K. I., Ourednik, J., Zurakowski, D., Langer, R., Snyder, E. Y. Functional recovery following traumatic spinal cord injury mediated by a unique polymer scaffold seeded with neural stem cells. *Proceedings of the National Academy of Sciences of the United States of America.* **2002**, *99*, 3024–3029.
132. Neubauer, M., Hacker, M., Bauer-Kreisel, P., Weiser, B., Fischbach, C., Schulz, M. B., Goepferich, A., Blunk, T. **2005**Adipose tissue engineering based on mesenchymal stem cells and basic fibroblast growth factor in vitro. *Tissue Engineering.* *11*, 1840–1851.

133. Bhang, S. H., Lim, J. S., Choi, C. Y., Kwon, Y. K., Kim, B. S. The behavior of neural stem cells on biodegradable synthetic polymers. *Journal of Biomaterial Science, Polymer Edition*. **2007**, *18*, 223–239.

134. Yoon, J. J., Nam, Y. S., Kim, J. H., Park, T. G. Surface immobilization of galactose onto aliphatic biodegradable polymers for hepatocyte culture. *Biotechnology and Bioengineering*. **2002**, *78*, 1–10.

135. Yoon, J. J., Song, S. H., Lee, D. S., Park, T. G. Immobilization of cell adhesive RGD peptide onto the surface of highly porous biodegradable polymer scaffolds fabricated by a gas foaming/salt leaching method. *Biomaterials*. **2004**, *25*, 5613–5620.

13

BioMEMS

L. James Lee

CONTENTS

Introduction

Miniaturization methods and materials are well developed in the integrated circuit industry. They have been used in other industries to produce microdevices, such as camera and watch components, printer heads, automotive sensors, micro-heat exchangers, micro-pumps, microreactors, etc., in the last 15 years.[1,2] These new processes are known as microelectro-mechanical systems (MEMSs), with a combined international market of over US\$ 15 billion in 1998.[3] In recent years, MEMS applications have also been extended to the optical communication and biomedical fields. The former are called micro-optic electro-mechanical systems (MOEMSs), while the latter are known as biomicroelectromechanical systems (bioMEMSs). Potential MOEMS structures include optical switches, connectors, grids, diffraction gratings, and miniature lenses and mirrors. Major potential and existing bioMEMS products are biochips/sensors, drug delivery systems, advanced tissue scaffolds, and miniature bioreactors.

Future markets for biomedical microdevices for human genome studies, drug discovery and delivery in the pharmaceutical industry, clinical diagnostics, and analytical chemistry are enormous (tens of billions of U.S. dollars).[4] In the following sections, major bioMEMS applications and microfluidics relevant to bioMEMS applications are briefly introduced. Because of the very large volume of publications on this subject, only selected papers or review articles are referenced in this chapter.

Biomems Applications

Biochips/Biosensors

Chip-based microsystems for genomic and proteomic analysis are the first bioMEMS products to have been commercialized. A large number of articles have been published in this field in recent years. Here, a brief introduction is given based on several recent review articles.[5–9] Biosensors are not necessarily microsystems.[10,11] MEMS techniques, however, may greatly enhance the performance of biosensors and reduce their cost. Microfabricated biosensors can be considered a division of biochips.[10–12]

Most molecular and biological assays and tests are very tedious, as shown in Figure 13.1. They include the following steps: 1) obtaining a cellular sample (e.g., blood or tissue); 2) separating the cellular material of interest; 3) lysing the cells to release the crude DNA, RNA, and protein; 4) purifying the crude lysate; 5) performing necessary enzymatic reactions, such as denaturing, cleaving, and amplifying of the lysate by polymerase chain reaction (PCR); 6) sequencing DNA/genes using gel or capillary electrophoresis; and finally, 7) detecting and analyzing data. This process requires skilled technicians working in well-equipped biomedical laboratories, for periods of time ranging from many hours to several days to analyze a single sample. Much of today's diagnostic equipment is costly and bulky. It has limited use in medical diagnostics and is unsuited for emergency response at sites of care. To improve public health services, there is a great need to develop efficient and affordable methods and devices that can simplify the diagnostic process and be used as portable units. In recent years, the concept of integrating many analysis systems into one microdevice has attracted a great deal of interest in industry and academia. Such devices are called "laboratories-on-chips." They combine a number of biological functions (such as enzymatic reactions, antigen–antibody conjugation, and DNA/gene probing) with proper microfluidic techniques (such as sample dilution, pumping, mixing, metering, incubation, separation, and detection in micrometer-sized channels and reservoirs) in a miniaturized device. The integration and automation involved can improve the reproducibility of results and eliminate the labor, time, and sample preparation errors that occur in the intermediate stages of an analytical procedure. The miniaturized devices also allow realization of

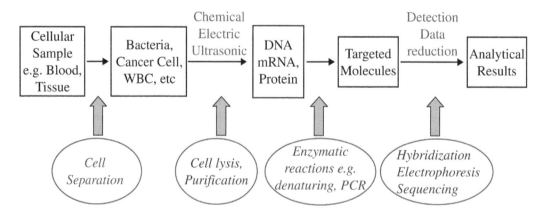

FIGURE 13.1
Schematic of molecular diagnostics. (*View this art in color at www.dekker.com.*)

low-energy and "point-of-care," parallel detection from a very small sample size, and easy data storage and transfer through computers and the Internet.

Biochips used for genomic analysis range from those used for separations for DNA sequencing, to those used in microvolume PCR, to complete analysis systems. Sequencing separations of single-stranded DNA fragments on a microchip follow the same principles as in conventional capillary electrophoresis. The process, however, is much faster because a higher electric field can be applied to the micrometer-sized separation channels without Joule heating problems. Automatic injection of a very small sample volume and parallel processing of a large number of separations can also be easily achieved. PCR allows amplification of a specific region of a DNA chain. PCR carried out on microchips is much more efficient than that on commercial PCR thermocyclers. In PCR, a sample solution is mixed, containing DNA, primers (synthesized short oligomers whose sequences flank the region of DNA to be amplified), a thermostable DNA polymerase enzyme, and the individual deoxyribonucleotides. Melting (or denaturization) of the double-stranded DNA molecules to single-stranded ones is done by heating the sample to ~95°C. The system is then quickly cooled to ~60°C for annealing; during this process, the added primers adhere to the single-stranded DNA. Finally, the sample is heated to ~72°C, at which temperature the polymerase is most active. During this extension period, complementary dinucleotide triphosphates are added to the growing strand using the target DNA as template. Each PCR cycle may double the amount of DNA of the required length. In the ideal case, 1 mol of targeted DNA fragments can be produced after 79 cycles. Practically, 20–50 cycles are needed to obtain a measurable quantity. Due to the high surface area/volume ratio associated with microdevices (it is important that PCR-friendly surfaces are produced in these devices), heat transfer is more efficient and temperature control is much easier than in large systems. PCR time can be easily reduced from hours to minutes, particularly in continuous-flow PCR chips.[6]

Miniaturized proteomic analysis devices include enzyme assays and immunoassays. The enzyme assay chip is mainly a sophisticated incubator and flow-through system. It can perform multiple functions typically required by the biochemist, namely, diluting substrate and buffer, mixing enzyme and substrate, incubating during conversion, and allowing for detection in a flow channel. Immunoassay chips are similar to enzyme assay chips, except that the main focus is antigen–antibody interaction for clinical diagnostics and drug discovery.

There are a small number of commercially available biochips in the market today. Most are microarray-based systems, with biomolecules such as DNA probes, enzymes, and antigens being immobilized on the chip surface (e.g., GenChip® from Affymetrix, NanoChip™ from Nanogen, Inc., and GeneXpert® from Cepheid), or simple microfluidic systems capable of DNA sequencing by either electrophoresis (e.g., LabChip® from Caliper, Inc., and LabCard™ from ACLARA BioSciences, Inc.) or PCR. DNA microarray chips and DNA sequencing microfluidic chips have contributed to the Human Genome Project.

In addition to commercially production, a great deal of research and development work on biochips has been going on both in industry and in academia. Genomic analysis of DNA and RNA continues to be the focus of interest, but more and more effort is being spent on proteomic analysis of proteins and peptides.[5–9] Several enzyme assays and immunoassays designed based on microarray-based systems with simple microfluidic control are close to commercialization. They can be a vital tool in clinic diagnostics, drug discovery, and biomedical research.

Completely integrated micro-total analysis systems (μ-TAS) that can perform all the functions mentioned in Figure 13.1 would be very valuable for high-throughput drug screening and personalized healthcare. However, only model systems have been proposed by research groups at present.[9] The mass production of such complicated systems at low cost is a challenging issue. Silicon and glass have been the most popular materials for fabricating microchips, but polymers are increasingly being used because of the availability of flexible, low-cost, high-throughput manufacturing methods for the micro-/nanoscale features needed for these types of applications.

Sensitive detection in microfluidic analytical devices is a challenge because of the extremely small detection volumes available. In conventional capillary electrophoresis, the most commonly employed detection method is UV absorption. In microscale biochips, laser-induced fluorescence (LIF) in conjunction with optical microscopy is currently the dominant detection technique because of its high sensitivity and noncontact nature. LIF microscopy, however, is costly, and the equipment size is quite large. To ensure wide application of the miniaturized biomedical devices, simple, portable, and low-cost detection methods are essential. Considerable efforts have been made lately to explore electrochemical methods, because the use of electrodes for detection leads to smaller instruments and cost reduction. Amperometry, conductimetry, and electro-chemiluminescence are also likely methods to complement fluorescence detection for on-chip analysis.[13]

Drug Delivery

Self-Regulated and Controllable Drug Delivery Systems

Most conventional drug delivery systems are based on polymers or lipid vesicles. Drug safety and efficacy can be greatly improved by encapsulating the drug inside or attaching it to a polymer or lipid. The three general mechanisms by which drugs are delivered from polymer or lipid systems are: 1) diffusion of the drug species through a polymer membrane; 2) a chemical or enzymatic reaction leading to cleavage of the drug from the system; and 3) solvent activation through swelling or osmosis of the system.[14] A major limitation of currently available delivery devices is that they release drugs at a predetermined rate. Certain disease states, such as diabetes, heart disease, hormonal disorders, and cancer, require drug administration either at a life-threatening moment or repeatedly at a certain critical time of day. Drug delivery technology can be taken to the next level by the fabrication of "smart" polymers or devices that are "responsive" to the individual patient's therapeutic requirements and deliver a certain amount of drug in response to a biological state. Given the miniature size of implantable devices, micromachining techniques will be essential for their manufacture. Currently, there are no commercial products based on the micromachined responsive drug delivery approach, and only some early research activity is seen in this direction.

The controlled release of drugs has been explored by adapting intelligent polymers, such as functional hydrogels, which respond to stimuli such as magnetic fields, ultrasound, electric current, temperature, and pH change.[15–17] These chemically synthesized materials are biocompatible and have good functionality. However, they often lack well-defined properties because of their inherent size and structure distribution resulting from chemical synthesis.[18] On the other hand, microfabrication technology developed for microelectronic applications is capable of mechanically creating devices with more precisely defined features, in a size range similar to that of polymeric and lipid materials.[19] Using hydrogels as switches or gates for controlled drug delivery and microfluidics has been explored recently by several researchers.[20,21] In a recent paper, Cao, Lai, and Lee[22] describe the design of a self-regulated drug delivery device based on the integration of both mechanical and chemical methods. A pH-sensitive hydrogel switch is used to regulate the drug release, while a constant release rate is achieved by carefully designing the shape of the drug reservoir.

Biocapsules, Membranes, and Engineered Particles for Drug Delivery

Immunoisolation is the protection of implanted cells from the host's immune system by the complete prevention of contact of immune molecules with the implanted cells, generally by the use of a semipermeable membrane. To achieve this without preventing nutrients from reaching the cells or waste from being removed, it is necessary to have an absolute pore size just below the minimum size needed to block out the smallest immune molecule, immunoglobulin G (IgG). The polymer and ceramic membranes used currently in biomedical devices possess nanopores with nonuniform size distributions, which makes it difficult to control the passage of drugs and immunoglobulins (~30–50 nm in size) through these membranes.[23,24] Nonuniform porosity also requires the use of long, torturous flow paths, necessitating the use of thick membranes. Nanoscale resistance to flow in such thick membranes is high, so that high applied pressures (~1–4 MPa) are needed, which further complicates use.[25] These membranes also show incomplete virus retention.

Ferrari and co-workers[19] examined the feasibility of using microfabricated silicon nanochannels for immunoisolation. A suspension of cells was placed between two microfabricated structures with nanoporous membranes to fabricate an immunoisolation biocapsule. Characterization of diffusion through the nanoporous membranes demonstrated that 18 nm channels did not completely block IgG but did provide adequate immunoprotection (immunoprotected cells remained functional in vitro in a medium containing immune factors for more than two weeks, while unprotected cells ceased to function within two days). A major application of biocapsules containing nanochannels is immunoisolation of transplanted cells for the treatment of hormonal and biochemical deficiency diseases, such as diabetes.

Polymer microparticles have attracted much attention for drug delivery applications. Traditional microparticle fabrication protocols, such as phase separation, emulsification, and spray drying, have been successfully used for the production of drug delivery microspheres.[26,27] However, due to the surface-driven manufacturing process of these methods, the structural complexity of the resulting particles is limited. These methods are also difficult to apply for producing a monodispersed particle size distribution. Size control of microparticles is an important factor, since there are many routes of drug administration. According to DeLuca et al.,[28] very large microparticles (>100 μm) with a broad particle size distribution are acceptable for embolization and drug delivery by implantation. Microparticles in the size range of 10–100 μm can be used for subcutaneous and intramuscular administration. Here, the particle size distribution is not a critical factor. Intravenous administration results in localization in the capillary vasculature and uptake by macrophages and phagocytes. Microparticles larger than 8 μm lodge predominately in the lung capillaries, whereas those smaller than 8 μm may clear the lung and be localized in the liver and spleen. Therefore, it is most important to control the size of the largest particle.

During inhalation administration to the lung, filtering of particles occurs in the upper airways by inertial impaction, with large particles (aerodynamic diameter $d_a > 5$ μm) being deposited in the mouth and the first few generations of airways. Very small particles ($d_a < 1$ μm) are dispersed by diffusion, and a large fraction of these particles remain suspended in the airflow and are exhaled. Microparticles with the optimal size range of 1–5 μm are deposited in the central and peripheral airways and in the alveolar lung region by a combination of inertial impaction and sedimentation.[29] Therefore, the size and distribution of microparticles for inhalation therapies must be closely controlled to achieve high efficiency. Inhalation is a noninvasive drug delivery route and has been used widely for the treatment of diseases such as asthma, cystic fibrosis, and chronic obstructive pulmonary disease. Potential applications of new inhalation products in the near future include the treatment of diseases such as diabetes, pain, and growth deficiency, where proteins and lipids-based drugs will be used. For these biomolecule-based drugs, processing conditions such as high temperature and long solvent contact time may result in drug denaturization, so particle formation methods must avoid such conditions. For certain envisioned functional features of drug delivery vehicles, such as targeted and controlled vector release on cancer tumors, "highly engineered" microparticles (i.e., each particle is essentially a microdevice) may be required. This is another limiting factor for the traditional microparticle fabrication methods.

Compared to conventional polymer microparticle fabrication methods, microfabrication offers greater control of particle features and geometries. The shape and size of the particles can be controlled tightly. Perhaps more importantly, the components and surface properties can be designed to achieve particular functions. Using soft lithography (this fabrication method is explained in a later section), Guan and Hansford[30] recently developed a simple method to fabricate nonspherical polymer microparticles of precise shape and size, which can serve as either drug delivery vessels or substrates for further processing to produce functional drug delivery devices. Figure 13.2 shows a micrograph of thin, platelike microparticles fabricated using this method. Combining the surface micropatterning and surface-tension self-assembly of autofolding,[31,32] well-defined 3D micropolyhedra, e.g., cubes and pyramids, can be fabricated from metals.[33] In our laboratory, similar micropolyhedra are currently being developed using functional polymers (e.g., biodegradable polymers and functional hydrogels). Large protein and gene molecules may be wrapped in such well-defined microstructures and delivered to targeted sites by either pulmonary delivery or intravenous administration.

Tissue Engineering

Tissue engineering is the regeneration, replacement, or restoration of human tissue function by combining synthetic and living molecules in appropriate configurations and environments.[34] The scaffold, the cells, and the cell–scaffold interactions are the three major components of any tissue-engineered construct. Although many tissue scaffold materials, such as foams and nonwoven fabrics, have been developed and used,[35,36] many challenges must be overcome for the promise of tissue engineering to become a reality. These include: 1) low-cost fabrication of well-defined 3D scaffold configurations at both micro- and nanoscale; 2) incorporation of appropriate biocompatibility, bioactivity, and biodegradability in the scaffolding construct to manipulate cellular and subcellular functions; and 3) active control of transport phenomena and cell growth kinetics to mimic microvasculature functions. Micro-/nanofabrication technology of polymers has tremendous potential in this field because it can achieve topographical, spatial, chemical, and immunological control over cells and thus create more functional tissue engineering constructs.[37]

An ideal tissue scaffolding process should be able to produce well-controlled pore sizes and porosity, provide high reproducibility, and use no toxic solvents. This is because these physical factors are associated with nutrient supply and vascularization of the cells in the implant as well as the development of a fibrous tissue layer that may impede nutrient access to the cells. Current processing methods used for polymer scaffolds include solvent

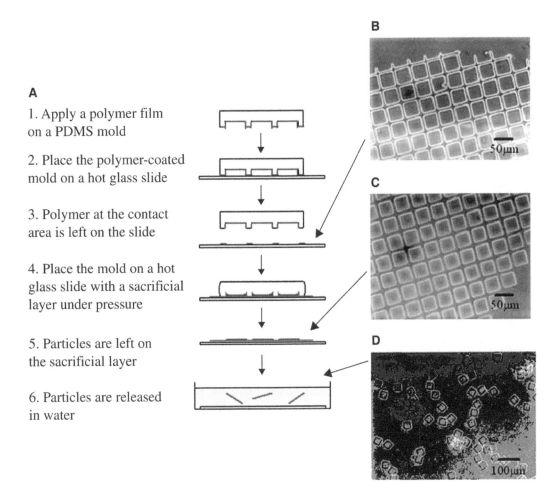

A

1. Apply a polymer film on a PDMS mold

2. Place the polymer-coated mold on a hot glass slide

3. Polymer at the contact area is left on the slide

4. Place the mold on a hot glass slide with a sacrificial layer under pressure

5. Particles are left on the sacrificial layer

6. Particles are released in water

FIGURE 13.2
Thin platelike microparticle fabricated by hot stamping. (A) The schematic of the procedure; (B) polymer at the contact area left on the glass slide; (C) polymeric microparticles left on the sacrificial layer after hot stamping; (D) and in water after release.

casting, plastic foaming, fiber bonding, and membrane lamination.[35] However, precise, reproducible features in the micrometer and nanometer range are difficult to attain using these methods. By combining living cells and microfabricated 2D and 3D scaffolds with carefully controlled surface chemistry, investigators have begun to address fundamental issues such as cell migration, growth, differentiation, apoptosis, orientation, and adhesion, as well as tissue integration and vascularization.

The functioning of tissues such as retinal, cardiac, and vascular tissue is dependent on the controlled orientation of multiple cell types. A key issue in the engineering of these tissues is control of the spatial distribution of cells in vitro to recreate a lifelike environment. The current approach to seeding cells is to allow cells to be randomly distributed in the scaffold. Microfabrication techniques, on the other hand, can produce short- and long-range surface patterns to mediate cell distribution and adhesion, biological interaction, and immune responses. Porous scaffolds without integrated blood supply rely solely on diffusion for mass transfer. They are limited to millimeters in size, while normal tissues leverage convection from blood vessels to enable oxygenation of large tissues.[38] Incorporating microfluidic networks in 3D tissue scaffolds for cell culture and implantation can be achieved by microfabrication techniques.[47] This new approach can provide the functional equivalents of microvasculature and enable scale-up of tissue engineering. Many cell-based bioreactors can be designed in a similar manner.[39]

Scientific and commercial work to date in tissue engineering has been largely devoted to clinical needs and focused on physiological aspects. There is a lack of low-cost, solventless, and mass-producible processing methods to fabricate scaffolds with well-defined micro- and nanostructures. In our laboratory, a manufacturing protocol is currently being developed for 3D tissue scaffolds of various shapes. The scaffold can be easily fabricated by combining micropatterned biodegradable polymers and supercritical CO_2 foaming technology.

Depending on the type of bioMEMS application, the polymers used can range from low-cost commodity plastics for disposable biochips to biodegradable and biofunctional polymers for drug delivery and tissue engineering. The feature size in these microsystems can be in either the micrometer or the nanometer scale. For instance, 10–100 µm is the desired microchannel size in microfluidic biochips. Below that, detection is too difficult, and, above that, mixing, heat transfer, and mass transfer are too slow. Particles used in drug delivery and the cell size in tissue scaffolds are also in the micrometer range. On the other hand, nanosized features are essential for immunoisolation in cell-based gene therapy and cell culture in tissue engineering. The enabling processing methods need to cover a broad size range, be mass producible and affordable, and be compatible with the polymers and biomolecules used in the process. Fluid transport in bioMEMS devices is crucial in many applications such as fast DNA sequencing, protein separation, drug delivery, and tissue generation.

Microfluidics

Microfluidics is the manipulation of fluids in channels, with at least two dimensions at the micrometer or submicrometer scale. This is a core technology in a number of miniaturized systems developed for chemical, biological, and medical applications. Both gases and liquids are used in micro-/nanofluidic applications,[40,41] and generally, low-Reynolds-number hydrodynamics is relevant to bioMEMS applications. Typical Reynolds numbers for biofluids flowing in microchannels with linear velocity up to 10 cm/s are less than 30.[42] Therefore, viscous forces dominate the response and the flow remains laminar.

Fluid motion in these small-scale systems can be driven by applied pressure difference, electric fields associated with charged Debye double layers (or electrical double layer—EDL)—common when ionic solutions are present, or capillary driving forces owing to wetting of surfaces by the fluid.[42] Pressure-driven flow is similar to the classic Poiseuille flow. Electrokinetic effects can result in either electro-osmotic flow (EOF) or electrophoretic responses. Electro-osmotic flow is a bulk flow driven by stresses induced in the thin EDL (i.e., 1–10 nm) near the channel walls, caused by an electric field imposed across the channel length. The velocity profile in the core of the channel is pluglike, even for a channel height as small as 24 nm.[43] Higher electrical permittivity of the fluid, imposed electric field strength, and zeta potential on the wall surface may all increase the flow rate.

Electrophoretic response, on the other hand, is the motion of charged molecules in a fluid caused by an electric field imposed across the channel length. Positively charged molecules move to the negative electrode, while negatively charged molecules move towards the positive electrode, leading to molecule separation. Electrophoresis is the most widely used separation method in the biotechnology field today. Typically, a buffer solution is chosen such that all biomolecules in the fluids, e.g., DNA/RNA fragments and proteins, are negatively charged. They all migrate from the sampling point to the detection point. Since DNA molecules have the same charge/mass ratio, separation is usually achieved by placing an immobilized gel or a mobile "gel" solution in the separation channel. Electro-osmotic flow may cause unwanted washout of the gel solution during electrophoresis, so some sort of channel coating may be necessary for EOF suppression if the channel wall has a high zeta potential (e.g., glass). On the other hand, undesirable electrophoretic separation may occur in EOF if the sample solution contains components with different charges. A high-ionic-strength-plugs method has been developed to facilitate sample transport. The use of solutions at different ionic strengths and therefore different electroosmotic mobility, however, creates a quite complex situation in microfluidics. Electrokinetic flows work very well in microchannels because of the large surface-to-volume ratio, which minimizes the Joule heating problem in this type of flow. Very high electric field strength (i.e., hundreds to thousands of volts per centimeter channel length) can be easily applied in microdevices to speed up the processing time from hours to minutes or even seconds. For nanosized channels, it has been found that very low electric power (e.g., several volts per micrometer channel length) can generate a volume flow rate that is practical for controlled drug delivery.[43]

Capillary separation is also highly favored in microfluidics. This method is simple and low-cost, but a gas–liquid interface must exist. It is mainly used for reagent loading and release in portable biochips and drug delivery systems. The velocity profile is similar to that in pressure-driven flow, but the flow is very sensitive to the surface tension of the fluid, solid surface energy and roughness, and channel shape.[44] Active control of surface tension forces to manipulate flows in microchannels can be achieved by forming gradients in interfacial tension on the channel surface[45] or by electrowetting.[46]

For most cases involving the flow of small-molecule liquids, such as buffer solutions, the standard continuum description of transport processes works very well, except that surface forces (surface tension, electrical effects, van der Waals interactions, and, in some cases, steric effects) play a more important role than usual. Although some discrepancies have been reported between pressure-driven flow measurements made in microchannels and calculations based on the Navier–Stokes equations, most have been found to be experimental errors.[42] This is because the pressure drop, as a function of flow rate, varies as the inverse fourth power of channel radius (or inverse third power of channel height), and a small change in the radius (or channel height) due to manufacturing imperfections or channel-wall contamination produces large changes in the flow. Since the volumetric flow rate varies linearly with channel radius (or height) for electrically driven flow,[43] EOF is a more reliable way than pressure-driven flow to verify microfluidic experiments with calculations. A recent study[43] shows that calculated flow rates from classical EOF analysis agree well with experimental data for channel heights in the range of 10–20 nm.

Retardation of flow of ionic liquids and solutions in microchannels, however, can be significant when the channel walls have either the same static charge[47] (e.g., glass surface is negatively charged) in pressure-driven flow or opposite charges in EOF.[43] In the former case, the flow causes charges inside the EDL to accumulate downstream, while charges on the solid channel wall remain immobile. Such excess charge creates a potential drop in the channel direction, causing a "backflow." For channel height in the range of 100 μm, this electroviscous effect (flow retardation is often counted as an increase in fluid viscosity) is small. But a retardation of 70% is observed when the glass channel diameter is in the range of several micrometers.[47] In the latter case, the backflow can be manipulated by surface micropatterning of opposite charges on the walls of the microchannel to achieve laminar chaotic mixing[48] or controllable membrane permeation.

In many bioMEMS applications, the sample fluid contains molecules and particles of various sizes. Small organic molecules are a few angstroms in size, typical protein molecules are about 2–5 nm, and large DNA molecules and cells are in the range of 1–10 μm. In some cases, the radii of near-spherical fluid droplets or gas bubbles are comparable to that of the channel, but in others, their lengths may be larger.[42] Non-Newtonian fluid and multiphase flow mechanics must be applied. In microchannels, the shear rate can be very high, e.g., 10^7 sec^{-1}, even though the Reynolds number is low. Rheological characterization of polymeric fluids and biofluids in such a flow field has recently been studied in our laboratory.[48] It was found that the standard rheological analysis used at the macroscale also works at the microscale. The high-shear Newtonian plateau can be easily observed. For solutions containing large polymer (or DNA) molecules, polymer degradation and wall slip are substantial when the flow rate is high. Rheology in microchannels needs to be studied further because many biofluids are highly non-Newtonian. One advantage of microfluidics is that a single biomolecule such as DNA can be isolated and analyzed on a biochip containing small channels or wells.[49,50] Since the molecule size is comparable to the channel (well) dimension, understanding and manipulating both the macroscopic and microscopic transport phenomena of the confined molecule undergoing flow is an active area of research.[42]

Conclusions

The miniaturization of biomedical and biochemical devices for bioMEMSs has gained a great deal of attention in recent years. Products include biochips/biosensors, drug delivery devices, tissue scaffolds, and bioreactors. In the past, MEMS devices have been fabricated almost exclusively in silicon, glass, or quartz because of the comparable technology available in the microelectronics industry. For applications in the biochemistry and biomedical field, polymeric materials are desirable because of their lower cost, good processability, and biocompatibility. Polymer microfabrication techniques, however, are still not well developed.

References

1. Madou, M.J. *Fundamentals of Microfabrication: The Science of Miniaturization*, 2nd Ed., CRC Press: Boca Raton, FL, 2002.
2. Jensen, K.F. Microchemical systems: status, challenges, and opportunities. *AIChE J.* **1999**, *45*, 2051.
3. Freemantle, M. Downsizing chemistry: chemical analysis and synthesis on microchips promise a variety of potential benefits. *Chem. Eng. News* **1999**, *77*, 27–36.
4. Snyder, M.R. Micromolding technology extends sub-gram part fabrication capability. *Mod. Plast.* **1999**, *76* (1), 85.
5. Bousse, L., Cohen, C., Nikiforov, T., Chow, A., Kopf-Sill, A.R., Dubrow, R., Parce, J.W. Electrokinetically controlled microfluidic analysis systems. *Annu. Rev. Biophys. Biomol. Struct.* **2000**, *29*, 155.

6. Sanders, G.H.W., Manz, A. Chip-based microsystems for genomic and proteomic analysis. *Trends Anal. Chem.* **2000**, *19* (6), 364.

7. Carrilho, E. DNA sequencing by capillary array electrophoresis and microfabricated array systems. *Electrophoresis* **2000**, *21*, 55.

8. Kricka, L.J. Microchips, microarrays, biochips and nanochips: personal laboratories for the 21st century. *Clin. Chim.* **2001**, *307*, 219.

9. Krishnan, M., Namasivayam, V., Lin, R., Pal, R., Burns, M.A. Microfabricated reaction and separation systems. *Curr. Opin. Biotechnol.* **2001**, *12*, 92.

10. Vo-Dinh, T., Cullum, B. Biosensors and biochips: advances in biological and medical diagnostics. *Fresenius J. Anal. Chem.* **2000**, *366*, 540.

11. Wang, J. Glucose biosensors: 40 years of advances and challenges. *Electroanalysis* **2001**, *13* (12), 983.

12. Lauks, I.R. Microfabricated biosensors and microanalytical systems for blood analysis. *Acc. Chem. Res.* **1998**, *31*, 317.

13. Schwarz, M.A., Hauser, P.C. Recent developments in detection methods for microfabricated analytical devices. Lab on a Chip Miniaturis. *Chem. Biol.* **2001**, *1* (1), 1.

14. Langer, R. Drug delivery and targeting. *Nature* **1998**, *392* (Suppl.), 5.

15. Lowman, A.M., Peppas, N.A. Analysis of the complexation/decomplexation phenomena in graft copolymer networks. *Macromolecules* **1997**, *30*, 4959.

16. Torres-Lugo, M., Peppas, N.A. Molecular design and in vitro studies of novel pH-sensitive hydrogels for the oral delivery of calcitonin. *Macromolecules* **1999**, *32*, 6646.

17. Traitel, T., Cohen, Y., Kost, J. Characterization of glucose-sensitive insulin release systems in simulated in vivo conditions. *Biomaterials* **2000**, *21*, 1679.

18. Lanza, R.P., Chick, W. Encapsulated cell therapy. *Sci. Am. Sci. Med.* **1995**, *2* (4), 16.

19. Desai, T.A., Hansford, D., Ferrari, M. Characterization of micromachined silicon membranes for immunoisolation and bioseparation applications. *J. Membrane Sci.* **1999**, *159*, 221.

20. Kaetsu, I., Uchida, K., Shindo, H., Gomi, S., Sutani, K. Intelligent type controlled release systems by radiation techniques. *Radiat. Phys. Chem.* **1999**, *55*, 193.

21. Liu, R.H., Yu, Q., Bauer, J.M., Jo, B.-H., Moore, J.S., Beebe, D.J. In-channel processing to create autonomous hydrogel microvalues. In *Micro Total Analysis systems 2000*, Proceedings of the 4th μ_{TAS} symposium, Enschede, Netherlands, May 14–18 2000; 45–48.

22. Cao, X., Lai, S., Lee, L.J. Design of a self-regulated drug delivery device. *Biomed. Microdev.* **2001**, *3* (2), 109.

23. Colton, C.K. Implantable biohybrid artificial organs. *Cell Transplant.* **1995**, *4* (4), 415.

24. Desai, T.A., Hansford, D.J., Kulinsky, L., Nashat, A.H., Rasi, G., Tu, J., Wang, Y., Zhang, M., Ferrari, M. Nanopore technology for biomedical applications. *Biomed. Microdev.* **2000**, *2* (1), 11.

25. Kim, K.J., Stevens, P.V. Hydraulic and surface characteristics of membranes with parallel cylindrical pores. *J. Membrane Sci.* **1997**, *123*, 303.

26. Jain, R.A. The manufacturing techniques of various drug loaded biodegradable poly(lactide-co-glycolide) (PLGA) devices. *Biomaterials* **2000**, *21*, 2475.

27. Langer, R. Biomaterials in drug delivery and tissue engineering: one laboratory's experience. *Acc. Chem. Res.* **2000**, *33*, 94.

28. DeLuca, P.P., Mehta, R.C., Hausberger, A.G., Thanoo, B.C. Biodegradable polyesters for drug and polypeptide delivery. In *Polymer Delivery Systems, Properties and Applications*, El-Nokaly, M.A., Piatt, D.M., Charpentier, B.A., Eds., ACS Symposium Series 520, Amercian Chemical Society: Washington, DC, 1993; 53–79 (Chapter 4).

29. Edwards, D.A. Delivery of biological agents by aerosols. *AIChE J.* **2002**, *48* (1), 2.

30. Guan, J., Lee, L.J., Hansford, D.J. Layered thin-film polymer microparticles fabricated by soft lithography. To be submitted to chemistry of marterials.

31. Green, P.W., Syms, R.R.A., Yeatman, E.M. Demonstration of three-dimensional microstructure self-assembly. *J. Microelectromech. Syst.* **1995**, *4* (4), 170.

32. Harsh, K.F., Bright, V.M., Lee, Y.C. Solder self-assembly for three-dimensional microelectromechanical systems. *Sensors Actuators* **1999**, *77*, 237.

33. Gracias, D.H., Kavthekar, V., Love, J.C., Paul, K.E., Whitesides, G.M. Fabrication of micrometer-scale, patterned polyhedra by self-assembly. *Adv. Mater.* **2002**, *14* (3), 235.

34. Langer, R., Vacanti, J.P. Tissue engineering: the design and fabrication of living replacement devices for surgical reconstruction and transplantation. *Lancet* **1999**, *354*, 23.

35. Mikos, A.G., Sarakinos, G., Leite, S.M., Vacanti, J.P., Langer, R. Laminated three-dimensional biodegradable foams for use in tissue engineering. *Biomaterials* **1993**, *14* (5), 323.

36. Li, Y., Yang, S.-T. Effects of three-dimensional scaffolds on cell organization. *Biotechnol. Bioprocess Eng.* **2001**, *6*, 311.

37. Desai, T.A. Micro- and nanoscale structures for tissue engineering constructs. *Med. Eng. Phys.* **2000**, *22*, 595.

38. Griffith, L.G., Noughton, G. Tissue engineering—current challenges and expanding opportunities. *Science* **2002**, *295* (5557), 1009.

39. King, K.R., Terai, H., Wang, C.C., Vacanti, J.P., Borenstein, J.T. Microfluidics for tissue engineering microvasculatuer: endothelial cell culture. In *Micrototal Analysis Systems*; 2001, Proceedings of the 5th μ_{TAS} 2001 Symposium, Monterey, CA, USA, October 21–25, 2001; 247–249.

40. Gad-el-Hak, M. The fluid mechanics of microdevices. *J. Fluids Eng.* **1999**, *121*, 5.

41. Giordano, N., Cheng, J.-T. Microfluid mechanics: progress and opportunities. *J. Phys. Condens. Matter* **2001**, *13*, R271.

42. Stone, H.A., Kim, S. Microfluidics: basic issues, applications, and challenges. *AIChE J.* **2001**, *47* (6), 1250.

43. Conlisk, A.T., McFerran, J., Zheng, Z., Hansford, D. Mass transfer and flow in electrically charged micro- and nanochannels. *Anal. Chem.* **2002**, *74*, 2139.

44. Kang, K., Lee, L.J., Koelling, K.W. High shear microfluidics and its application in rheological measurements. *Experiments in Fluids* **2005**, *38*, 222–232.

45. Gallardo, B., Gupta, V.K., Eagerton, F.D., Jong, L.I., Craig, V.S., Shah, R.R., Abbott, N.L. Electrochemical principles for active control of liquids on submillimeter scales. *Science* **1999**, *283*, 57.

46. Pollack, M.G., Fair, R.B., Shenderov, A.D. Electrowetting-based actuation of liquid droplets for microfluidic applications. *Appl. Phys. Lett.* **2000**, *77* (11), 1725.

47. Kulinsky, L., Wang, Y., Ferrari, M. Electroviscous effects in microchannels. *SPIE Proc.* **1999**, *3606*, 158.

48. Stroock, A.D., Weck, M., Chiu, D.T., Huck, W.T.S., Kenis, P.J.A., Ismagilov, R.F., Whitesides, G.M. Patterning electro-osmotic flow with patterned surface charge. *Phys. Rev. Lett.* **2000**, *84* (15), 3314.

49. Smith, D.E., Babcock, H.P., Chu, S. Single-polymer dynamics in steady shear flow. *Science* **1999**, *283*, 1724.

50. Shrewsbury, P.J., Muller, S.J., Liepmann, D. Effect of flow on complex biological macromolecules in microfluidic devices. *Biomed. Microdev.* **2001**, *3* (3), 225.

14

Immobilized Enzyme Technology

Charles G. Hill, Jr., Cristina Otero, and Hugo S. Garcia

CONTENTS

Introduction

Enzymes are proteins employed by Mother Nature to catalyze the chemical reactions necessary to sustain life in plants and animals. As catalysts, enzymes may influence the rates and/or the directions of chemical reactions involving an enormous range of substrates (reactants). Enzymes function by combining with substrates to form enzyme-substrate complexes (reaction intermediates) that subsequently react further to yield products while regenerating the free enzyme.

The region of the enzyme that interacts with substrates is referred to as the active site. For reaction to occur there must be an appropriate fit between the three-dimensional structure of this site and the geometry of the reactant molecule so that an enzyme-substrate complex may form (Emil Fischer's "lock and key" hypothesis). Enzymes are relatively labile species and when subjected to unfavorable conditions of temperature, pH, pressure, chemical environment, etc., they can lose their catalytic activity. In these situations, deactivation of the enzyme can usually be attributed to changes in the geometric configuration of the active site.

Enzymes are characterized by unusual specific activities and remarkably high selectivities. They are effective catalysts at relatively low temperatures and ambient pressure. The primary driving force for efforts to develop immobilized forms of these biocatalysts is cost, especially when one is comparing process alternatives involving either conventional inorganic catalysts or soluble enzymes. Immobilization can permit conversion of labile enzymes into forms appropriate for use as catalysts in industrial processes—production of sweeteners, pharmaceutical intermediates, and fine chemicals— or as biosensors in analytical applications. Because of their high specificities, immobilized versions of enzymes are potentially useful in situations where it is necessary to obtain high yields of the desired product to minimize downstream processing costs and the environmental impact of a process.

Because the costs of isolation and purification of soluble enzymes are high and it is often both technically difficult and costly to recover an active form of the enzyme from product mixtures when the reaction of interest is completed, soluble enzymes are normally employed only in batch operations in which the enzymes are removed from the liquid product by precipitation. Thermal deactivation may be used instead to destroy the catalytic activity of the enzyme. Immobilization of the enzyme circumvents these difficulties because the solid phase containing the enzyme is easily recovered from the product mixture. Use of immobilized enzymes makes it possible to conduct the process in a continuous flow mode, thereby facilitating process control via manipulation of the flow rate of the process stream. One can offset losses in enzyme activity as time elapses by reducing the flow rate to maintain a constant product composition. Operation in this mode permits one to obtain more product per unit of enzyme employed.

Techniques for Immobilization of Enzymes

A variety of physical and chemical methods have evolved for immobilizing enzymes on or within solid supports. Kennedy and Cabral employed a variation of the scheme in Figure 14.1 to classify techniques for immobilization of enzymes.[1] Judicious choice of the support is essential not only for the stability of immobilized enzymes, but also for the operational characteristics of the device containing the immobilized enzyme and the economic viability of the intended application. The discussion below and the information in Table 14.1 indicate some of the criteria employed in selecting a mode of immobilization.

Enzyme Carriers—Particulate Supports

Industrial scale processes involving immobilized enzymes are normally carried out in fixed-bed reactors. Hence, the desired characteristics of the catalyst support are closely akin to those for the heterogenous catalysts commonly employed in the chemical industry, namely:

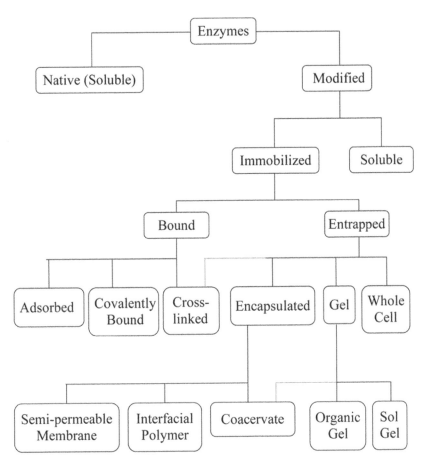

FIGURE 14.1
Schematic representation of modes of immobilization of enzymes. (Dotted lines indicate potential alternative uses/classifications.)

- Chemical, mechanical, and thermal stability.
- Resistance to both microbial degradation and swelling or dissolution in the reaction medium.
- High permeability to reactant and product species (and to the enzyme during the immobilization process) (pore diameters in the 10–50 nm range and porosities of ~50% or better).
- The requisite hydrophobicity or hydrophilicity for the intended application.
- A moderately high specific surface area (~50 m^2/g) and a pore size distribution that provides adequate capacity for adsorption of the enzyme.
- A suitable shape and particle size (to minimize pressure drop), typically greater than 0.5 mm in diameter.
- Relatively low cost.
- Regenerability.

TABLE 14.1

Comparison of Different Modes of Immobilization of Enzymes

Characteristic	Modes of Immobilization							
	Physical Adsorption	Ionic Binding	Chelation	Covalent Bonding	Cross-Linking	Physical Entrapment	Membrane Entrapment	Whole Cells
Ease of preparation	Simple	Simple	Simple	Difficult	Intermediate	Difficult	Simple	Intermediate
Binding force	Weak	Intermediate	Intermediate	Strong	Strong	Intermediate	Weak	Intermediate
Fraction of original enzyme activity	Intermediate	High	High	High	Low	Small	High	High
Ease of regeneration	Possible	Possible	Possible	Rare	Impossible	Impossible	Possible	Impossible
Cost of immobilization	Low	Low	Intermediate	High	Intermediate	Intermediate	Intermediate	Intermediate
Stability	Low	Intermediate	Intermediate	High	High	High	Intermediate	Intermediate
General applicability	Yes	Yes	Yes	No	No	Yes	Yes	Some
Protection of enzyme from microbial attack	None	None	None	None	Good	Some	Some	Some

- Inertness with respect to both the enzyme-mediated reactions of interest and reactions leading to deactivation of the enzyme.

Because naturally occurring materials do not meet the morphological specifications, most carrier materials are synthesized via routes that produce the desired characteristics. Boller, Meier, and Menzler have indicated that although immobilization of enzymes on solid supports has been studied for half a century, there are no generally applicable rules for selecting the proper support for a specific application.[2] Nonetheless, they also indicate that microporous and mesoporous epoxy-activated acrylic beads with particle diameters in the 100–250 μm range (pore radii in the 10–100 nm range) are popular supports for the preparation of multiton quantities of catalysts for industrial biotransformations.

Methods of Immobilization

Several physical and chemical methods can be utilized to immobilize an enzyme on a solid support (Figure 14.1).

Physical Adsorption

The simplest method of immobilization is physical adsorption of the enzyme on the carrier. The procedure consists of contacting a solution of the enzyme with the support material under appropriate conditions, and after allowing sufficient time to elapse, separating the solution from the now insoluble enzyme preparation by filtration, centrifugation, or other means. Because no chemical interactions are involved, there is little or no conformational change in the enzyme. Thus, the impact on the geometry of the site at which the biocatalyst interacts with substrates is minimal. The forces binding the enzyme to the carrier are relatively weak and may involve hydrogen bonding and hydrophobic interactions in addition to conventional van der Waals forces. The extents of adsorption and retention of activity are dependent on experimental parameters such as the pH of the solution/suspension, temperature, ionic strength, species concentrations, and the chemical nature of the solvent.

It is important to allow sufficient time for diffusion of the enzyme into the pore structure of the support to maximize the extent of physical adsorption of the enzyme. Surface coverages of the support by adsorbed enzymes may range from a fraction of a monolayer to multiple layers. Because the forces binding the enzyme to the support are relatively weak, enzymes immobilized by physical adsorption are susceptible to desorption during use. Shifts in microenvironmental conditions, such as changes in **pH**, ionic strength, temperature, composition of the solvent, etc., can lead to desorption with concomitant apparent loss of activity of the biocatalyst. On the other hand, these characteristics can sometimes be advantageous in protocols used to regenerate a fixed bed of immobilized enzyme, once it has lost a significant fraction of its original activity. In some cases, it may be appropriate to use a multifunctional cross-linking agent such as glutaraldehyde to chemically bind the adsorbed protein molecules to one another or to the underlying surface of the support. This approach minimizes the potential for loss of activity by desorption of the enzyme, but renders the task of regeneration of the activity of the biocatalyst much more difficult.

Ionic Bonding and Chelation

A useful variation of the physical adsorption method involves adsorption of the enzyme on carriers whose structures contain anion or cation exchange residues. Unless the **pH** of the system corresponds to the isoelectric point of the enzyme, the interactions of the net charge on the protein with opposite fixed charges on the solid support enhance the strength of adsorption of the enzyme on the support. In practice, both ionic bonding and physical adsorption occur simultaneously; the main difference is that when ionic forces are present, the strength of the interaction is greater. An alternative to employing charge bearing carriers is to enhance the strength of the adsorbate-adsorbent interaction by taking advantage of the ability of some transition metal compounds to form chelate structures with enzymes.

Covalent Bonding

Stronger bonds between the enzyme and the carrier can be formed when covalent bonding is employed. Enzymes are copolymers (proteins) composed of a variety of amino acid monomers. They possess a number of reactive side chains that can be utilized for the purpose of forming covalent bonds with solid supports, whose surfaces contain appropriate functional groups. The functionalities present in the side chains of the protein typically include amino, carboxylic acid, sulfydryl, hydroxyl, imidazole, disulfide, indole, and phenol groups. The particular chemistry employed in forming covalent bonds between the enzyme and the carrier is selected on the basis of the chemical nature of the support and the intended application. The range of potential coupling processes is vast for both inorganic and organic supports, and methodologies based on several different types of chemical reactions are described in Kennedy and Cabral[1], Bickerstaff[3], Scouten[4], Woodward.[5] A sample of some of the most commonly employed types of reactions is presented in Table 14.2.

If the enzyme is to retain significant catalytic activity after being covalently bound to the support, immobilization should occur via functional groups that are not associated with the active site of the enzyme. This limitation may be difficult to surmount, and enzymes immobilized via covalent bonding may thus suffer significant losses of activity relative to the activity of the soluble precursor. Nonetheless, the wide variety of supports with functional groups capable of reacting (or susceptible to appropriate functionalization) via a range of chemistries with enzymes makes covalent bonding a generally applicable route for immobilization of enzymes. Enzymes bound to supports in this manner are not susceptible to desorption from the surface during use.

Cross-Linking

Enzymes can be readily cross-linked using a bi- or multifunctional reagent, such as glutaraldehyde, bisdiazobenzidine-2,2-disulfonic acid, or toluene diisocyanate, which can react with free amino or carboxyl groups or with other functional groups that might be present in the enzyme. Cross-linking of proteins results in an insoluble polymer that may not possess appropriate physical properties for the intended application. Cross-linking is often used in combination with other techniques (especially physical entrapment and physical adsorption) to obtain a material that has enhanced stability.

Physical Entrapment of Enzymes

Enzymes can also be immobilized by physical entrapment, either in a solid matrix or encapsulated by a membrane that is permeable to low-molecular-weight species, but not to high-molecular-weight species. These membranes may be fabricated from polymers or formed by interfacial polymerization of appropriate monomers. The membranes may also be parts of living cells that have the capacity for bringing about the desired chemical

TABLE 14.2

Examples of Reactions Commonly Used for Covalent Bonding of Enzymes to Solid Surfaces

Functional Group of the Support	Functional Group of the Enzyme	Coupling Reagent
Amine, $-NH_2$	$-NH_2$	Glutaraldehyde
Hydroxyl, $-OH$	$-NH_2$	$(CH_3O)_3$ Si $(CH_2)_3$ NH_2 + glutaraldehyde
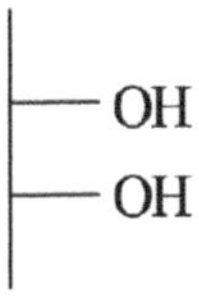	$-NH_2$	Cyanogen bromide or tresyl chloride

| ─OH
├─OH | | |

Aldehyde, $-CH\!=\!O$	$-NH_2$	None
Acid anhydride, 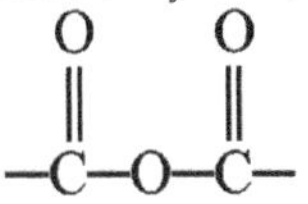 $\overset{O}{\underset{\|}{}}\ \ \overset{O}{\underset{\|}{}}$ $-C-O-C-$	$-NH_2$	None
Imidocarbonate 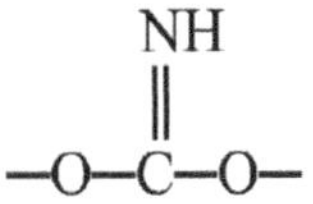 $\overset{NH}{\underset{\|}{}}$ $-O-C-O-$	$-NH_2$	None
Cylic carbonate, 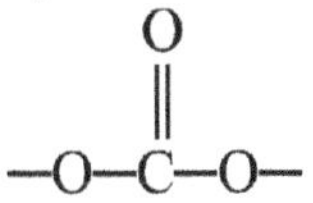 $\overset{O}{\underset{\|}{}}$ $-O-C-O-$	$-NH_2$	None
Triazinyl	$-NH_2$	None

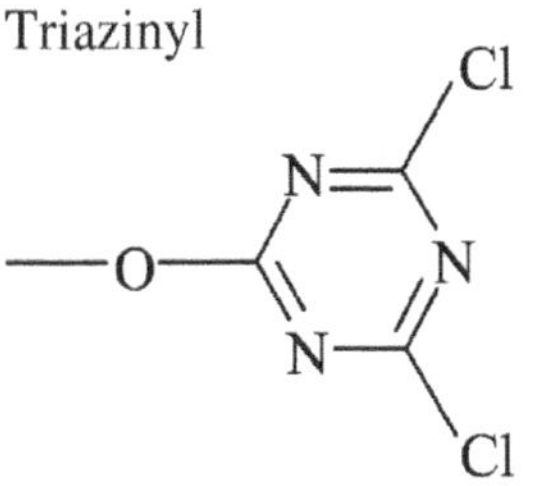

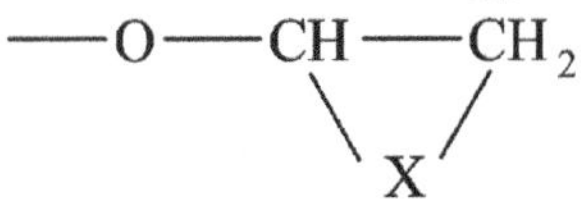

(X = NH, O, S)

| (X=NH, O, S) | NH_2 $-OH$, $-SH$ | None |
| $-COOH$ | $-NH_2$ | Carbodiimide (RN=C=NR) |

transformations as part of the metabolic processes required for their existence. In addition, the membranes can consist of the walls of cells subjected to lysis.

Gel Formation

Entrapment of enzymes in solid matrices composed of synthetic or natural polymers or inorganic gels is a relatively simple process. The basic technique involves occlusion of the enzyme within the lattice of a solid matrix as the matrix is formed by polymerization, precipitation, or coacervation. Gels formed from polysaccharides, especially those derived from cellulose and algae, have often been used in combination with cross-linking agents to immobilize enzymes. Both anionic polymers (e.g., carrageenan, carboxymethylcellulose, and sodium alginate) and cationic polymers (e.g., chitosan) have been used to form ionotropic gels within which enzymes or whole cells can be entrapped. This approach leads to a product that is not robust with respect to changes in ionic strength or **pH**, but it is usually the method of choice for immobilizing whole cells. Gels formed by polymerization of acrylic and methacrylic acids can be activated using a soluble carbodiimide as a precursor to production of covalently bound enzymes. In addition to the organic gels noted above, it is also possible to employ sol-gel techniques based on hydrolysis of metal alkoxides to produce inorganic polymeric structures within which enzymes can be physically entrapped.

A major disadvantage of the gel entrapment route to immobilization is the potential for physical loss of the enzyme as time elapses. To circumvent this problem, cross-linking agents, such as N,N''-methylene-bis-acrylamide or glutaraldehyde, may be used to more firmly immobilize the enzyme or to provide mechanical stability. However, the more rigid the matrix the greater is the possibility that diffusional resistance to transport of reactants (substrates) to the site of the enzyme and of products out of the gel will limit the reaction rate.

Encapsulation

Immobilization of enzymes by encapsulation within semipermeable structures dates back to the 1970s.[6] There are three fundamental variations of this approach. In coacervation, aqueous microdroplets containing the enzyme are suspended in a water-immiscible solvent containing a polymer, such as cellulose nitrate, polyvinylacetate, or polyethylene. A solid film of polymer can be induced to form at the interface between the two phases, thereby producing a microcapsule containing the enzyme. A second approach involves interfacial polymerization in which an aqueous solution of the enzyme and a monomer are dispersed in an immiscible solvent with the aid of a surfactant. A second (hydrophobic) monomer is then added to the solvent and condensation polymerization is allowed to proceed. This approach has been used extensively with nylons, but is also applicable to polyurethanes, other polyesters, and polyureas.

A third general approach to encapsulation involves the use of permselective membrane devices of the types employed in ultrafiltration and nanofiltration of aqueous solutions, especially those devices that employ the membrane in the form of hollow fibres. In effect, the enzymes are retained within a macrocapsule. An aqueous solution of the soluble enzyme or whole cells is contained on the retentate side of the membrane, while a solution containing the substrates is supplied to the permeate side of the membrane. The reactants are transported across the membrane to the retentate side where they undergo reaction. The products then diffuse out of the retentate zone across the membrane to the permeate side where they can be removed by convective transport. Throughout, the enzyme is retained on the retentate side provided that one employs a membrane whose molecular cutoff value is significantly below the molecular weight of the enzyme. An advantage of this approach is that this apparatus permits periodic replacement of the biocatalyst.

Reactions mediated by enzymes immobilized by coacervation, interfacial polymerization, retention by semipermeable membranes, or gelation are particularly susceptible to mass transfer/diffusional limitations on the rate.

Immobilized Cells

Instead of immobilizing individual enzymes in molecular form, one can elect to immobilize whole (living) cells containing the enzymes of interest. Many of the techniques employed for immobilization of enzymes are readily extended to immobilization of whole cells, especially those methods involving physical entrapment of the enzymes. In essence, immobilization of whole cells is just another means of physically encapsulating the enzyme(s) of interest. Furthermore, immobilization of whole cells circumvents the need for the multiplicity of processing steps involved in isolating and purifying intracellular enzymes with concomitant reductions in cost. The stability of the desired enzyme(s) is usually enhanced by virtue of the fact that its natural environment is maintained during both immobilization and use. This advantage is of particular benefit in the case of membrane-bound enzymes and for enzyme-mediated reactions involving the participation of either cofactors or multiple enzymes. The necessity for purification of multiple enzymes is avoided and the optimal spatial location of these enzymes within various compartments of the cell remains intact, as do the sites for regeneration of cofactors. Hence, the structural integrity of the catalytic complex is retained. Moreover, the enzymes present in immobilized cells are much more robust with respect to local perturbations in **pH**, temperature, ionic strength, and the presence of substances that cause deactivation of the enzyme (e.g., toxic metal ions).

On the other hand, use of whole cells as the vehicle for immobilization of enzymes is not without problems. These disadvantages include the susceptibility to mass transfer/diffusional limitations on reaction rates and possible losses in the yield of the desired product as a consequence of unwanted side reactions. In addition, there are potential problems associated with maintaining the integrity of the immobilized cells—supplying the nutrients, energy sources, or cofactors necessary to maintain the cells in a sufficiently viable condition to mediate the reaction(s) of interest.

Comparison of Immobilization Techniques

Table 14.1, an extension of the work of Kennedy and Cabral, is a concise summary of important characteristics of different immobilization techniques.[7]

Applications of Immobilized Enzymes

Industrial Applications

In spite of the high expectations generated by immobilized enzyme technology in the last third of the 20th century, only a limited number of reductions to industrial practice have been accomplished. Very few large-scale immobilized enzyme processes can successfully compete with processes based on either free enzymes or more conventional catalysts. Some of these are indicated below.

Applications in the Food Industry

Immobilized Glucose Isomerase for the Production of High-Fructose Corn Syrup

A major shift in the technology employed for the production of sweeteners began in the early 1960s with the introduction of soluble enzymatic methods for the production of dextrose syrups from starch. (In commercial practice, the term dextrose is used instead of glucose.) This seminal change was followed about a decade later by utilization of immobilized glucose isomerase for the production of high-fructose corn syrups (HFCS) for use as sweeteners in the manufacture of soft drinks and other foods and beverages. Glucose and fructose have the same molecular formula ($C_6H_{12}O_6$), but differ in geometric configuration (Figure 14.2) and sweetening power. Schenck has reviewed the technology for production of high-fructose syrups.[8]

Starch, a polymer formed from glucose monomers, is the principal storage carbohydrate of plants and commercially is obtained primarily from corn. The industrial process involves, first, acid or enzyme (soluble bacterial α-amylase) catalyzed hydrolysis of aqueous suspensions of gelatinized starch to obtain a product with a low dextrose equivalent (DE) of 5–12 and sugars such as glucose and maltose. (The DE is the percentage of the dry matter that consists of reducing sugars expressed as dextrose. This parameter indicates the percentage of the glycosidic linkages in the starch precursor that have been cleaved by hydrolysis.) This hydrolysis dissolves the starch. Further hydrolysis with soluble α-amylase yields a product with a DE in the range of 8–15. This product is then saccharified using one or more soluble enzymes. Fungal α-amylase and fungal glucoamylase (GA) are utilized either separately or in combination (and sometimes in combination with pullulanase) to produce a dextrose syrup with a typical DE of 42. Soluble enzyme preparations from *Aspergillus niger, A. oryzae,* or *Rhizopus oryzae* are then employed for additional saccharification to obtain the 95–98 DE feedstock necessary for the production of HFCS. The resulting products are sweeteners whose compositions and applications depend on the extent of hydrolysis mediated by the enzymes in question. Yields of glucose may be as high as 95–97%. However, the sweetening power of glucose suffers by comparison to that of fructose, and it is the subsequent conversion of glucose to fructose for which immobilized glucose isomerase (xylose isomerase) is an effective biocatalyst. This isomerization reaction constitutes the heart of the technology that brought about a revolution in the manufacture of sweeteners. The resulting syrups compete successfully with sucrose (cane sugar) in many food applications. Virtually all manufacturers of soft drinks use HFCSs in their formulations.

Production of fructose from glucose became commercially viable only after adequate procedures for immobilization of glucose isomerase were developed, so that the same quantity of enzyme could isomerize large quantities of substrate in a packed-bed reactor fed continuously with a solution of maltodextrins. Process conditions (typically 55–65°C and pH 7.5–8.5) depend on the particular form of the immobilized enzyme. Reactor diameters are typically between 0.6 and 1.5 m, with corresponding heights of 2–5 m. Initial residence times are less than 1 hr, but to compensate for the loss of enzyme activity as time

FIGURE 14.2
Isomerization of glucose to fructose over an immobilized glucose isomerase.

onstream elapses it is necessary to increase the residence time by reducing the flow rate of the feed stream so as to maintain the composition of the effluent constant. Manufacturers frequently employ large numbers of reactors in tandem to maintain constant production rates. After several months when the activity of a particular packed bed decreases to about 10% of its initial value, that bed can be removed from the reactor network and the biocatalyst replaced by a new charge of enzyme.

Because glucose isomerase is formed intracellularly in many bacterial strains of commercial interest, some industrial processes have utilized immobilized cells, rather than isolated enzymes, in this application. In whole-cell processes the microbial cells are recovered from the fermentation broth and treated to maintain both enzymatic activity and particle integrity. For the biocatalysts derived from soluble enzymes, the enzyme is separated from the cells and purified prior to immobilization. Relatively few organisms (*Actinoplanes missouriensis, Bacillus coagulans, Streptomyces rubiginosus, S. olivochromogenes, S. murinus, and Microbacterium arborescens*) have been used to generate the glucose isomerase used in commercial operations. Although the equilibrium yield of fructose for the typical operating conditions cited above corresponds to 50–55% on a dry basis (db), the reactors are normally operated to obtain yields of 42–45% db to circumvent the limitations on the rate imposed by the approach to equilibrium and to obtain an economically viable reactor size. The effluent from the reactor is then polished to remove color and salts using a combination of activated carbon and ion exchange resins and then concentrated to ca. 71% (w/w) solids. This 42% HFCS product can be used directly in formulating some food products, but manufacturers of soft drinks who desire to effect complete replacement of sucrose in their formulations require that the 42% HFCS be fractionated (for example, by a continuous chromatographic technique) to obtain a product enriched in fructose and a raffinate enriched in glucose. The latter can then be recycled to the immobilized enzyme reactor (IMER) for further conversion to fructose. One possible mode of operation to produce the three fructose-rich syrups of commercial interest and crystalline fructose is shown in Figure 14.3. Data provided by the USDA Economic Research Service indicates that in 2002 U.S. production of HFCS corresponded to 9.3 million short tons (db) of fructose.

Other Food-Related Applications

Swaisgood has reviewed applications of immobilized enzymes in the food industry during the past 40 yr.[9] He discusses not only the HFCS application, but also a variety of others, some of which are no longer employed commercially. Table 14.3 contains a summary of these applications.

Applications Involving Fine Chemicals-Chiral Synthesis and Chiral Separations

In the mid-1980s researchers in the pharmaceutical industry demonstrated that single enantiomers of pharmaceutical compounds often functioned better as therapeutic agents than racemic mixtures. There are many examples for which one particular chiral form (enantiomer) of a compound demonstrated therapeutic efficacy, while the other chiral form was ineffective or produced deleterious effects. (For example, dextromethorpan is commonly used as a cough suppressant, while its enantiomer, levomethorphan, is a powerful narcotic.) By 2002 annual sales of the top 10 single-enantiomer drugs totaled $34.2 billion.[10]

In the pharmaceutical industry, immobilized enzymes (especially lipases) are used to mediate reactions of two general types: reactions involving prochiral substrates and

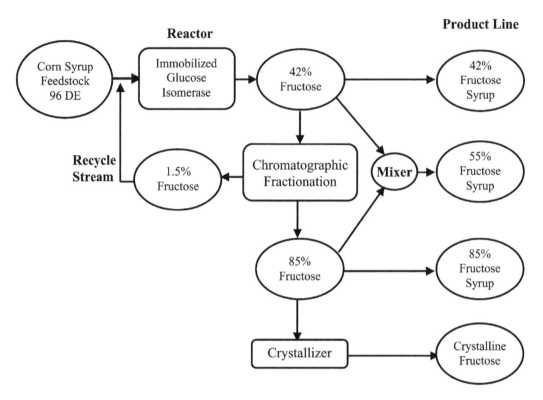

FIGURE 14.3
Flow diagram for production of HFCS products.

TABLE 14.3

Applications of Immobilized Enzymes in the Food Industry

Product	Immobilized Enzyme (Application)	Countries	Half-Life of Enzyme	Productivity (wt/wt of Enzyme)
Whey hydrolysate	β-Galactosidase (hydrolysis of lactose)	Finland, Norway, France, and U.K.	20 mo	2000
L-Amino acids	Aminoacylase (resolution of optical isomers)	Japan	65 days	n.a.
L-Phenylalanine	Transaminases (production of aspartame)	Japan	>8 mo	n.a.
5′-Ribonucleotides	5′-Phosphodiesterase (chemical synthesis)	Germany	>500 days	n.a.
Isomaltulose	Isomaltulose synthase (chemical conversion)	U.K., Germany, Japan	1 yr	1500
Invert sugar	Invertase (hydrolysis of sucrose)	Europe	n. a.	6000
Modified fats and oils	1,3-Specific lipases (migration of acyl groups)	Japan, U.K.	n. a.	1100

n.a., not available.

kinetic resolution of racemates. Although prochiral or chiral alcohols and carboxylic acid esters initially served as the primary classes of substrates, compounds susceptible to processing via these two routes now encompass diols, α- and β-hydroxy acids, cyanohydrins, chlorohydrins, diesters, lactones, amines, diamines, amino alcohols, and α- and β-amino acid derivatives. Gotor and Arroyo have reviewed the use of biocatalysts for the preparation of pharmaceutical intermediates and fine chemicals.[11,12] Some specific examples are indicated below.

Penicillins and cephalosporins are characterized by β-lactam structures and are the antibiotics that have traditionally been those most commonly used in the treatment of infections. Pharmaceutical companies have synthesized a variety of semisynthetic β-lactam compounds for use as oral antibiotics, for example, ampicillin and amoxicillin. These penicillin derivatives are prepared by acylation of 6-amino-penicillanic acid (6-APA) derived from penicillin G (benzyl penicillin) or penicillin V (phenoxymethyl penicillin). An immobilized penicillin amidase (penicillin acylase) from *Escherichia coli* or *Bacillus megaterium* is used to prepare the 6-APA in nearly quantitative yield (Figure 14.4). This substance is used as the starting material for the production of a number of other penicillins. The immobilized enzyme can be reused more than 600 times in the batch reactor used to accomplish this transformation.

A similar reaction scheme can be used to produce derivatives of cephalosporin via acylation of 7-amino-cephalosporamic acid or 7-amino-desacetoxycephalosporamic acid. These compounds could be produced from a natural cephalosporin using an immobilized cephalosporin acylase, but alternative routes to these compounds are more cost-effective.

Immobilized forms or reticulated crystals of *Candida antarctica* lipase are effective biocatalysts for the synthesis of pure enantiomers utilized as anti-inflammatory agents. For example, one route for production of the *S*-isomers of 2-aryl propionic acids (ibuprofen, naproxen, ketoprofen, and flurbprofen) involves enantioselective hydrolysis of the corresponding racemic esters. Arroyo has indicated that an immobilized form of *C. antarctica* lipase (fraction B) is used to mediate the selective acetylation of a diol to form the *S*-enantiomer of a monoacetate (Figure 14.5), which is further processed to obtain an antifungal agent.[12]

Analytical chromatographic separation of enantiomeric acids and N-substituted amino acids has been effected using immobilized α-chymotrypsin supported on activated silica.[13] Similarly, cellulase immobilized on silica gel can bring about chromatographic separation of enantiomers of propanolol.[14] This approach can be employed in industrial separation of racemic mixtures using simulated moving-bed chromatography as in the production of naproxen, warfarin, propanolol, and ephedrine.

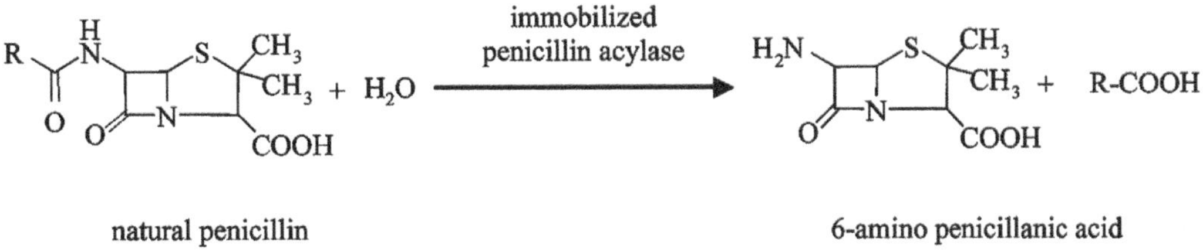

natural penicillin   6-amino penicillanic acid

FIGURE 14.4
Conversion of natural penicillin to 6-APA over an immobilized penicillin acylase.

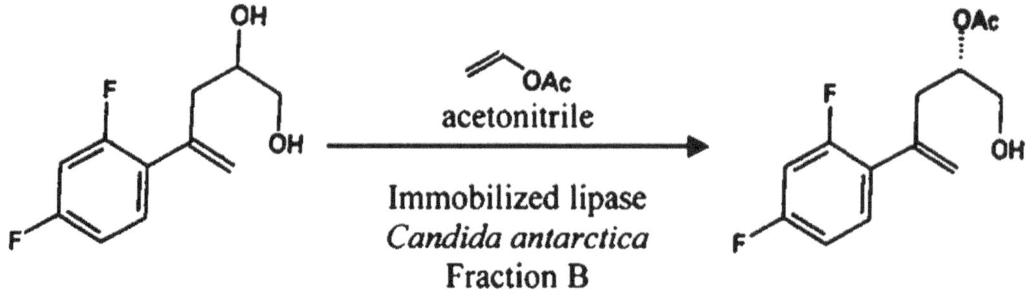

diol   S-monoacetate

FIGURE 14.5
Generation of the *S*-enantiomer of the monoacetate precursor of an antifungal agent from a diol as mediated by an immobilized lipase from *Candida antarctica*, fraction B.

Biosensors and IMERs

A biosensor is a sensing device consisting of a recognition element (enzyme or cell) in intimate contact with an appropriate transducer that is able to convert the concentration of a species involved in the recognition reaction into a measurable electronic signal. Biosensors based on immobilized enzymes have been employed for decades in analyses for solutes present in either aqueous solution or biological fluids. Fabrication of these biosensors frequently involves immobilization of an enzyme on a membrane electrode that is capable of donating electrons to (or accepting electrons from) species participating in an enzyme-mediated reaction. The substrate to be quantified diffuses to this surface where the biocatalytic reaction occurs. Electrochemical (potentiometric or amperometric) measurements are then employed to monitor changes in the concentration of the analyte of interest. In many applications, the transducer element is located downstream from an IMER that converts the substrate into a chemical form that stimulates the response of the transducer.

Analytical Applications

Schuhmann has reviewed amperometric biosensors and indicated that these sensors can be categorized as devices employing: 1) direct electron transfer between redox proteins and electrodes modified with self-assembled monolayers; 2) anisotropic orientation of redox proteins at monolayer-modified electrodes; 3) electron transfer cascades via redox hydrogels; and 4) electron transfer via conducting polymers.[15] These biosensors are employed to quantify the concentrations of a wide variety of substrates. For these determinations, the choice of the support material and the method of immobilization are tailored to the particular application. Gupta and Mattiasson have described several unique applications of enzymes in bioanalytical systems.[16]

Physical entrapment of an enzyme in a gel or polymeric matrix is often employed in analytical applications of biosensors because much of the original activity of the free enzyme is retained. However, this approach is limited to relatively small analytes that can readily penetrate the solid matrix. For in vivo measurements, heparin may also be employed to coat the sensor to create a biocompatible nonthrombogenic surface. A technique for reversible binding of enzymes in biosensors involves use of antibodies raised against enzymes (antigens). For reversible immobilization of glycoenzymes one can utilize lectins (carbohydrate binding proteins). Novel protein structures (sequences) including one partner of an affinity pair can be constructed using recombinant DNA technology. These tagged enzymes can subsequently be immobilized to complementary ligands bound to the support for use in enzyme purification and in ELISA.

Analytical protocols based on the use of immobilized enzyme-mediated reactions that are highly specific to the analyte require little, if any, manipulation of the sample (extraction, addition of reagents, dialysis, filtration, etc.). The sensing element is normally reusable and has the great advantage that it does not require consumption of reagents during the reaction. Thus, a single biosensor may often be utilized to make hundreds or thousands of measurements so that the resulting cost per assay is low. In other cases, the biosensor involves a dipstick that undergoes a color change as the enzymatic reaction proceeds.

The most widely employed types of biosensors are those that employ an oxidase to generate hydrogen peroxide. A classic example is the electrode containing an immobilized glucose oxidase that generates an amperometric signal related to the concentration of glucose present in the sample. However, biosensors of these types are often susceptible to interference from other electrochemically active solutes in the sample. A wide variety of techniques have been developed to circumvent or minimize this problem, for example, application of a semipermeable membrane above the enzyme matrix, use of enzyme field effect transistors, etc. For analytes that are not susceptible to quantitation by simple electrochemical methods, it may be necessary to employ coupled enzyme reactions or optical modes of detection, perhaps taking advantage of optical fibers. Other biosensors may utilize field effect transistors, thermistors, bioluminescence, or chemiluminescence for detection of the analyte.

The IMER approach does not require that the enzyme be placed in close proximity to the detector if the transducer signal is generated by a soluble product or cosubstrate of the enzymatic reaction. In the latter case, a variety of flow systems and postreactor detectors can be utilized to produce simultaneous determinations of the concentrations of several analytes. For example, an IMER can be combined with a high-performance liquid chromatography (HPLC) instrument (perhaps also in combination with mass spectroscopy) for purposes of both qualitative and quantitative analysis. The chemo-, stereo-, and regio-selectivities of enzymes facilitate separation and/or identification of analytes that may be present as different isomers (e.g., in peptide analysis based on use of peptidase IMERs in combination with these techniques to obtain structural information about the sequence of amino acids in peptides).

In some cases, coimmobilization of multiple enzymes is required, especially when successive use of two or more enzymes is employed to convert the original substrate to a final product that is easier to detect or that permits one to drive an unfavorable equilibrium situation to completion by consumption of the initial product in a detection reaction yielding the product that serves as the actual analyte.

Two important general categories of biosensor applications are discussed below.

Detection and Analysis of Sugars

Because the refractive index detectors used in HPLC are not very selective and are characterized by low sensitivities for different sugars and their various isomers, reactors containing immobilized enzymes are often coupled to the corresponding chromatographic columns to obtain significant improvements in selectivity and sensitivity. Examples include:

1. Separation and determination of glucose and lactose in a penicillin fermentation broth using a glucose dehydrogenase (GDH) to catalyze oxidation of the β-anomeric form of aldoses to lactones in the presence of NAD^+. The analysis is based on amperometric detection of the NADH formed. The enzymatic reactor is packed with silanized porous glass beads activated with glutaraldehyde as a support for immobilization of GDH from *Bacillus megaterium*.[17]

2. Determination and quantification of sugars in the discharge from the sulfite pulping of lignocellulose using coimmobilization of a mutarotase (MT), from porcine kidney that converts all of the aldoses to their active β-anomers, with a xylose isomerase from *Streptomyces* sp. (that produces the detectable ketoses), a galactose dehydrogenase (from a recombinant *Escherichia coli*), and GDH from *Bacillus megaterium*.[18]

3. Detection of glucose and malto-oligomers (G2-G10) in corn syrup. This assay employs the selective hydrolysis of α-(1,4) and α-(1,6) linkages of malto-oligosaccharides to glucose by GA with amperometric detection at a gold electrode. Glucoamylase from *Aspergillus niger* is supported on Nucleosil 300 previously silanized with (γ-glycidoxypropyl) trimethoxysilane and activated with 1,1'-carbonyldiimidazole.[19]

4. Detection of oligosaccharides (e.g., stachyose, raffinose, sucrose, and fructose) in a soybean extract using invertase hydrolysis of β-D-fructofructoside to fructose, and further oxidation of this sugar by hexacyanoferrate (III) ion in the presence of fructose dehydrogenase (FDH). This analysis is based on a coimmobilization of invertase from *Candida utilis* and FDH from *Gluconobacter* on poly(vinyl alcohol) (PVA) beads and coulometric quantification of the hexacyanoferrate(II) ions formed.

5. Analysis of malto-oligosaccharides using an IMER containing GA, MT, and GDH coimmobilized on an aminated porous silica matrix. This analysis has been used for quantification of sugars in soft drinks, beer, and fermentation broth containing *Penicillium* and *Fusarium sysporum s*.

Analysis of Amino Acids

Differentiation of the L- and D-forms of amino acids is essential because they differ in their biological and physiological properties. Although chromatographic columns that effect separations of chiral compounds could be used for analysis of solutions of these acids, a combination of a reactor containing a stereoselective immobilized enzyme and a chromatographic system provides the necessary selectivity for such analyses.

1. Immobilized L-amino acid oxidase catalyzes the oxidation of L-amino acids to 2-oxo acids. Detection of the hydrogen peroxide product is accomplished using a fluorometric or chemiluminescent assay subsequent to the reaction in the packed bed. L-Amino acids can also be quantified amperometrically using a platinum electrode Ag/AgCl on which the enzyme is immobilized. L-Lys, L-His, L-Cys, L-Arg, L-Met, L-Leu, L-Ile, L-Tyr, L-Phe, and L-Trp can be quantified by this procedure.

2. Branched-chain L-amino acids can be analyzed using leucine dehydrogenase immobilized on aminated PVA activated with glutaraldehyde. This enzyme catalyzes the deamination of L-Leu, L-Ile, and L-Val to 2-oxo acids in the presence of NAD^+.

There are many instances where it is helpful to use biosensors for the detection of contaminants and monitoring of air and water quality. In some cases, appropriate biosensors exist. Generally speaking, development of such biosensors is a demanding task because of the wide range of potential substrates (including such hazardous substances as chemical and biological warfare agents) and the associated problems of interference effects, the necessity for unattended operation, and the need for robust sensors in harsh environments. Analyses for aldehydes produced in industrial plants, incinerators, automobile exhausts, or foodstuffs (aroma and storage controls) are based on oxidation of aldehydes to carboxylic acids by immobilized aldehyde dehydrogenase in the presence of a thiol, K^+, and NAD^+. Determinations of some metals (e.g., zinc) via IMERs are sensitive, specific, fast (minutes), and do not require chromatographic separations.

Medical/Clinical Applications

Liang, Li, and Yang have reviewed biomedical applications of immobilized enzymes with emphasis on the use of biosensors for the diagnosis of disease states.[20] Electrodes containing immobilized enzymes constitute the primary technology used in this application. Table 14.4, adapted from Liang, Li, and Yang, contains a summary of applications of enzyme-based sensors in clinical diagnosis.[20]

Diagnosis of renal problems, xanthinuria, and toxemia of pregnancy via determination of the ratio of hypoxanthine to xanthine in plasma is facilitated by the use of biosensors. Xanthine oxidase immobilized on aminopropyl-CPG (controlled pore glass) activated with glutaraldehyde oxidizes hypoxanthine first to xanthine and then to uric acid. Use of an IMER with biosensors for hypoxanthine, xanthine, and uric acid provides the necessary data. Pre- or postcolumn enzymatic reactions catalyzed by creatinine deiminase, urease, alkaline phosphatase, ATPase, inorganic pyrophosphatase, or arylsufatase facilitate analysis of uremic toxins (simultaneous detection of electrolytes, serum urea, uric acid, creatinine, and methylguanidine).

TABLE 14.4

Applications of Biosensors Containing Immobilized Enzymes for Clinical Analyses

Substrate (Analyte)	Immobilized Enzyme(s)	Linear Range (Approximate)
Glucose	Glucose oxidase/GDH	50 mM
Lactate	Lactate oxidase	27 mM
Oxalate	Oxalate oxidase	1 mM
Urea	Urease	100 mM
Glutamate	Glutamate oxidase	200 µM
Carnitine	Carnitine dehydrogenase and diaphorase	1 nM
Theophylline	Theophylline oxidase	30 µM
Creatine and creatinine	Creatininase, creatinase, and sarcosine oxidase	30 mM
Cholesterol	Cholesterol oxidase	3 mM
Amino acids	Amino acid oxidase	10 mM
Acetylcholine and choline	Acetylcholine esterase and choline oxidase	100 µM
Bilirubin	Hemoglobin and glucose oxidase	
γ-Aminobutyric acid	Catalase and γ-glutamate oxidase	10 nM

Analysis of steroid hormones for control of endocrine functions by radioinmmunoassay fails for homologous steroids. Instead, one can employ highly specific steroid dehydrogenases to mediate oxidation of the hydroxy functions of hydroxysteroids. Then, highly sensitive fluorescence detection is used to determine specific positions of OH groups and α and β configurations with HPLC systems combined with an appropriate postcolumn enzymatic reactor. Similarly, one can determine serum bile acid concentrations as an indicator of liver disease by combining an IMER containing 3α-hydroxysteroid dehydrogenase with gas or liquid chromatography.

In addition to the steroid hormones, the steroids utilized for pharmaceutical applications can be selectively determined using appropriate immobilized enzymes in an IMER. Use of α-, β-, and/or stereoespecific dehydrogenases permits one to enhance the selective detection and chiral resolution of steroids. Detection of anabolic steroids in urine by these methods is employed in sports medicine as a means of controlling illicit use of these drugs by athletes.

Flow injection analysis (FIA) for ethanol can be utilized for blood alcohol determination in drunken driving situations.

One can also envision applications involving the use of encapsulated or entrapped enzymes in bioreactors for therapeutic applications involving detoxification of deleterious substances or correction of metabolic deficiencies. In these applications, the enzymes could be contained within artificial cells [e.g., modified red blood cells (erythrocytes) or liposomes]. Liang, Li, and Yang have reviewed biomedical applications of immobilized enzyme bioreactors.[20]

Summary

Techniques for immobilization of enzymes and applications of immobilized enzymes are discussed.

Both chemical and physical methods may be used to immobilize biocatalysts while retaining or modifying their activity, selectivity, or stability. Among the techniques used for immobilization of enzymes are physical adsorption, covalent bonding, ionic binding, chelation, cross-linking, physical entrapment, microencapsulation, and retention in permselective membrane reactors. The mode of immobilization employed for a particular application depends not only on the specific choice of enzyme and support, but also on the constraints imposed by the microenvironment associated with the application.

Commercial uses of immobilized enzyme technology are limited in scope, but encompass industrial production of HFCS, biosensors, clinical diagnostic procedures, chemical analyses, chiral syntheses, and therapeutic applications.

Conclusions

Immobilized enzyme technology is not a stagnant technology. It has evolved in recent decades to the point where it can be employed for select industrial processes and, more importantly, for rapid analyses of significant import in both clinical and analytical situations. The analytical applications also have important implications for monitoring air and water quality in support of environmental regulations, as well as in analyses of process effluents and industrial wastewaters. Biosensors based on immobilized enzyme technology offer significant commercial potential for detecting food spoilage and chemical and biological warfare agents, as well as for monitoring food storage conditions.

One of the areas in which immobilized enzymes are expected to have major impact is in the biocatalysis of reactions in organic media, for example, in the synthesis of chiral compounds as intermediates in the manufacture of pharmaceuticals and in the modification of naturally occurring fats and oils to produce value-added products targeted at the nutraceuticals market.

Use of recombinant DNA technology, other means of genetic engineering, and enzymes obtained from thermophilic and halophilic organisms can be expected to produce novel enzymes with enhanced selectivity, activity, or stability. These novel enzymes may be utilized to effect reactions at elevated temperature, in organic media, and in other harsh environments where stringent requirements must be met. In addition, further advances in permselective membrane technology and/or affinity separation media may facilitate improvements in the biosensors employed in analytical applications as a result of immobilization of both enzymes and cofactors in a manner that minimizes leakage problems.

Advances in the understanding of structure-activity/selectivity relations for enzymes evolving from the use of x-ray, NMR, and other instrumental methods for characterization of enzyme structures should contribute to the development of improved immobilized enzyme systems for both analytical and industrial applications. Immobilized enzyme technology has enormous potential, but significant advances on several fronts are necessary prior to widespread industrial use of this technology. Katchalski-Katzir has discussed this problem in a review of past successes and failures in efforts to employ immobilized enzymes in the food, pharmaceutical, and chemicals industries.[21]

From an industrial perspective, economic considerations are paramount when deciding whether or not to adopt a new technology. A crucial consideration for future development of large-scale industrial processes employing immobilized enzyme technology is that they be cost-effective relative to alternative technologies. To date, immobilized enzyme technology has, in most instances, failed to pass the test of economic viability for chemical, pharmaceutical, and food processing applications. On the other hand, this technology has found increasing numbers of applications in clinical and analytical applications, where it proves to be cost-effective. Indications are that in the future, novel and better commercial IMERs will be utilized more extensively and routinely in FIA systems in analytical applications.

Acknowledgments

A sabbatical grant from the Spanish Ministerio de Educacion, Cultura, y Deportes and financial support from the U.S. National Science Foundation (BES-00 77524) for Charles G. Hill, Jr. are gratefully acknowledged.

References

1. Kennedy, J.F., Cabral, J.M.S. Enzyme immobilization. In *Biotechnology, Vol. 7a, Enzyme Technology*; Rehm, H.J., Reed, G., Eds.; VCH: Weinheim, Germany, 1987; 347–404.
2. Boller, T., Meier, C., Menzler, S. EUPERGIT oxirane acrylic beads; how to make enzymes fit for biocatalysis. *Org. Process Res. Dev.* **2002**, *6*, 509–519.
3. Bickerstaff, G.F. Immobilization of enzymes and cells. In *Methods in Biotechnology*; Humana Press: Totowa, NJ, 1997; Vol. 1.
4. Scouten, W.H. A survey of enzyme coupling techniques. *Methods Enzymol.* **1987**, *135*, 30–65.
5. Woodward, J. Immobilized enzymes: adsorption and covalent coupling. In *Immobilized Cells and Enzymes: A Practical Approach*, Woodward, J., Ed.; JRL: Oxford, 1985; , 3–17.
6. Chang, T.M.S. Microencapsulation of enzymes and biologicals. In *Methods in Enzymology*; Mosbach, K., Ed., Academic Press: New York, 1976; Vol. 44, 201–218.
7. Kennedy, J.F., Cabral, J.M.S. Enzyme immobilization. In *Biotechnology, Vol. 7a, Enzyme Technology*, Rehm, H.J., Reed, G., Eds.; VCH: Weinheim, Germany, 1987; 393
8. Schenck, F.W. High fructose syrups—a review. *Int. Sugar J.* **2000**, *102*, 285–288.
9. Swaisgood, H.E. Use of immobilized enzymes in the food industry. In *Handbook of Food Enzymology*, Whitaker, J.R., Voragen, A.G.J., Wong, D.W.S., Eds., Marcel Dekker, Inc.: New York, 2002; 359–366.
10. Rouhi, A.M. Chiral business. *Chem. Eng. News* **2003**, *81* (18), 45–55.
11. Gotor, V. Biocatalysis applied for the preparation of pharmaceuticals. *Org. Proc. Res. Dev.* **2002**, *6*, 420–426.
12. Arroyo, M. Empleo de biocatalizadores en la sintesis de compuestos de interes farmaceutico. *Rev. R. Acad. Ciens. Exact. Fis. Nat.* (Esp.) **2000**, *94*, 131–142.
13. Marle, I., Karlsson, A., Pettersson, C. Separation of enantiomers using α-chymotrypsin-silica as a chiral stationary phase. *J. Chromatogr.* **1992**, *604*, 185–196.
14. Fornstedt, T., Sajonz, P., Guiochon, G. Thermodynamic study of an unusual chiral separation.Propanolol enantiomers on an immobilized cellulase. *J. Am. Chem. Soc.* **1997**, *119*, 1254–1264.
15. Schuhmann, W. Amperometric enzyme biosensors based on optimized electron-transfer pathways and non-manual immobilization procedures. *Rev. Mol. Biotechnol.* **2002**, *82*, 425–441.
16. Gupta, M.N., Mattiasson, B. Unique applications of immobilized proteins in bioanalytical systems. Bioanal. *Appl. Enzymes* **1992**, *36*, 1–34.
17. Marko-Varga, G. High-performance liquid chromatographic separation of some mono- and disaccharides with detection by a post-column enzyme reactor and a chemically modified electrode. *J. Chromatogr.* **1987**, *408*, 157–170.
18. Marko-Varga, G., Dominguez, E., Hahn-Hagerdahl, B., Gorton, L. Selective post-column liquid chromatographic determination of sugars in spent sulfite liquor with two enzymic electrochemical detectors in parallel. *J. Chromatogr.* **1990**, *506*, 423–441.
19. Larew, L.A., Johnson, D.C. Quantitation of chromatographically separated maltooligosaccharides with a single calibration curve using a postcolumn enzyme reactor and pulsed amperometric detection. *Anal. Chem.* **1988**, *60*, 1867–1872.
20. Liang, J.F., Li, Y.T., Yang, V.C. Biomedical application of immobilized enzymes. *J. Pharm. Sci.* **2000**, *89*, 979–990.
21. Katchalski-Katzir, E. Immobilized enzymes—learning from past successes and failures. *Tibtech* **1993**, *11*, 471–478.

Bibliography

Godfrey, T., West, S., Eds. *Industrial Enzymology*, 2nd Ed., Macmillan Press: London, 1996.

Kress-Rogers, E., Ed., *Handbook of Biosensors and Electronic Noses, Medicine, Food, and the Environment*, CRC Press: Boca Raton, FL, 1997.

Lam, S., Mallikin, G. Eds., *Analytical Applications of Immobilized Enzyme Reactors*, Blackie Academic and Professional: Glasgow, U.K., 1994.

Liese, A., Seelbach, K., Wandrey, C. *Industrial Biotransformations*, John Wiley & Sons-VCH: Weinheim, Germany, 2000.

Rehm, H.-J., Reed, G. *Biotechnology: A Comprehensive Treatise in 8 Volumes*, Kennedy, J.F., Ed., VCH Verlagsgesellschaft mbH: Weinheim, Germany, 1987; Vol. 7a.

Schmid, A., Dordick, J.S., Hauer, B., Kieners, A., Wubbolts, M., Witholt, B. Industrial biocatalysis today and tomorrow. *Nature* **2001**, *409*, 258–268.

Uhlig, H. *Industrial Enzymes and Their Applications*, Linsmaier-Bednar, E.M., Ed., John Wiley & Sons: New York, 1998.

15

Biocompatibility-Imparting Nanofilm Coating

Hirotsugu Yasuda

CONTENTS

Introduction

A material placed in or contacting with a biological system causes various extent of interfacial interaction with biological components that constitute the contacting biological surface. The extent of the interfacial interaction determines the overall perturbation to the host biological system. If the extent is greater than the tolerance limit of the host biological system, it is generally conceived that the material is not biocompatible. If the level of perturbation caused by the interfacial interaction is within the tolerance limit, the material could be tolerated by the host biological system[1]; in such a situation, the material could be viewed, in practical sense, as "biocompatible."

The perturbation to a host biological system caused by creating a new contact with an artificial material does not solely depend on the chemical nature of the surface but also depends on the size and shape of the material and specific location within the whole biological system where the new contact is induced. It is important to recognize that any biological system depends on the overall balance of numerous concurrent transport processes such as pressure-driven mass flow (e.g., blood flow), diffusive transport within a cell, transport of energy, supply of nutrients, and removal of metabolites, etc., which are readily examined in the cornea of eye. Thus, the overall perturbation caused by the insertion of a material (implant) is highly dependent on the inadvertent interruption of transport processes, and the biocompatibility of an implant highly depends on the specific location where the implant is used. Accordingly, the searching for or trying to develop generic biocompatible materials by chemical synthesis of a new material is infertile effort.

The biocompatibility heavily depends on the type and level of interfacial interaction with the host biological system, but the main objective of placing an implant in a specific part of biological system (body) is dependent on other bulk characteristics of the material, which cannot be sacrificed in order to gain biocompatibility. Consequently, it is nearly mandatory to employ the surface coating that provides a biocompatible surface without altering the bulk properties and the functional capability of the implant.

This chapter describes the very unique nanofilm coating technology; magneto-luminous polymerization (MLP), which could be also termed as magneto-luminous chemical vapor deposition. The process utilizes the unique magneto luminous gas phase, which is the closest kin to the gas phase generally described as low-pressure plasma or low-temperature plasma but is distinctively different from those. Some important distinctions and advantageous features are also presented in this chapter.

Factors that Control Biocompatibility

Man-made materials, in general, are not highly compatible with a biological system or components of the biological system, and it is necessary to create or improve the biocompatibility by tailoring the surface of materials. In this effort, it is mandatory to understand the following fundamental factors that control so-called "surface properties," particularly dealing with organic polymers. Those factors should be realized by the distinctions between two similar terminology; terms specifically applicable to an interface (the factor that influence biocompatibility) are underlined: 1) surface vs. interface; 2) bulk-state vs. surface-state of a solid; 3) molecular configuration vs. surface configuration; 4) bulk properties vs. surface properties; 5) surface properties vs. interfacial properties; 6) surface dynamics vs. interfacial dynamics; and 7) means to minimize interfacial interaction.

"Surface" is the end of the bulk phase of condensed matters; liquid and solid phases. A surface always exists in contact with a different contacting phase, which can be gas, liquid, or solid, however, the influence of contacting gas phase, e.g., ambient air or absence of it (vacuum), is generally ignored, and surface properties are represented by the "interfacial properties" observed in contact with ambient air or in vacuum. In a strict sense, a surface always exists as an "interface" with a contacting medium. The interfacial properties require the description of the contacting medium.

The "bulk-state" of a solid is the majority of a macroscopic solid state, in which atom–atom or molecule–molecule interaction is more or less evenly balanced. In the region near the surface, however, the balance of interactive forces cannot be maintained and the excessive interactive force remains in the "surface state." Consequently, the surface properties, which are governed by the properties of the surface state, generally differ significantly from those of the bulk state. This difference could also explain the difference in molecular configuration and surface configuration described below.

The "molecular configuration" is the description of what, where, and how atoms occupy in a molecule. The presence of a hydrophilic moiety, such as hydroxyl group –OH, makes a molecule hydrophilic, and some polymer molecules, such as poly(vinyl alcohol), water-soluble. However, the surface of poly(vinyl alcohol) film is not as hydrophilic as one might expect from its molecular configuration. This discrepancy is because not all –OH groups are aligned at the surface facing outward; in reality, more –OH groups are facing inward (away from the surface) in a poly(vinyl alcohol) film cast in air and kept in air. Thus, the surface configuration refers to the molecular configuration in the surface state. The surface configuration is the special mode of arrangement of functional moieties in the surface state without altering the configuration of the molecules in consideration.

The bulk properties of materials, such as mechanical strength, rigidity, flexibility, and sustainability or decay characteristics of those bulk properties when the material is used in contact with biological system, are vitally important factors that cannot be substituted or traded for any surface characteristic requirement. On the other hand, the surface characteristics are the major factor that determines the biocompatibility of the whole material under consideration. The most serious and practical dilemma is how to obtain two groups of requirements which, in many cases, are incompatible. If the effort to change surface properties alters the bulk properties of the materials in consideration, it is defeating the purpose of the use of artificial materials in biological system. This is unfortunately the reality dealing with biocompatibility of artificial materials.

"Surface properties" are, in a strict sense, misleading hypothetical properties of materials as far as the use of artificial materials in contact with biology or medicine, since the crucial factor that controls the biocompatibility of materials is the interfacial properties. On the other hand, the interfacial properties require identification of the contacting medium, i.e., biological environment and the characterization of the system with respect to the interfacial characteristics with the material in consideration. The interfacial characteristics therefore cannot be dealt with a generic manner either.

"Surface dynamics" refers to the time-dependent change of the interfacial properties under a set of environmental conditions. The rate of surface dynamic change is dependent on the parameters involved in the contacting medium such as humidity in contacting air, the osmolality of contacting medium, etc. In other words, "interfacial dynamics with well-defined contacting medium" is the key factor that determines the longevity of biocompatibility.

Factors described above clearly point out that alteration of interfacial characteristics by applying a coating without influencing bulk and surface characteristics of the substrate material is an extremely difficult task. It might be best considered that it cannot be done by conventional coatings, because whatever is being applied must face the same problems that the substrate material could not overcome. Thus, it is necessary to employ totally new methods and significantly different materials in order to impart biocompatibility to adequately selected materials. MLP coating is a unique method to achieve such a goal by creating a nanofilm that does not interact with any contacting biological system. The following figures illustrate the factors described above.

Figure 15.1 shows plots of the advancing sessile droplet contact angles of water on the surface of a gelatin hydrogel against the water droplet volume measured under a constant

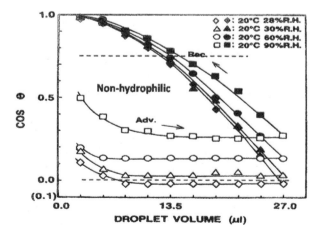

FIGURE 15.1
Sessile droplet contact angle of water on surfaces of gelatin hydrogel (water content > 95%) measured at 20°C and various relative humidity of air. cos < 0.0: hydrophobic domain; cos θ > 0.75: hydrophilic domain; 0.75 > cos θ > 0.0: amphoteric domain.

temperature (film, water, and air) of 20°C but under different relative humidity of air. This figure shows, first of all, that the surface of a highly hydrated hydrogel of gelatin is surprisingly hydrophobic. These plots also clearly show that the sessile droplet contact angle, which is an interfacial parameter, is dependent on the relative humidity of air, which is the parameter of the contacting medium. These plots also show that the advancing contact angle and the receding contact angle do not follow the same dependency on the droplet volume. This discrepancy is due to the change of interfacial configuration of gelatin under the water droplet.

Figure 15.2 shows the contact areas under a water droplet on the advancing and receding contact angle measurement processes shown in Figure 15.1. This figure tells that the contact area of water droplet increases appreciably on the receding contact angle measurement from the last contact area of the advancing contact angle measurement, That is, the advancing contact angle and the receding contact angle represent two different phenomena. The increasing contact area during the process of decreasing the droplet volume tells us an important fact that the interfacial configuration of gelatin under the water droplet has changed by contacting liquid water, and the strong interfacial interaction between liquid water and the surface of gelatin (under the water droplet) not only prevents the receding of contact area but also increases the contact area during the receding contact angle measurement (i.e., the change of the interfacial configuration causes increase of wetting of surface). Since the contact area does not decrease with decreasing water droplet volume, the receding contact angle merely represents the flattening of the water droplet rather than receding of whole droplet with identifiable contact angle. The flattening of water droplet is caused by the decrease of droplet volume and the increasing contact area, which clearly shows that interfacial configuration of gelatin, in air, had been changed by the influence of the contacting liquid water.

Figure 15.3 shows similar plots shown in Figure 15.1 for a hydrogel of agar–agar with similar hydration (above 95% water content). The striking difference between Figures 15.1

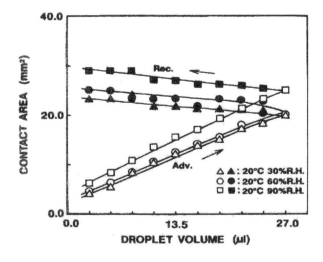

FIGURE 15.2
Contact area of water droplet observed the sessile droplet contact angle measurement shown in Figure 15.1 as a function of water droplet volume in microliter. On gelatin surface, the contact area does not decrease on the receding contact angle measurement but slightly increases.

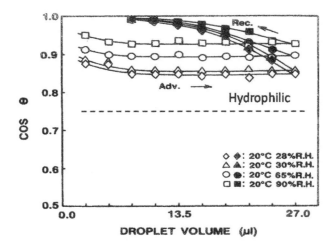

FIGURE 15.3
Sessile droplet contact angle of water on the surface of agar–agar hydrogel (water content > 95%) measured under the identical conditions used in the experiments shown in Figure 15.1. The surface of agar–agar hydrogel remains in hydrophilic domain but shows the influence of air humidity and slight increase of the interfacial configuration under the water droplet to increase hydrophilicity (decrease of advancing contact angle).

and 15.3 is that the interface of agar–agar hydrogel/air is highly hydrophilic but the interface of gelatin hydrogel/air is in the hydrophobic end of amphoteric domain. The reason why the surface of gelatin hydrogel that imbibe large amount of water can be hydrophobic and the surface of agar–agar hydrogel that also imbibe the same amount of water is highly hydrophilic lies in the difference of the molecular configuration of both polymers and their ease of changing interfacial configuration.

Gelatin consists of degenerated protein molecules, which takes random coil configurations with high degrees of mobility. When the gelatin hydrogel is taken out of water for contact angle measurement, rearrangement of the surface configuration occurs instantaneously and nearly all hydrophilic moieties turn toward the bulk of hydrogel, which contains a large amount of water, making the surface populated with highly hydrophobic moieties. This postulation is confirmed by the measurement of contact angle by the air bubble insertion method (an air bubble is introduced below the hydrogel surface in water), which showed a very hydrophilic surface.[2] Thus, the difference of Figures 15.1 and 15.3 demonstrates the importance of the interfacial configuration change caused by the contacting medium.

In contrast to the rather loose random coil configuration of gelatin molecules, the repeating unit of an agar–agar molecule is di-saccharide, as depicted in Figure 15.4. The repeating unit is in a planar configuration and the center linkage that combines two sugar units is rigid, with no rotational freedom, and the equal amount of –OH groups are located on both sides of planer rings. With such a molecular configuration, the quick and significant change of interfacial configuration, which requires redistribution of functional groups, cannot occur.

The influence of contacting medium on the interfacial configuration of a polymer is not limited to contact angle measurement. Figure 15.5 depicts the change of interfacial configuration as a function of immersion time observed with a copolymer of ethylene/vinyl alcohol in which oxygen atoms exist only as –OH groups. The O_{1s}/C_{1s} signal ratio profile in the surface-state domain of a film obtained by the angular dependence of X-ray

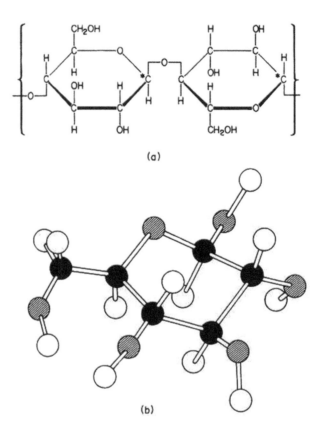

FIGURE 15.4
The molecular configuration of the repeating unit of agar–agar molecule. Due to the equally distributed hydrophilic moieties on both sides of a planar structure, the rotation of molecule does not change overall distribution patterns of hydrophilic groups.

photoelectron spectroscopy (XPS) measurement changes with the immersion time. In order to avoid the change of surface configuration during the process of XPS measurement, water-immersed samples were freeze-dried in the removing process of the mobile liquid water from samples.

The dry sample (water immersion time = 0) shows the lowest oxygen content at the surface. The distribution pattern changes with water immersion time of film sample. The oxygen content at the surface increases with water immersion time, and the highest value is observed with the sample kept in water for the longest time. These plots clearly show that the (–OH) groups buried in the deeper sections from the top surface are brought out towards the top surface by water immersion of the film. This change does not occur instantaneously such as the case of gelatin hydrogel, and the time factor of the change due to contacting environment becomes evident. This delayed surface configuration change could be related to the in vivo durability of the biocompatibility of an implant with polymer surface.

The interfacial interaction does not solely depend on the localized motion of functional groups that cause the interfacial interaction with biological environment. The rotational movement of a certain length of polymer segments could also contribute to the interfacial interaction or the absence of it. Figure 15.6 depicts molecular models of a short segment (7–8 carbons) of poly(oxymethylene) $[(CH_2-O)_n]$ and poly(oxyethylene) $[(CH_2-CH_2-O)_n]$. The

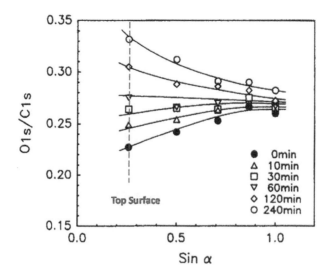

FIGURE 15.5

Depth profile change of oxygen atoms in copolymer of ethylene/vinyl alcohol due to the immersion of sample film in water measured by incident angle (α) dependent of O_{1s} signal plotted as the ratio of O_{1s}/C_{1s} against sin α. In dry film, more –OH groups are buried in the deeper section from the top surface. Water immersion of film brings out the buried –OH groups to the top surface of the film (interfacial configuration change).

ratio of O/C for poly(oxymethylene) (POM) is 1.0, and the same ratio for poly(oxyethylene) (POE) is 0.5. According to the O/C ratio, one might anticipate that POM is more hydrophilic than POE. However, the reality is the opposite; that is, POM is hydrophobic polymer, and PEO is highly hydrophilic water soluble polymer. Figure 15.6 depicts the consequence of simply rotating the short segments. The top view of POM changes completely by rotating the segment 180 degree. On the other hand, the top view of POE is essentially identical to the bottom view after the segment is rotated 180 degree, although the location of the unit segments shifts laterally.

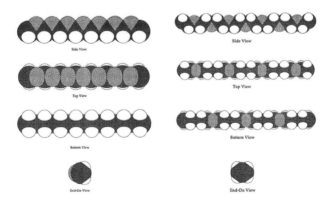

FIGURE 15.6

Comparison of the influence of rotation of molecular model of the repeating segment of poly(oxymethylene), *left*, and of poly(oxyethylene). The *top view* and *bottom view* of POM is totally different, but the *top view* and *bottom view* of POE is essentially identical, which makes the change of interfacial configuration very small to none.

It has been generally considered that agar–agar and POE are biocompatible. Both polymers have symmetrical distribution of functional groups or atom (O) and, consequently, the chain conformation change has minimal effect on the interfacial configuration of polymers, which supports the concept that the absence of interfacial configuration change is important in biocompatibility of polymers because the configuration change caused by the influence of interfacial contact triggers the interfacial interactions, which is the key element of biocompatibility.

The factors that influence interfacial interaction can be expressed by Gibbs free energy of interfacial interaction, ΔG_i:

$$\Delta G_i = (\Delta \gamma + \Delta H_\chi)_i - T\Delta S_i$$

Enthalpy term, ΔH_i, involves $\Delta \gamma$, interfacial tension (γ-interaction) and ΔH_χ, enthalpy of physico-chemical interaction (χ-interaction). Entropy term, $T\Delta S_i$, involves the entropy of interfacial configuration change.

The above equation indicates that, in order to have good biocompatibility of an implant, the following requirements should be met: 1) the interfacial tension should be minimal; 2) chemical interaction should be also minimal; and 3) no interfacial configuration change should occur. These requirements also indicate that chemical surface modification has a very slim chance to get the sustainable biocompatibility. Furthermore, chemical modification of a surface could alter the bulk properties of the material, especially polymers, which makes the situation worse. Another tempting approach is the synthesis of a new polymer; however, we should be aware of the fundamental principle that the molecular configuration cannot be used to control the interfacial configuration in a specific interface as described in the above sections; that is, the of molecular configuration could be used to alter bulk properties of a polymer, but not to change the interfacial properties as schematically illustrated in Figure 15.7.

Consideration of all factors described above leads to the concept that the only realistic mean to provide biocompatibility to an implant is to apply a coating that changes the

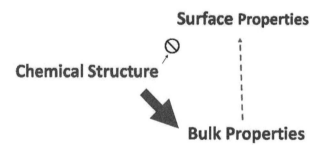

FIGURE 15.7
Schematic representation of the role of molecular configuration to bulk properties and surface properties. The major influence of molecular configuration is on the bulk properties of polymer, and the effect on surface properties are the secondary effect based on the bulk properties.

surface state that contact with a biological system. This approach comes with very stringent requirements including: 1) superb adhesion to the substrate; 2) no or minimal interfacial interaction with the contacting biological system; 3) no uptake of matters from the contacting medium; 4) no leaching out of matters to the contacting medium; 5) superb dimensional stability; and 6) sterilization by steam autoclaving should not change any function of whole implant. (Other sterilization methods, e.g., γ-radiation and gas sterilizations, have been used when autoclaving cannot be used; however, those sterilization could chemically alter polymeric surface and the fact that autoclaving cannot be used is often an indication of poor biocompatibility of the material.)

Magneto-Luminous Polymerization

The deposition of an amorphous carbon nanofilm from methane by MLP fulfills all of the abovementioned stringent requirements. In order to understand the uniqueness of the method, the knowledge of what kind of gaseous species are involved in the low-pressure deposition of a nanofilm to be used in imparting biocompatibility is necessary. Nature of the gaseous species created from a simple organic gas, such as methane, entirely depends on the mechanism by which a luminous gas phase is created, i.e., the mechanism of the gas-phase breakdown or the transformation of gas phase from the dielectric dark gas phase to the electrically conducting luminous gas phase (glow). This crucially important step of the process, however, remains as the least elucidated factor in the process generally known as "plasma polymerization" or "plasma enhanced chemical vapor deposition."

Traditional physics of low-pressure plasma deals mainly with the ionization process and mechanisms of ionization of atomic gases. It has been widely considered that "plasma" is synonymous to "ionized gas." This concept is applicable to the high-temperature plasma phase in which numbers of ions and electrons could be equal. However, in "low-temperature plasma" or "low-pressure plasma," the situation is quite different. First of all, the energy associated with ions and electrons are much lower, and the number of ions and the number of electrons are not equal because ions are created by electron impact reactions.

An electron impact ionization of Ar to create Ar^+ and e^- leaves two electrons (one created by the ionization and another on that caused the ionization reaction) and one ion. The number of ions in low-temperature "plasma" in vacuum is less than 1% of gas atoms. Such a gas system cannot be dealt by the same terminology of "plasma," which is ionized gas. The concentration of ionized species in the low pressure typically used in plasma polymerization is generally in the range of 1.0–0.001%. The lower the system pressure, the concentration of ionized species approaches to the lower end.

With (mono) atomic gas such as argon, which does not participate in chemical reaction, the nature of the main species in low-temperature plasma, which is over 99% of mass in the gas phase, is of the secondary importance or an issue of no serious concern. Dealing with organic molecules in gas phase, which participate chemical reactions (i.e., plasma polymerization), the situation reverses completely. The nature of mass that constitute over 99% in the luminous gas phase of organic molecules and the mechanisms of creation of chemically reactive species are far more important issues than the mechanism of how ions, which are minor components in such a system, are created.

Due to widespread use of the vague term "plasma" and the concept that "plasma is ionized gas," it has been assumed (without careful examinations) that the ionization causes the formation of chemically reactive species from an organic molecule in low pressure. According to the traditional concept of plasma inception in low pressure, the primary electrons and ions of gas atoms are created by the impact of naturally occurring (unspecified) high-energy radiation in the (dark) gas phase, and the bombardment of the primary ions on the cathode surface causes the emission of the secondary electrons, which are accelerated in the electric field. When the energy of electrons being accelerated in the electric field reaches a sufficiently high level, the ionization of gas atoms occurs and creates a luminous gas phase that is conceived as a "plasma phase."

While the mechanism of the formation of chemically reactive species out of simple non-reactive gas molecules, such as methane and benzene at the onset of luminous gas phase, plays the key role in determining the physical and chemical nature of the deposition products, it has not been treated as the key issue. That is, it has been assumed that the glow developed at the onset of glow is caused by the ionization of gas molecules.

The distinction between the electron impact ionization of atomic gas and the electron impact dissociation of organic molecules has been first recognized by the discovery of the presence of the *dissociation glow* of organic molecules that can be clearly distinguished from the *ionization glow* of atomic gas in the observation of the initial step of direct current (DC) discharge of an organic molecule [trimethylsilane (TMS)] in 2003 and published in 2004.[3,4]

Figure 15.8 compares the ionization glow of Ar and the dissociation glow of TMS under identical experimental condition, in which a plane plate cathode is placed in the center of two anode plates which is equipped with magnetic field enhancement. (The use of a magnetic field on the anode plates was for investigating other factors involved in the gas-phase breakdown, but does not influence the breakdown of the two gases.) The mechanism of

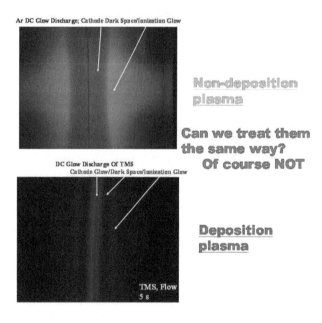

FIGURE 15.8
Comparison of the initial glow develops at the onset of gas-phase breakdown with a non-deposition gas (Ar) and deposition gas (TMS). In Ar discharge, the initial glow (ionization glow) develops an appreciable distance away from the cathode surface. In TMS discharge, the initial glow (dissociation glow) develops right at the surface of cathode.

gas-phase breakdown should be able to explain these obviously different phenomena. The ionization mechanism cannot explain the phenomenon shown in the lower picture of Figure 15.8, since there is no cathode fall dark space in which acceleration of electrons necessary for gas ionization should occur.

As shown in the bottom picture of Figure 15.8, the dissociation of organic molecules occurs first; that is, the dissociation glow that adheres to the cathode surface (without a dark space in between) develops before the ionization glow, which is separated by the cathode fall dark space from the cathode surface, become visible. The emission spectroscopy analysis of the dissociation glow and the ionization glow showed that the ionization glow is mainly ions of hydrogen, which is the dissociation product of TMS. Furthermore, plasma polymerization in DC discharge is controlled by the current density near the cathode, and the DC voltage, which is the key factor in creating glow of Ar, has no influence on DC plasma polymerization.[6,7] Accordingly, the mechanisms of gas-phase breakdown in DC discharge of gases, particularly organic molecules, should be investigated with broader viewpoint than investigation by DC voltage influence, which is based on ionization principle.

Electrons that cause the breakdown of the gas phase, mono-atomic gas (Ar) as well as molecular organic gas (TMS or methane) are pulled out of metal surface by the applied electric field[8], not by the bombardment of accelerated ions, as the classical view postulated. The electrons emanating from the cathode surface are accelerated in the dark space by the electric field. Electrons must be accelerated in the dark space to gain enough energy to ionize atoms; consequently, the ionization glow appears at the characteristic distance away from the cathode surface, which is recognized as the cathode fall dark space. Figures 15.9 and 15.10 schematically illustrate the processes of causing ionization glow and dissociation glow.

In contrast to the ionization glow, the dissociation glow appears very close to the cathode surface without visible cathode fall dark space, indicating that the electrons emanating from the cathode surface have sufficient energy to dissociate organic molecules, which is nearly an order of magnitude smaller than the ionization energy of Ar.

The role of electrons that causes ionization glow (upper picture, Figure 15.8) and dissociation glow (lower picture, Figure 15.8) can be explained by examination of the dark spaces in both cases. Figure 15.9 depicts the dark space that causes the ionization glow of Ar, and

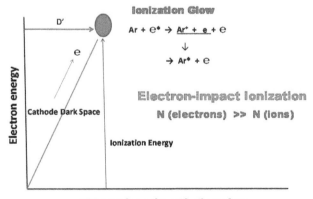

FIGURE 15.9

Schematic presentation of the acceleration of electrons in the cathode dark space to gain sufficient energy to cause ionization of Ar atoms. The electron emanating from the cathode is shown by $\ominus$, and the electron created by ionization is shown by e. Energetic species are shown with *.

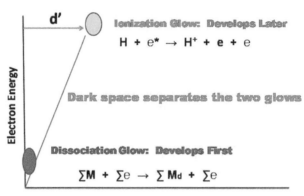

Distance from the cathode surface

FIGURE 15.10
Schematic presentation of electron impact dissociation of organic molecules (TMS) and the ionization of H atoms, which is the product of dissociation of TMS, by electron impact ionization. The electron emanating from the cathode surface is shown by Θ, and the electron created ionization is shown by e. Energetic species are shown with *.

Figure 15.10 depicts the dark space with dissociation glow of an organic molecule (TMS) and the subsequently developed ionization glow of H atoms. In both cases, electrons are pulled out from the cathode surface by the electric field developed under the breakdown voltage, V_b.

In DC discharge of Ar, the energy carried by the electrons emanating from the cathode surface is not enough to cause excitation or ionization of Ar, though electrons inevitably collide with Ar atoms. The electrons are accelerated under the electric field, and when electrons gain sufficient energy, the collision of electrons causes the ionization of Ar atom. It is likely that the majority of excited species of argon, Ar^*, are caused by the recombination of the ion and electrons; $Ar^+ + e^- = Ar^*$, since excitation requires precise energy with a narrow band, and the probability of Ar atom colliding with an electron at the precise moment when the electron gains the precise energy for excitation is deemed to be very slim.

In the DC discharge of TMS, the electrons emanating from the cathode surface have sufficient energy to dissociate an organic molecule, mainly splitting of Si–H and C–H bonds, causing the dissociation glow. The detached hydrogen atoms recombine to form H_2 molecules. The recombination of hydrogen atoms lowers the overall net dissociation energy lower, which makes the dissociation of organic molecules quite feasible at low levels of electron energy. Electrons emanating from the cathode surface, some of which have lost energy by causing the dissociation of TMS molecules, are accelerated under the electric field, exactly the same process as the case of Ar discharge. At the onset of gas-phase breakdown, the intensity of the ionization glow is faint, and sometimes it is difficult to identify the presence of the ionization glow as the second glow, simply because hydrogen is created by the dissociation glow (there is no hydrogen atom in the feed gas).

The phenomenon of gas-phase breakdown in DC discharge of Ar, as a function of the applied voltage, V, is explained in Figure 15.11. When a very low voltage is applied, very low dark current (leak current) is observed, but gas the phase is dark. As the applied voltage is slowly increased, the dark current does not increase and the gas phase remains dark. Although the meaning of an electric field without the current flowing through the gas phase is questionable, the calculated electric field, (V/d, where d is the distance between

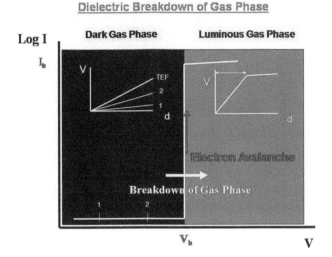

FIGURE 15.11
Breakdown process of Ar in DC discharge expressed by the plot of current (in logarithmic scale) and voltage.

cathode and anode), increases proportional to the applied voltage, as shown by the insert in the dark gas phase domain, which is a straight line connecting cathode and anode.

As the applied voltage reaches the critical value, designated as the breakdown voltage, V_b, the discharge current, I_b, develops and the gas phase is converted to the luminous gas, or glowing gas phase. In practical terms, this transition occurs within a short period of time and the electron avalanche from the cathode to anode occurs. V_b and I_b, in this context, represent for the sustainable electrical discharge. When the luminous gas phase is developed, the electric field profile changes significantly and the electric field in the luminous gas phase is no longer the straight line from the cathode to anode as depicted in the luminous gas phase domain in Figure 15.11.

The electric field that caused the gas-phase breakdown can be given by V_b/d', which is greater than V_b/d. The parameter, d', is the width of dark gas phase existing between the cathode surface and the edge of glow. The width of the cathode dark space and the width of the luminous gas phase (glow) at the onset of gas-phase breakdown are dependent on the system pressure, as depicted in Figure 15.12. The cathode dark space shrinks, and the intensity of glow adjacent to the dark space increases at the expense of the width of glow, as seen in the figure, with pressure. These phenomena are reversible and reproducible. When the system pressure of a broken-down gas phase is decreased, the cathode dark space increases, and the glow width also increases. When the glow touches the anode surface, the glow extinguishes, which confirms that the gas-phase breakdown to create luminous gas phase can be dealt with as a function of V_b and I_b, in the context described above, in all practical purposes. Thus, gas-phase breakdown should be investigated as a function of V_b and I_b, and the parameters that can be derived from V_b and I_b.

The conventional approach in investigation of gas-phase breakdown assumes that low temperature plasma is created by ionization of gas. Accordingly in the study of gas-phase breakdown, only breakdown voltage, V_b, has been measured as a function of pressure and the separation distance, d, of cathode and anode in DC discharge. In recent studies, however, considering all factors described above, breakdown process was investigated as a function of breakdown voltage, V_b, current, I_b, and many derived parameter from V_b and

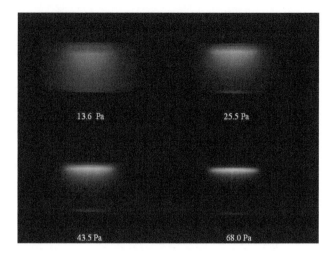

FIGURE 15.12
The dependence of dark space width, d', and the width of glow on the system pressure. *Top surface*: cathode; *bottom surface*: anode.

I_b, e.g., conductivity of broken down gas phase (I/V), discharge energy ($W = V * I$), and the energy transfer parameter, which is given by energy per mass of the feed gas (W/FM in J/kg, where F is molar flow rate, and M is the molecular weight of gas).[9]

The value of V_b (of a simple mono atomic gas; Ar) is dependent on experimental parameters, such as the size of electrodes, separation distance of cathode and anode, and the overall size of reactor, which changes the flow rate of gas and the pumping rate to maintain a specific system pressure. It was found that the variance of V_b due to types of gas (with a fixed reactor parameters) is marginal compared to that of operational parameters. In contrast to this situation, the dependence of I_b on types of gas is remarkably strong. Since variance in V_b is small, all derived parameters follow the same pattern of dependence of I_b on the system pressure. In other words, the investigation of gas-phase breakdown of various gases by V_b alone cannot provide necessary information for the mechanism of gas-phase breakdown. Only the investigation of gas-phase breakdown in terms of V_b and I_b, or derived parameters from them, can provide the necessary information for comprehending the gas-phase breakdown phenomena with various types of gases.

Since values of V_b and I_b can be obtained only in DC and alternating polarity in kHz range, but not in higher-frequency domain, e.g., radio frequency discharge at 13.5 MHz, in which most deposition work has been carried out, gas-phase breakdown characteristics of various gases can be best compared by plots of energy per mass (J/kg) vs. system pressure, as shown in Figure 15.13. The breakdown energy expressed in the units of J/kg is also an indication of how much energy can be transported through the conductive luminous gas phase. The energy of the luminous gas phase increases with the system pressure only above the transition point pressure. Below the transition point pressure, θ, unstable discharge with high voltage/low current develops. The transition point pressure is the minimum pressure to create the practically utilizable luminous gas phase. Above the transition point pressure, the discharge could be characterized as low voltage/high current. The transition point pressure is by and large the same regardless of the types of gas. The major differences due to the types of gas are seen in the slope of the plots.

The increase of discharge power is highly dependent on the conductivity of the broken down gas phase, which is dependent on the dissociation of molecules and the

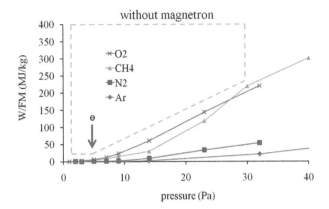

FIGURE 15.13
Gas-phase breakdown phenomena expressed by the plots of breakdown power, W/FM in MJ/kg against system pressure. Gas-phase breakdown occurs only in the domain above the threshold pressure, Θ. The zone shown by *red line* is the domain in which gas-phase breakdown cannot occur.

electronegativity of atoms in molecules. Thus, the gas-phase breakdown can be viewed as creation of excited species with high electron conductivity rather than ionization of gas. Only in absence of dissociation, e.g., in the case of Ar, the ionization and subsequent excitation become the major mechanisms to create luminous gas phase. However, even in such a case, the majority of the gas phase in the low-pressure domain is not comprised of ions because ions recombine with electrons to form photon-emitting excited neutral species.

Figure 15.14 depicts the change due to the presence of a magnetic field in the same mode of the data shown in Figure 15.13. The remarkable changes depicted in the plots shown in Figures 15.13 and 15.14 indicate the magnitude of the influence of magnetic field on the mechanisms of gas-phase breakdown of gases investigated.

The most significant change is the shift of the domain of the gas-phase breakdown. With magnetic field, the main domains shift to the no-breakdown domain without magnetic field. As a consequence of this shift, exceptionally high currents are caused by very low

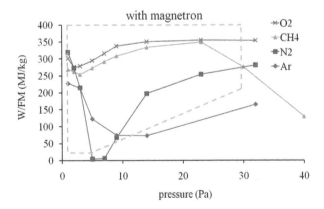

FIGURE 15.14
Gas-phase breakdown under the influence of super imposed magnetic field on the cathode (magnetron). The domain of gas-phase breakdown shifts to the no-breakdown zone observed without magnetic field. Very high current discharge occurs in very low pressure, below the threshold pressure observed without magnetic field.

voltage applied in the very low pressure domain. Without a magnetic field in this domain, very high voltage is necessary to cause very low current, yielding an unstable discharge that cannot be used in coating process. Significant differences in the mechanisms of gas-phase breakdown due to the types of gases are more pronounced with magnetic field, i.e., mono atomic gas vs. deposition gas, molecular gas with high electronegative atom (e.g., O) vs. low electronegative atom (e.g., N).

Coating of Amorphous Carbon Nanofilm by Means of MLP

The magneto luminous gas phase is created by superimposing a magnetic field on the cathode consisting of nonmagnetic metal. For coating of nanofilm, audio frequency (1–50 kHz) discharge, rather than DC discharge, is used to create uniform coating on both sides of a substrate placed in the center of the electrodes separation distance. Audio frequency discharge is essentially alternating polarity DC discharge, and all unique features of DC discharge are retained. Although numerous ways of coupling the magnetic field to the cathode could be used, the simplest coupling of magnetic field to the cathode has been used in MLP in laboratory reactors, which can be also used in industrial scale operation with minor modification to accommodate the change of mode of operational conditions for the final coated products. In industrial scale continuous operation, substrates must be transferred by linear (vertical or horizontal) motion.

Figure 15.15 is a pictorial view of the magnets assembly; a central circular iron plate and a circular iron plate ring are bridged by eight bar magnets maintaining the same polarity of all bar magnets. The center circular plate becomes one magnetic pole, e.g., the south pole, and the circular ring plate becomes the opposite magnetic pole, i.e., the north pole. This assembly is attached to the back side of an electrode (nonmagnetic metal plate, e.g., titanium), and audio frequency power is applied to both electrodes separated with predetermined distance. In order to maintain the symmetry of the luminous gas phase, a symmetrical power supply, a power supply with floating power outlets, is used. Without the magnetic field, the cathode glow develops, covering the entire surface of an electrode. With the magnetic field, the toroidal glow, as shown in Figure 15.16, develops in the gas phase near the electrode surface.

The electrode surface beneath the toroidal glow (toroidal glow surface) has no deposition of material under properly selected operational conditions, while the remaining electrode surface received varying degree of deposition and characteristics. Since the toroidal glow surface remains deposition free, MLP can be operated continuously for extended periods of time, e.g., continuous operation for a month, with continuous feeding of substrates and gas. The characteristics of the coating do not change in the entire span of 1 mo of continuous coating operation. This is a very unique feature of MLP, which cannot be achieved by any other mode of low-pressure plasma deposition processes because of the deposition on electrodes (energy input surface).[11]

In most cases, small amounts of oxygen are added to the feeding gas. The main objective of adding oxygen to the methane gas flow is enhancing the deposition-free "toroidal glow surface" and slowing down deposition rate by consuming carbon atoms produced by the dissociation of methane and removing from the vacuum system. The addition of oxygen into low-pressure plasma polymerization, in general, does not contribute to making the surface of depositing film hydrophilic by the same principle; that is, added oxygen does

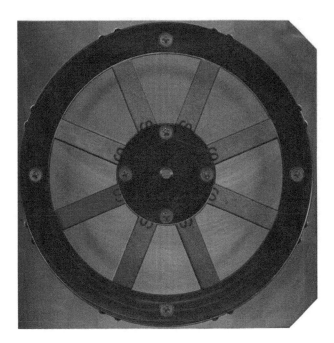

FIGURE 15.15
Pictorial view of magnetic field arrangement; the circular center (iron) plate becomes the south pole and the outer circular ring plate becomes the north pole of the circular planar magnet, which will be placed behind a titanium plate electrode.

not remain in the film deposition: nitrogen in and oxygen out (iN–Out) rule.[9] Without feeding O_2, the surface of the nanofilm is moderately hydrophilic, because the trapped free radicals on the top surface of nanofilm react with oxygen when coated substrates are taken out of the vacuum reactor. The O contents, measured by XPS, of coatings with and without O_2 feed are essentially the same. Amphoteric hydrophilicity–hydrophobicity (not highly hydrophilic surface) seems to be one of the key requirements for biocompatibility by the minimum perturbation concept, because biological systems contain both hydrophilic and hydrophobic components.

The gas phase in between two toroidal glows is filled with a less intense but more uniform luminous gas phase (with respect to the toroidal glow), which is the main luminous gas phase of alternating magneto DC discharge. Since the reactive species are created in the toroidal glow, the distribution of reactive species within the luminous gas phase depends on the distance from the toroidal glow. The substrates to be coated are placed in the middle portion of two electrodes by means of a rotating sample holder, and the substrates move in and out of the luminous gas phase, which normalize the distribution pattern of reactive species in the luminous gas phase. (The more detailed description and explanation of the process can be seen in Yasuda[10]).

In the preceding section, the unique differences of MLP from conventional low-pressure plasma processes are described. How those differences reflect on the characteristics of nanofilm are described below, particularly with respect to the biocompatibility of the surface created by the method.

Figure 15.17 schematically depicts the scheme of how a polymeric nanofilm is formed from a simple molecule of methane, CH_4, by the repeating step growth polymerization (RSGP) mechanism. Free radicals are produced mainly by the electron impact dissociation

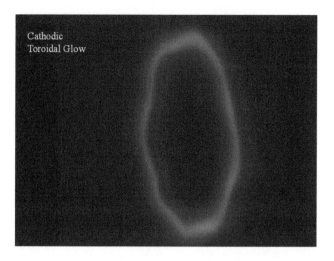

FIGURE 15.16
Pictorial view of the toroidal glow develops by a magnetic field enhanced electrode by DC discharge. The toroidal glow develops near the surface where the magnetic field line is parallel to the electrode surface. The glow develops only in the toroidal glow and the remaining surface of electrode is not participating with the creation of luminous gas phase.

of methane in the toroidal glow, i.e., homolytic splitting of C–H bonds and carbine formation. The formation of a stable molecule, H_2, as a product reduces the overall reaction energy significantly. The recombination of two free radicals yields a species with larger molecular weight, which is subjected to repeat the same cycle. The RSGP mechanism becomes much more complicated if monomers of free radical polymerization or molecules with various functional groups are used. The surface of deposited material would have ability to interact with contacting medium (e.g., biological system), which defeats the objective of producing a non-interacting surface for biocompatibility.

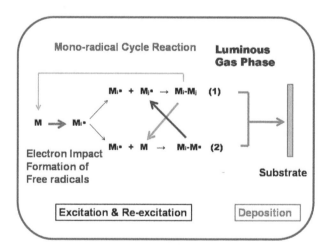

FIGURE 15.17
Schematic presentation of the mechanism of forming polymeric nanofilm from simple molecule of methane by RSGP.

In a gas phase contained in a vessel, there are two fundamentally important collisions that influence material formation and the subsequent characteristics of the deposited surface, namely: 1) gas–gas collision and 2) gas–surface collision. The surfaces that collide with gas species are the surfaces of the reactor in various forms and the surface of substrate to be coated. The ratio of gas–gas collision/gas–surface collision is dependent on the total surface area (reactor surface and substrate surface), size of gas phase (reactor volume) and gas pressure.

The adhesion of depositing material mainly depends on gas–surface collision. On the other hand, the propagation of polymer formation mainly depends on the frequency of gas–gas collision, although the deposited species reacting with gas species contribute significantly to the overall polymerization reactions. Figure 15.18 schematically depicts, in a simplified manner, the influence of gas–surface collision and gas–gas collision onto the characteristics of the nanofilm. Gas–gas collision in the luminous gas phase increases the size of species that eventually deposit on the substrate surface. The increase of size of depositing species increases the mass of depositing species per contact area of the deposition and substrate surface, which reduces the adhesion strength of nanofilm. At lower pressure, the deposition process would yield: 1) assembly of smaller size depositing species; 2) increased adhesion of overall nanofilm; 3) increased smoothness; and 4) higher barrier characteristics for larger permeating molecules (lower permeation cutoff). (Some details of MLP reactors could be seen in Yasuda[2] and Yasuda[10].)

Evaluation of Biocompatibility

The very unique and advantageous features of the MLPs of methane are: 1) capability to be operated at a lower pressure than the transition point pressure of DC discharge; and 2) exceptionally high deposition rate in the low-pressure regime. The polymeric material thus produced can be best described as amorphous carbon with minor content of hydrogen

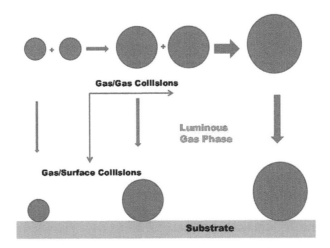

FIGURE 15.18
Schematic presentation of competitive processes of deposition (gas–surface collision) and gas phase growth (gas–gas collision) that determine the factors important to biocompatibility of nanofilm.

atoms; a-(CH_x). The most significant aspect of the MLP of methane (MLP-Me), with respect to the biocompatibility of the nanofilm, is the polymerization of carbon atoms, and the surface-state of nanofilm has no chemically or physico-chemically active (functional) group.

The biocompatibility should be examined from the viewpoint of the interfacial interaction; that is, if the interfacial interaction is high between an artificial material (implant) and the biological system that contact with the implant, the biocompatibility cannot be anticipated. Thus, the biocompatibility is inversely proportional to the level of interfacial interaction, and it is necessary to reduce the interfacial interaction in order to make an implant biocompatible.

The level of interfacial interaction between an implant, I, and the biological system, B, that contact with I, can be expressed by the interacting capabilities of respective phases: the interfacial interaction potential, Φ. The overall interfacial interaction could be expressed by the products of interaction capabilities of two phases, i.e., interfacial interaction potential of the implant surface, Φ_I, and that of the biological system, Φ_B. Thus, the interfacial interaction between the surface of an implant and biological system could be interfacial interaction potential of the implant surface, Φ_I, and that of the biological system, Φ_B:

$$\text{Interfacial interaction} = F(\Phi_I \times \Phi_B)$$

It is important to recognize that there is no way to reduce Φ_B, but there are possibilities of reducing Φ_I. As described in earlier section, the value of Φ_I for gelatin hydrogel surface is much greater than Φ_I for agar–agar gel surface due to the difference in the rotating capability of functional groups within a molecule. The value of Φ_I for poly(ethylene oxide) is low by virtue of the molecular configuration. It has been known (without reasons, such as described here) that those polymers, which have lower value of Φ_I, belong to polymers with better biocompatibility.

If one could reduce the value of Φ_I to zero, then the product of $(0 \times \Phi_B)$ becomes zero. If $(\Phi_I \times \Phi_B) = 0$, the overall interfacial interaction becomes zero regardless of the value of Φ_B, and such a material becomes close to a generic biocompatible material regardless of the contacting biological system, within the framework of the interfacial interaction between two surfaces. If a material surface cannot cause interfacial interaction with a biological system, the presence of material is ignored by the biological system. This is the fundamental principle of the minimum perturbation concept for biocompatibility, and that is the case for the surface coated with a nanofilm of amorphous carbon, because there is no chemical functional group on the surface that could interact with a biological system.

Experimental results indeed showed that the interfacial interaction of the amorphous carbon film, deposited on various materials, with various biological systems are remarkably low. It should be cautioned, however, that if implant is placed in an obvious flow system, e.g., artery or vein, in a body, the perturbing of the flow system by the implant, which largely depends on the size and shape of the implant, becomes as an additional important factor. However, the additional factor would mainly influence on the Φ_B factor, if the value of Φ_I is small enough as the case of the amorphous carbon nanofilm. Consequently, imparting biocompatibility to stent, which will be placed in the blood stream, could be possible as described in the following section.

Quantitative examination of the interfacial interaction with biological systems is virtually impossible, but the general trends that we anticipate from the general characteristics of the coating can be examined by comparing samples with and without nanofilm of amorphous carbon as shown in the following cases. If the surface does not interact with the surrounding biological system, the presence of the implanted material is ignored by the biological system, which is the basic principle of the neutral approach by means of imperturbable surface state. Comparative studies of adsorption of protein and bacteria on the amorphous carbon film surface, comparing with a standard reference surface, seem to be a good method for preliminary evaluation of biocompatibility of the amorphous carbon surface.

In the evaluation shown in Figures 15.19 and 15.20, the reference probe with gold surface and a reference probe coated with MLP-Me were placed in the flow of a test solution side by side, and adsorption was monitored with time.[11] At the end of the preset adsorption time of 250 min, the flow of a test solution was stopped and switched to a cleaning solution under a preset flow rate, and the change of adsorbed materials was followed with time to check if the adsorbed materials were adhering to the surface.

Figure 15.19 shows the adsorption kinetics of albumin on the gold surface (upper line) and on the MLP-Me-coated gold surface. The time-dependent adsorption of a protein, bovine serum albumin, in a surface plasma resonance (SPR) measurement setup using 10 µg/ml phosphate buffered saline (pH 7.4) standard phosphate buffer solutions, were measured at a flow rate of 50 µl/min. The adsorption experiment was stopped at 250 min and switched to the washing cycle at the increased flow rate of 250 µl/min (five times of the flow rate in adsorption cycle). (The upper line A: gold reference surface, the lower line B: the amorphous carbon nanofilm coated on the gold surface.)

Figure 15.20 shows the time-dependent unspecific bacterial adsorption in the identical SPR measurement setup with 1×10^8 *Enteroccocus faecalis* in Luria–Bertani media (flow rate of 50 µl/min); the experiment was stopped after 250 min; however, in this case, the flow rate of washing was not increased (flow rate of 50 /min). (The upper line A: gold reference surface; the lower line B: the amorphous carbon nanofilm coated on the gold surface.)

Both figures show the following consistent trends: 1) MLP-Me-coated surface showed significantly lower adsorption than the uncoated gold surface; and 2) adsorbed materials do not adhere to the surfaces (gold and coated gold). These findings indicate that the application of a nanofilm of MLP-Me is a viable process to impart biocompatibility to surfaces of various implants. Some experimental data previously available also supported this assessment.

Figure 15.21 shows a summary of the blood coagulation tests with stainless steel stents in pig model. The nanofilm encapsulation of all surfaces of metallic stent, by MLP-Me,

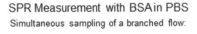

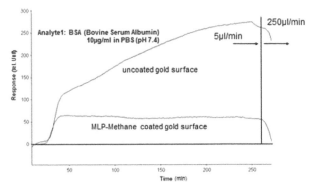

FIGURE 15.19
Comparison of adsorption dynamics of protein on surfaces in a flow system; two surfaces are placed in split apparel flow tubes.

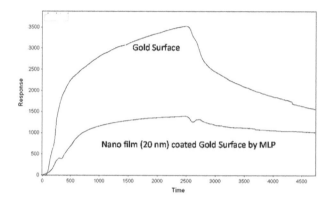

Adsorption of bacteria on surfaces in a flow system

FIGURE 15.20
Comparison of adsorption dynamics of bacteria on surfaces in a flow system; two surfaces are placed in split apparel flow tubes.

yield no closure of all five stents tested, while all of uncoated stents, without coagulation preventive drug in the blood stream, were closed by the clots of blood.[12] (The nature of the coating was not disclosed, and the reference coating as polymer-coated stents.)

Figure 15.22 shows comparison of the amount of protein adhered on the surface of extended wear contact lenses after a contact lense was worn for a predetermined time (days) continuously without removing the lens from the eye. (Identification of contact lenses are changed to samples A and B in order to avoid potential commercial implication.) Sample A is an extended wear contact lens coated with MLP-Me, and sample B is another extended wear contact lens with a different surface treatment by a different contact lens maker. Sample A showed three orders of magnitude lesser adsorption of protein, in spite of four times longer continuous wear time, compared to the adsorption observed with sample B.[13] Virtually, no adhesion of proteins occurs onto the MLP-Me-coated contact lens surface.

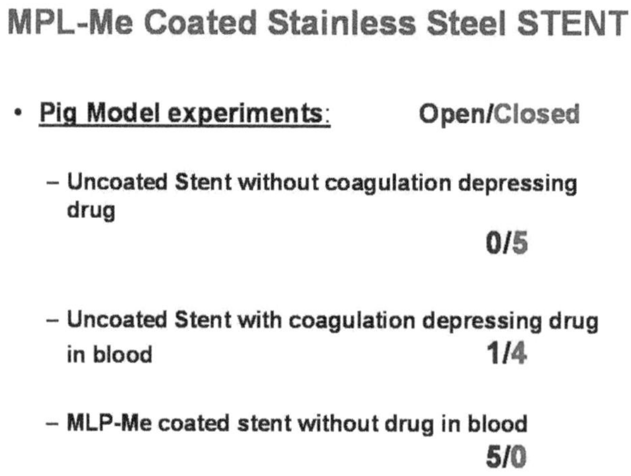

MPL-Me Coated Stainless Steel STENT

- **Pig Model experiments**:   **Open/Closed**

    – Uncoated Stent without coagulation depressing
      drug
                                                        0/5

    – Uncoated Stent with coagulation depressing drug
      in blood   **1/4**

    – MLP-Me coated stent without drug in blood
                                                        5/0

FIGURE 15.21
Summary of in vivo blood coagulation tests (pig model); the details of MLP-Me coating was not disclosed, and the coating was described as a polymer coating in the reference.

Protein Deposition - EW

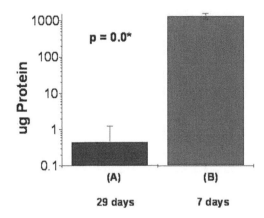

FIGURE 15.22
Comparison of protein adsorption by (A) MLP-Me-coated extended wear contact lens and (B) similar lens (different brand) coated with other method.

The first step of creating adhesion of a biological element, e.g., bacteria, on an artificial surface is the interfacial interaction between the surface and the biological element. If the surface has very low interfacial interaction potential, such as a surface coated with nanofilm of amorphous carbon, it can be predicted that the bacteria could not adhere to the surface and hence cannot grow on the surface. Figure 15.23 shows a vivid display of

Antibacterial Effect of the Coating

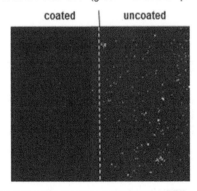

E. coli XL1 with GFP (green fluorescent protein)

- Fluorescent microscope detecting the GFP
- Samples incubated for 24 h at 37 °C with 100 rpm
- Washing to remove non adherent bacteria from the surface

FIGURE 15.23
Comparison of bacterial adhesion and growth on uncoated surface and coated surface with a nanofilm of MLP-Me.

the predicted phenomena. One-half of a polystyrene thin plate was coated with nanofilm of amorphous carbon and the remaining half was left uncoated, and the adhesion and growth of *Escherichia coli* was examined after the plate was placed in a solution of *E. coli*, which were treated to have luminescence under ultraviolet light. After incubation for 24 hr, the surface was washed to remove non-adhering *E. coli* from the surface before observing under ultraviolet light. The coated surface has no *E. coli*, whereas the uncoated surface shows significant amount of *E. coli* adhering to the surface.[14]

The use of artificial material in contact with a biological system causes the perturbation of the host biological system. If the level of the perturbation is within the tolerance limit of the biological system, the material could remain in or in contact with the biological system, and the material can be considered as biocompatible, which is the minimum perturbation principle of biocompatibility.[1]

All data shown in Figures 15.19 through 15.23 strongly indicate that the imperturbable surface-state of MLP-Me nanofilm yield the following remarkable advantages: 1) reduction of the adsorption of biological components; 2) the adsorbed biological components do not adhere to the surface, which would not cause the response of biological systems; and 3) accordingly, the presence of the surface would be ignored by the biological system.

Those positive trends strongly support the minimum perturbation principle of biocompatibility, and the application of nanofilm coating of amorphous carbon prepared by the MLP-Me has been proven to be an excellent means to impart biocompatibility to material surfaces in various shapes and sizes to be in contact with biological systems.

References

1. Yasuda, H. *Macromolecular Bioscience*, **2006**, *6*, 121.
2. Yasuda, H., *Luminous Chemical Vapor Deposition and Interface Engineering*, Marcel Dekker, New York, NY, 2004.
3. Yasuda, H. & Yu, Q. *Plasma Chemistry and Plasma Process*, **2004**, *24*, 325.
4. Yasuda, H. & Yu, Q. *Journal of Vacuum Science & Technology A: Vacuum, Surfaces, and Films*, **2004**, *22*(3), 472.
5. Yu, Q. S., Huang, C., & Yasuda, H. K. *Journal of Polymer Science, Part A: Polymer Chemistry* **2004**, *42*, 1042.
6. Yasuda, H. K., & Yu, Q. S. *Journal of Vacuum Science & Technology A: Vacuum, Surfaces, and Films* **2001**, *19*(3), 773.
7. Yu, Q., & Yasuda, H. *Plasmas and Polymers*, **2002**, *7*, 41–55.
8. Yasuda, H., Ledernez, L., Olcaytug, F., & Urban, G. *Pure and Applied Chemistry*, **2008**, *80*(9), 1883.
9. Yasuda, H. *Plasma Polymerization*, Academic Press, San Diego, CA, 1985.
10. Yasuda, H. *Magneto Luminous Chemical Vapor Deposition*, CRC Press, Boca Raton, FL, 2011.
11. Dame, G. and coworkers, data to be published.
12. van der Giessen, W. J., vanBeusekom, H. M. M., vanHouten, C. D., vanWoerkens, L. J., Verdouw, P. D., & Serruys, P. W. *Coronary Artery Diseases*, **1992**, *3*, 631.
13. Matsuzawa, Y., and Winterton, L., Presentation at International Round Table on Plasma Interface Engineering, Surface Science and Plasma Technology, University of Missouri, Columbia, MO, USA, 2002.
14. Yasuda, H., Ledernez, L., Olcaytug, F., Dame, G., & Bergmann, M. Biocompatible nanofilm coating by magneto-luminous polymerization of methane, *Special Issue of Progress in Organic Coatings*, for publication, **2011**.

Index

Page numbers followed by f and t indicate figures and tables, respectively.